Springer Theses

Recognizing Outstanding Ph.D. Research

Aims and Scope

The series "Springer Theses" brings together a selection of the very best Ph.D. theses from around the world and across the physical sciences. Nominated and endorsed by two recognized specialists, each published volume has been selected for its scientific excellence and the high impact of its contents for the pertinent field of research. For greater accessibility to non-specialists, the published versions include an extended introduction, as well as a foreword by the student's supervisor explaining the special relevance of the work for the field. As a whole, the series will provide a valuable resource both for newcomers to the research fields described, and for other scientists seeking detailed background information on special questions. Finally, it provides an accredited documentation of the valuable contributions made by today's younger generation of scientists.

Theses may be nominated for publication in this series by heads of department at internationally leading universities or institutes and should fulfill all of the following criteria

- They must be written in good English.
- The topic should fall within the confines of Chemistry, Physics, Earth Sciences, Engineering and related interdisciplinary fields such as Materials, Nanoscience, Chemical Engineering, Complex Systems and Biophysics.
- The work reported in the thesis must represent a significant scientific advance.
- If the thesis includes previously published material, permission to reproduce this must be gained from the respective copyright holder (a maximum 30% of the thesis should be a verbatim reproduction from the author's previous publications).
- They must have been examined and passed during the 12 months prior to nomination.
- Each thesis should include a foreword by the supervisor outlining the significance of its content.
- The theses should have a clearly defined structure including an introduction accessible to new PhD students and scientists not expert in the relevant field.

Indexed by zbMATH.

Václav Březina

Supramolecular Complexes of Oxoporphyrinogens with Organic Molecules

Doctoral Thesis accepted by
Charles University, Prague, Czech Republic

Springer

Author
Dr. Václav Březina
Department of Macromolecular Physics
Charles University
Prague, Czech Republic

Supervisor
Dr. Lenka Hanyková
Department of Macromolecular Physics
Charles University
Prague, Czech Republic

ISSN 2190-5053 ISSN 2190-5061 (electronic)
Springer Theses
ISBN 978-3-031-68645-0 ISBN 978-3-031-68646-7 (eBook)
https://doi.org/10.1007/978-3-031-68646-7

This Springer imprint is published by the registered company Springer Nature Switzerland AG
The registered company address is: Gewerbestrasse 11, 6330 Cham, Switzerland

Supervisor's Foreword

The alteration in color of a chromophore resulting from protonation or specific interactions with particular atoms, ions, or molecules serves as the fundamental principle underpinning all pH indicators and colorimetric sensors. These indicators and sensors find extensive utility in industry, scientific research, and health care by facilitating the identification of certain substances solely through observable color changes. The development of such indicators and sensors necessitates a comprehensive understanding of the mechanisms governing color changes and their practical limitations. The dissertation of Václav Březina deals with characterization of oxoporphyrinogen molecules, whose structure and properties are derived from macrocyclic porphyrin. Oxoporphyrinogens exhibit pH-dependent color variations in non-polar environments as well as a wide range of dynamic processes at millisecond timescale. The potential use of oxoporphyrinogens as colorimetric pH sensors renders this dissertation not only scientifically relevant but also opens up practical applications.

The first section of the dissertation provides a comprehensive account of the colorimetric response exhibited by the studied molecules upon interacting with organic acids and the subsequent formation of supramolecular complexes, employing UV/vis spectroscopy in combination with the singular value decomposition method. The subsequent section in the dissertation delves into a thorough examination of the dynamic behavior of oxoporphyrinogen molecules by the analysis of chemical exchange in NMR spectra. A pivotal outcome of this investigation is the development of an analytically derived model for multistate chemical exchange. Notably, Václav's presentation style is outstanding, ensuring that readers, even those new to the subject, can comprehend the intricacies of supramolecular bonds and the principles of employed methods (NMR and UV/vis absorption spectroscopies).

Václav Březina's dissertation is a meticulously crafted and comprehensive resource, poised to serve as an engaging and valuable guide for those seeking to navigate the realm of colorimetric pH sensors and dynamic molecular systems. It not only offers a strong foundation for understanding this topic but also provides a springboard for further exploration and research in this field.

Prague, Czech Republic
December 2023

Dr. Lenka Hanyková

The original version of the book has been revised: The author's name has been updated in xml only.
A correction to this book can be found at https://doi.org/10.1007/978-3-031-68646-7_9

Abstract

Oxoporphyrinogens are flat macrocyclic molecules possessing binding and protonation sites, and capable of light absorption in the visible region. These properties are prerequisites for a colorimetric molecular sensor, i.e. a specific detector of other molecules in the sample. In this work, we studied chromic properties of three oxoporphyrinogens, **OxP** and its partially (**Bz_2OxP**) and fully (**Bz_4OxP**) N-benzylated derivatives. Their colorimetric response to organic acids is caused by protonation and subsequent formation of supramolecular host–guest complex. We have shown that colorimetric sensitivity is highest for **OxP** and gradually weakens for **Bz_2OxP** and **Bz_4OxP** since the N-benzylation blocks the central binding sites, decreasing binding affinity of the oxoporphyrinogens. Furthermore, solvatochromic response of the oxoporphyrinogens to varying solvent polarity showed similar sensitivity decrease in **Bz_2OxP** and **Bz_4OxP**. The chromic and binding properties were studied by UV/vis and NMR spectroscopy, host–guest binding models were applied to describe the formation of oxoporphyrinogen-acid complexes.

In NMR spectra of protonated **OxP** and **Bz_2OxP**, we detected several dynamic processes, manifested as chemical exchange, such as prototropic tautomerization (i.e. change of the protonation site) or rotation of bulky side groups in **Bz_2OxP**. These processes were observed at different temperatures and acid concentrations, the corresponding transition rate coefficients were obtained by fitting of the NMR exchange lineshapes. In protonated **Bz_2OxP**, the effect of temperature obeys the Eyring equation, and the effect of acid concentration has been modeled using a kinetic model derived from competitive host–guest binding.

This thesis presents a detailed mechanism of color changes in **OxP**, **Bz_2OxP**, and **Bz_4OxP** and confirms potential application as colorimetric sensors in nonpolar media. Our NMR study of dynamic processes in the oxoporphyrinogens contributes to understanding the concentration dependence of transition rate coefficients, which is always system-specific. The presented models can be applied to a wide range of dynamic systems.

Acknowledgments

I would like to thank to my supervisor Lenka Hanyková for her support, patience, useful advice, and help with NMR measurements. Many thanks belong to my consultant Jan Labuta for endless discussions about data processing, interpretation of the results and for arranging my internship at the National Institute for Materials Science (NIMS) in Tsukuba, Japan. Revisions of the text of this thesis by my supervisor and consultant are greatly appreciated. I am also grateful to Jonatan Hill from NIMS for providing the oxoporphyrinogen samples. Last but not least, I thank to my family and my girlfriend Kát'a for moral support.

Contents

Abbreviations

COSY	Correlation spectroscopy
CSA	Camphorsulfonic acid
CW	Continuous wave
DFA	Difluoroacetic acid
DMF	Dimethylformamide
DSIS	Decomposition of spectra into individual species
EXSY	Exchange spectroscopy
FID	Free inductions decay
FT	Fourier transform
FTIR	Fourier transform infrared (spectroscopy)
HSQC	Heteronuclear single quantum coherence (spectroscopy)
IR	Infrared
NMR	Nuclear magnetic resonance (spectroscopy)
OxP	Oxoporphyrinogen
PC	Principal component
PCA	Principal component analysis
RF	Radiofrequency
RSD	Residual standard deviation
SVD	Singular value decomposition
TFA	Trifluoroacetic acid
TMS	Tetramethylsilan
TPP	Tetraphenylporphyrin
UV/vis	Ultraviolet/visible (spectroscopy)

Chapter 1
Introduction

Compounds capable of light absorption in the visible region as well as host-guest complexation with variety of ligands are perfect candidates for molecular sensors. These conditions are fulfilled by oxoporphyrinogen (**OxP**) and its derivatives, which are studied in this work. Oxoporphyrinogens are flat macrocyclic molecules possessing multiple binding and protonation sites. They have been previously investigated from the perspective of colorimetric detection of various molecules, anions and acids in nonpolar media. Here, we extend the current knowledge and use nontrivial analyses of UV/vis spectra to qualitatively and quantitatively explain the mechanism of color change of oxoporphyrinogens due to the presence of organic acids. Solvatochromic properties are also studied.

Several dynamic molecular processes have been previously observed in the system of oxoporphyrinogen + acid. This topic is also covered in this thesis in the analysis of chemical exchange in NMR spectroscopy. Rates of the processes are characterized as functions of temperature and concentrations, a new model is devised to qualitatively explain the observed data. This work is mainly based on our publications [1, 2].

In this first introductory chapter, we put oxoporphyrinogens in a broader context of similar natural and synthetic compounds. An overview of applications and recent literature about oxoporphyrinogens will be given as well as the concrete goals of this work.

1.1 Overview of Tetrapyrroles

Compounds of our interest belong to a wide class called *tetrapyrroles*, containing four pyrrole rings typically joined by single-atom links at the pyrrole α-positions [3]. Scheme in Fig. 1.1 shows simplified classification of tetrapyrroles, presenting example molecules for each group (names of depicted compounds are highlighted

V. Březina, *Supramolecular Complexes of Oxoporphyrinogens with Organic Molecules*, Springer Theses, https://doi.org/10.1007/978-3-031-68646-7_1

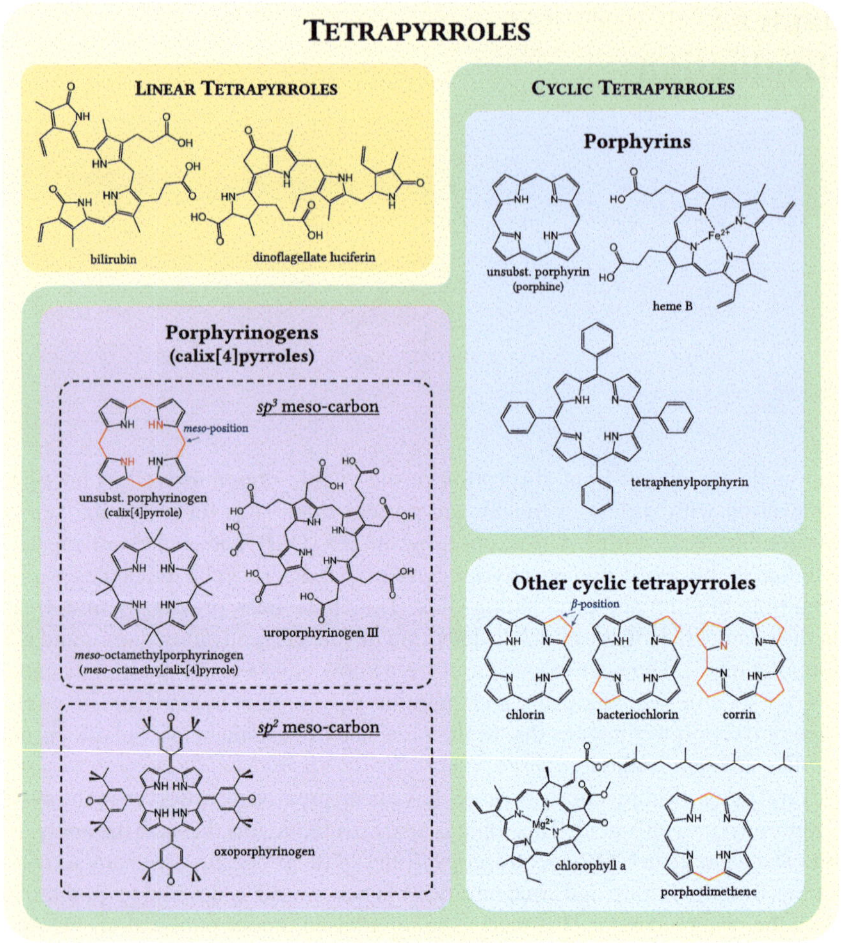

Fig. 1.1 Simplified structure-based classification of tetrapyrroles. Several tetrapyrrole subclasses are omitted. Red color denotes differences of the unsubstituted rings from porphine

by boldface in the text below). Basic division comprises linear (open) tetrapyrroles and cyclic (closed) tetrapyrroles. Both groups represent essential natural dyes and metabolites [4, 5] as well as widely-studied artificially synthesized molecules [6].

Linear tetrapyrroles include bile pigments such as **bilirubin** or biliverdin, breakdown products of heme [5]. Another example is **dinoflagellate luciferin**, which produces bioluminiscence ($\lambda_{max} = 475$ nm) after oxidation catalyzed by luciferase enzyme [7, 8].

Cyclic tetrapyrroles are classified according to the type of macrocycle conjugation. Many of them are capable of coordinating a metal atom in their molecular center, which is denoted as 'chelate' or by a prefix 'metallo-'. Prototypes of the tetrapyrrole macrocycle are fully conjugated *porphyrins*, whose unsubstituted form is called **porphine**. Their name is derived from a Greek word for 'purple'. A well known natural metalloporphyrin is heme (most abundant form **heme B**), coordinating Fe^{2+} atom. Heme is located in red blood cells, bound to hemoglobin protein as a prosthetic group, and serves for transport of oxygen [9]. Man-made porphyrins have many applications in both chelated and free-base forms as discussed below. A common example is **tetraphenylporphyrin** (TPP), that has already appeared in thousands of scientific publications [10]. Interestingly, TPP is the first molecule, whose vibrational modes have been experimentally imaged [11].

Another group of cyclic tetrapyrroles are *porphyrinogens*, also called *calix[4]pyrroles* [12]. Their central nitrogens are bound by single bonds, and unsubstituted porphyrinogens contain four NH groups. While free-base (i.e. non-chelated) porphyrins such as TPP possess flat macrocycle, porphyrinogen center is nonplanar. In this work, we distinguish two types of porphyrinogens based on their *meso*-position (shown in Fig. 1.1): (**i**) "true" porphyrinogens with sp^3 hybridization on *meso*-carbons, (**ii**) *oxoporphyrinogens* with the same formal ring structure as (i) but with sp^2 *meso*-carbon hybridization. Structure of unsubstituted **porphyrinogen** shows the presence of two additional NH groups and interruption of conjugation in contrast to porphine (denoted by red color in Fig. 1.1). The incomplete conjugation renders the "true" porphyrinogens colorless. In heme metabolism, porphyrinogens are precursors to porphyrins, hence the name 'porphyrinogen'. Examples of metabolic porphyrinogens are **uroporphyrinogen III**, coproporphyrinogen III and protoporphyrinogen IX [5]. A frequently investigated synthetic porphyrinogen is ***meso*-octamethylcalix[4]pyrrole** [12]. The other group based on porphyrinogen structure and also the subject of this work are *oxoporphyrinogens*, possessing sp^2 *meso*-carbon hybridization. **Oxoporphyrinogen** (**OxP**) has been first synthesized by Milgrom [13]. Oxoporphyrinogens are colored in contrast to "true" porphyrinogens due to the complete conjugation of the oxoporphyrinogen ring. On the other hand, both types of porphyrinogen share anion binding capabilities owing to the central NH groups.

There are other possible modifications of the porphyrin macrocycle illustrated in Fig. 1.1 (differences from the porphyrin ring are denoted by red color), with different degree of saturation at *meso*- or β-positions (marked in Fig. 1.1). Many of them constitute essential natural compounds. Reduced porphyrins (addition of hydrogens at the β-carbons) **chlorin** and **bacteriochlorin** constitute the core of chlorophylls (plants and algae) and bacteriochlorophylls[1] (photosyntetic bacteria) [14, 15]. These compounds are magnesium chelates and include many variants, for example, **chlorophyll a** and chlorophyll b are the green plant pigments. Chlorophylls and bacteriochlorophylls serve for light harvesting, energy transfer and also electron transfer in natural photosystems [15]. Another macrocycle, **corrin**, is substantially saturated and lacks one of the methine link. A member of the corrins is cobalamin,[2] also known as

[1] Some bacteriochlorophylls are based on the bacteriochlorin but others on the chlorin ring.

[2] There are several forms of cobalamin, e.g. hydroxycobalamin or cyanocobalamin [16].

vitamin B_{12} [4]. As the name suggests, it is a chelate of cobalt. Interestingly, the already mentioned uroporphyrinogen III is common biosynthetic precursor to hemes, chlorophylls and also cobalamin [4, 15]. The last shown example of tetrapyrrole macrocycle is **porphodimethene**.

Applications of artificial cyclic tetrapyrroles utilize their binding properties (central NH binding sites), electronic structure or the ability to form π–π stacked structures [17]. Porphyrin-based compounds can serve as molecular probes for detection of cations or anions [18, 19]. Similarly, "true" porphyrinogens serve as probes[3] of anions or neutral species [6, 12, 21]. TPP has wide applications in photochemistry, for example, as photosensitizer in light-induced oxidation[4] [22]. Protonated TPP has the capability of forming helical J-aggregates for chiral recognition [23]. Further application include, for example, photoacoustic imaging probes [24, 25] or photodynamic therapy [26, 27].

1.2 Properties and Applications of Oxoporphyrinogens

Oxoporphyrinogen (**OxP**) and its N-alkylated derivatives[5] N_{21}, N_{23}-bis(4-bromobenzyl)oxoporphyrinogen (**Bz$_2$OxP**) and also N_{21}, N_{22}, N_{23}, N_{24}-tetrakis(4-bromobenzyl)oxoporphyrinogen (**Bz$_4$OxP**) are the subject of the current work. Their chemical and X-ray crystal structures are given in Fig. 1.2a–c. The **OxP** molecule possesses two calix[4]pyrrole-type binding sites (shortly NH/amine binding sites, black arrows in Fig. 1.2) and four protonation sites at the carbonyl goups (C = O groups, green arrows). N-benzylation at the macrocyclic nitrogen atoms of oxoporphyrinogen leads to stepwise elimination of the two NH binding sites (red crosses in **Bz$_2$OxP** and **Bz$_4$OxP** X-ray structures) and subsequent modification of the chemical properties. Oxoporphyrinogens share the extensive conjugated system and consequent light absorbing capability with porphyrins thanks to the sp^2 *meso*-carbon. On the other hand, they share the nonplanar macrocycle and central calix[4]pyrrole-type binding sites (see Fig. 1.2) with (colorless) porphyrinogens. Hence, the advantages of porphyrins and porphyrinogens are conveniently combined in oxoporphyrinogens. Oxoporphyrinogens are insoluble in water, thus, their applications require nonpolar solvents.

Atoms in the central macrocycle are numbered in the same way as in porphine [3], see Fig. 1.2d. Methine links numbered 5, 10, 15 and 20 are denoted as *meso* positions. Labeling of α (1, 4, 6, 9, ...) and β (2, 3, 7, 8, ...) positions is derived from that

[3] For UV/vis probing, a chromophore is attached to colorless porphyrinogens or displacement assay technique is used (weakly bound colored anion is displaced by more strongly binding analyte, which leads to color change of the colored anion [20]).

[4] High triplet-state energy of tetraphenylporphyrin (142.3 kJ/mol) enables excitation of triplet oxygen to its singlet state (94.2 kJ/mol) [22].

[5] The 4-bromobenzyl derivatives are more soluble in chloroform than the corresponding non-brominated derivatives.

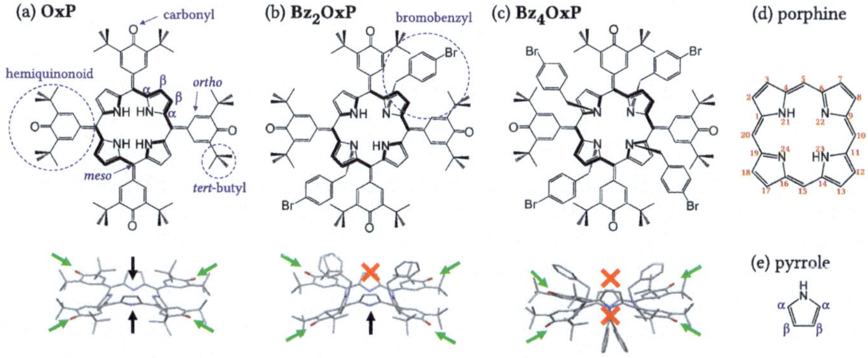

Fig. 1.2 Oxoporphyrinogens, notation of important sites and groups. **a–c** Oxoporphyrinogens investigated in this work, chemical structures (top) and X-ray crystal structures (bottom). X-ray crystal structure in panels (**b**, **c**) are of the non-brominated derivatives [30]. Black arrows: accessible calix[4]pyrrole-type binding sites, red crosses: blocked binding sites, green arrows: protonation carbonyl sites. **d** Labeling of pyrrole positions transferable to other cyclic tetrapyrroles, **e** labeling of pyrrole positions

of pyrrole in Fig. 1.2a, e. The bulky side groups are denoted as hemiquinonoid or cyclohexadienilidene, the corresponding hydrogens at *ortho*-position will be referred to as *ortho*-H. Carbonyl C = O sites can be protonated by an acid, the central NH groups serve as binding site for anions or neutral molecules. Next to the carbonyl, two *tert*-butyl groups are located, consisting of an sp^3 carbon with three methyls. Central nitrogen atoms are N-alkylated with bromobenzyl groups, see Fig. 1.2b, c. Noteworthy, chlorin and porphyrin macrocycles show NH tautomerism, where the two central hydrogens "jump" among the four nitrogen atoms [28, 29]. Such scenario is, however, impossible in **Bz$_2$OxP** since the N-alkylation prevents any rearrangements.

Various aspects of **OxP** and **Bz$_2$OxP** were studied before mainly by Jonathan Hill, Jan Labuta and coworkers in the National Institute of Materials Science (NIMS) in Tsukuba, Japan [30–45]. The bromobenzylated **Bz$_4$OxP** derivative was described in our publication [1] for the first time.

Colorimetric detection of anions is possible due to hydrogen bonding in 1:1 (**Bz$_2$OxP**) or 1:2 (**OxP**) stoichiometry [31, 33, 44]. Complex formation was also confirmed by ^{1}H NMR measurements, especially NH resonances are very sensitive to binding at the center. Hydrogen bonding of water is also possible, resulting to the shift of UV/vis absorption bands even due to small amounts of water. Hence, **Bz$_2$OxP** can be used, for instance, for colorimetric detection of trace amounts of water ($\approx$50 ppm) in tetrahydrofuran [35]. Similarly, oxoporphyrinogens form hydrogen bonds with solvent molecules, thus causing solvatochromism [31], i.e. color change of the solution induced by different solvents. Due to this property, nanocomposite film with **OxP** can provide naked-eye discrimination of methanol from ethanol [36]. Solvatochromic effects will also be addressed in this work.

Oxoporphyrinogens act as acidity sensors in nonpolar environment [32]. Carbonyl groups shown in Fig. 1.2a–c (green arrows) can be protonated and the acid anion is attached by hydrogen bond at the central NH binding site [32]. Acidity sensing will be one of the main topic of this thesis, we will focus on quantitative analysis of the color changes. Basic molecules can also form hydrogen bonds with oxoporphyrinogens bases [32]. The presence of a base might cause deprotonation of two central NH groups in **OxP** [32].

Another aspect is observation of dynamic processes present in complexed oxoporphyrinogens, such as prototropic tautomerization in **OxP** + acid systems [37, 38]. Dynamic processes, observable in NMR chemical exchange, are interesting from theoretical point of view, however, there is a practical application in NMR determination of enantiomeric excess of chiral analytes. This novel approach exploits the symmetry of oxoporphyrinogen molecules and fast dynamic exchange of bound analytes. In particular, a NMR resonance of the chirality sensor splits linearly with enantiomeric excess of the bound chiral analyte [38–43]. This method of chirality sensing is described in detail in Sect. 3.4.3. Apart from **OxP** and **Bz$_2$OxP**, also TPP [40] and other compounds [46, 47] can be used as chirality sensors. Dynamic processes are the second main topic of this work. We focus on quantitative description of previously known dynamic processes, and also discover new interesting dynamic phenomena.

Apart from N-alkylation, the oxoporphyrinogen structure can also be modified by β-substitution [33, 34]. These derivatives can be used e.g. for generation of singlet oxygen when incorporated into porous polymer [48–50].

1.3 Aims and Structure of This Work

This work is based on our publications [1, 2] including figures and parts of text. It is also a continuation of the author's master's thesis[6] [51]. Here, we investigate oxoporphyrinogen (**OxP**) and two N-alkylated derivatives denoted as **Bz$_2$OxP** and **Bz$_4$OxP** and their interaction with organic acids. In particular, chiral camphorsulfonic acid (CSA), achiral difluoroacetic acid (DFA) and trifluoroacetic acid (TFA) are used. The effect of chirality on NMR spectra has been taken into account, especially in fast exchange processes. Furthermore, solvatochromic color changes of the oxoporphyrinogens are observed in the mixture of polar dimethylformamid (DMF) and nonpolar chloroform. Our aims are as follows:

- Determination of the effect of N-alkylation on color changes. Color changes of **OxP**, **Bz$_2$OxP** and **Bz$_4$OxP** induced by CSA of DFA will be studied in chloroform solvent.

[6] Available at https://dspace.cuni.cz/handle/20.500.11956/67480.

- Analysis of UV/vis spectral changes by singular value decomposition. This type of analysis has not been used before in our systems, it allows determination of the exact number of absorbing species, their UV/vis spectra (and the corresponding degree of protonation) and the corresponding populations.
- Determination of binding constants. Binding isotherms for UV/vis and NMR titration experiments will be constructed and fitted by appropriate host-guest binding models.
- Investigation of solvatochromism. Color changes induced by solvent polarity change or hydrogen bonding with the oxoporphyrinogents will be studied in the DMF/chloroform mixture.
- Identification of dynamic molecular processes such as rotation of side groups or tautomerization. These processes will be observed in NMR spectra in a form of so-called chemical exchange.
- Derivation of analytical lineshape for three-state chemical exchange. The lineshape will be calculated from the Bloch-McConnell equations and will be used to retrieve the transition rate coefficients from the exchange spectra by fitting.
- Determination of transition rate coefficients and energy barriers as obtained from lineshape fitting of experimentally obtained exchange spectra. Concentration dependence of the transition rate coefficients will be described by host-guest binding models, and temperature dependence of the transition rate coefficients will be analyzed in terms of Eyring equation.

This work consists of theoretical part (Chaps. 1–5) and experimental part (Chaps. 6 and 7). Extensive appendices provide an overview of binding models, NMR lineshapes, details about experimental measurements, data analyses and measured NMR spectra. Codes for implementation in *Mathematica* software are also included.

In particular, structure of this work is as follows: The current Chap. 1 introduces the investigated oxoporphyrinogens and gives an overview of the wider class of tetrapyrroles. Chapter 2 provides theory of supramolecular binding and chemical kinetics. Chapter 3 introduces the NMR spectroscopy with focus on chemical exchange. Analytical three-state exchange lineshape, derived in our publication [2], is also shown in this chapter. Basics of UV/vis spectroscopy are given in Chap. 4. In Chap. 5, comprehensive description of singular value decomposition and its application to UV/vis spectra is presented. In Chap. 6, chromic and binding properties of the oxoporphyrinogens are studied experimentally with UV/vis and NMR spectroscopy. Results from our publication [1] are presented here in an extended form. In the next experimental chapter, Chap. 7, dynamic processes in **OxP** and **Bz$_2$OxP** are studied. Part of this research is published in our article [2], nevertheless, large portion of these results remains unpublished. Results of this work and further applications are concisely summarized in Chap. 8.

References

1. Březina V, Ishihara S, Lang J, Hanyková L, Ariga K, Hill JP, Labuta J (2018) Structural modulation of chromic response: effects of binding-site blocking in a conjugated calix[4]pyrrole chromophore. ChemistryOpen 7:323–335
2. Březina V, Hanyková L, Velychkivska N, Hill JP, Labuta J (2022) NMR lineshape analysis using analytical solutions of multi-state chemical exchange with applications to kinetics of host-guest systems. Sci Rep 12:17369
3. Moss GP (1987) Nomenclature of tetrapyrroles (Recommendations 1986). Pure Appl Chem 59:779–832
4. Meganathan R (2010) Comprehensive natural products II: chemistry and biology. In: Mander L, Liu H-W (eds) vol 7. Elsevier
5. Ortiz de Montellano PR (2008) Hemes in biology. In: Wiley encyclopedia of chemical biology. American Cancer Society, pp 1–10
6. Rather IA, Wagay SA, Hasnain MS, Ali R (2019) New dimensions in calix[4]pyrrole: the land of opportunity in supramolecular chemistry. RSC Adv 9:38309–38344
7. Schultz LW, Liu L, Cegielski M, Hastings JW (2005) Crystal structure of a pH-regulated luciferase catalyzing the bioluminescent oxidation of an open tetrapyrrole. PNAS 102:1378–1383
8. Valiadi M, Iglesias-Rodriguez D (2013) Understanding bioluminescence in dinoflagellates—how far have we come? Microorganisms 1:3–25
9. Hsia CC (1998) Respiratory function of hemoglobin. N Engl J Med 338:239–248
10. Taniguchi M, Lindsey JS, Bocian DF, Holten D (2021) Comprehensive review of photophysical parameters (ϵ, Φ_f, τ_s) of tetraphenylporphyrin (H_2TPP) and zinc tetraphenylporphyrin (ZnTPP)—critical benchmark molecules in photochemistry and photosynthesis. J Photochem Photobiol C Photochem Rev 46:100401
11. Lee J, Crampton KT, Tallarida N, Apkarian VA (2019) Visualizing vibrational normal modes of a single molecule with atomically confined light. Nature 568:78–82
12. Gale PA, Sessler JL, Král V (1998) Calixpyrroles. Chem Commun 1–8
13. Milgrom LR (1983) The facile aerial oxidation of a porphyrin. Tetrahedron 39:3895–3898
14. Castelfranco PA, Beale SI (1983) Chlorophyll biosynthesis: recent advances and areas of current interest. Annu Rev Plant Physiol 34:241–276
15. Senge MO, Ryan AA, Letchford KA, MacGowan SA, Mielke T (2014) Chlorophylls, symmetry, chirality, and photosynthesis. Symmetry 6:781–843
16. Eitenmiller RR, Landen Jr WO (2010) Vitamin analysis for the health and food sciences. CRC Press
17. Monti D, Nardis S, Stefanelli M, Paolesse R, Di Natale C, D'Amico A (2009) Porphyrin-based nanostructures for sensing applications. J Sens 2009:e856053
18. Lee H, Hong K-I, Jang W-D (2018) Design and applications of molecular probes containing porphyrin derivatives. Recent progress on fluorescent probes. Coord Chem Rev 354:46–73
19. Ishihara S, Labuta J, Rossom WV, Ishikawa D, Minami K, Hill JP, Ariga K (2014) Porphyrin-based sensor nanoarchitectonics in diverse physical detection modes. Phys Chem Chem Phys 16:9713–9746
20. Sedgwick AC, Brewster TT, Wu T, Feng X, Bull SD, Qian X, Sessler JT, James TD, Anslyn EV, Sun X (2021) Indicator displacement assays (IDAs): the past, present and future. Chem Soc Rev 50:9–38
21. Gale PA, Anzenbacher P Jr, Sessler JL (2001) Calixpyrroles II. Coord Chem Rev 222:57–102
22. Pfoertner K-H, Oppenländer T (2012) Photochemistry. In: Ullmann's encyclopedia of industrial chemistry. American Cancer Society
23. Watarai H, Kurahashi Y (2016) Chiral recognition of 2-alkylalcohols with magnetic circular dichroism measurement of porphyrin J-aggregate on silica gel plate. Anal Chem 88:4619–4623
24. Merkes JM, Rueping M, Kiessling F, Banala S (2019) Photoacoustic detection of superoxide using oxoporphyrinogen and porphyrin. ACS Sens 4:2001–2008

25. Merkes JM, Zhu L, Bahukhandi SB, Rueping M, Kiessling F, Banala S (2020) Photoacoustic imaging probes based on tetrapyrroles and related compounds. Int J Mol Sci 21:3082
26. Kou J, Dou D, Yang L (2017) Porphyrin photosensitizers in photodynamic therapy and its applications. Oncotarget 8:81591–81603
27. Tian J, Huang B, Nawaz MH, Zhang W (2020) Recent advances of multi-dimensional porphyrin-based functional materials in photodynamic therapy. Coord Chem Rev 420:213410
28. Helaja J, Stapelbroek-Möllmann M, Kilpeläinen I, Hynninen PH (2000) NH tautomerism in the natural chlorin derivatives. J Org Chem 65:3700–3707
29. Storm CB, Teklu Y (1972) Nitrogen-hydrogen tautomerism in porphyrines and chlorines. J Am Chem Soc 94:1745–1747
30. Hill JP, Hewitt IJ, Anson CE, Powell AK, McCarty AL, Karr PA, Zandler ME, D'Souza F (2004) Highly nonplanar, electron deficient, N-substituted tetra-oxocyclohexadienylidene porphyrinogens: structural, computational, and electrochemical investigations. J Org Chem 69:5861–5869
31. Hill JP, Schumacher AL, D'Souza F, Labuta J, Redshaw C, Elsegood MRJ, Aoyagi M, Nakanishi T, Ariga K (2006) Chromogenic indicator for anion reporting based on an N-substituted oxoporphyrinogen. Inorg Chem 45:8288–8296
32. Shundo A, Ishihara S, Labuta J, Onuma Y, Sakai H, Abe M, Ariga K, Hill JP (2013) Colorimetric visualization of acid-base equilibria in non-polar solvent. Chem Commun 49:6870–6872
33. Chahal MK, Labuta J, Březina V, Karr PA, Matsushita Y, Webre WA, Payne DT, Ariga K, D'Souza F, Hill JP (2019) Knock-on synthesis of tritopic calix[4]pyrrole host for enhanced anion interactions. Dalton Trans 48:15583–15596
34. Chahal MK, Velychkivska N, Webre WA, Labuta J, Ishihara S, Ariga K, D'Souza F, Hill JP (2019) Increasing the complexity of oxoporphyrinogen colorimetric sensing chromophores: N-alkylation and β-substitution. J Porphyr Phthalocyanines 23:1184–1194
35. Ishihara S, Labuta J, Šikorský T, Burda JV, Okamoto N, Abe H, Ariga K, Hill JP (2012) Colorimetric detection of trace water in tetrahydrofuran using N, N′-substituted oxoporphyrinogens. Chem Commun 48:3933–3935
36. Ishihara S, Iyi N, Labuta J, Deguchi K, Ohki S, Tansho M, Shimizu T, Yamauchi Y, Sahoo P, Naito M, Abe H, Hill JP, Ariga K (2013) Naked-eye discrimination of methanol from ethanol using composite film of oxoporphyrinogen and layered double hydroxide. ACS Appl Mater Interfaces 5:5927–5930
37. Labuta J, Futera Z, Ishihara S, Kouřilová H, Tateyama Y, Ariga K, Hill JP (2014) Chiral guest binding as a probe of macrocycle dynamics and tautomerism in a conjugated tetrapyrrole. J Am Chem Soc 136:2112–2118
38. Labuta J, Ishihara S, Ariga K, Hill JP (2014) Dynamic processes in prochiral solvating agents (pro-CSAs) studied by NMR spectroscopy. Symmetry 6:345–367
39. Shundo A, Labuta J, Hill JP, Ishihara S, Ariga K (2009) Nuclear magnetic resonance signaling of molecular chiral information using an achiral reagent. J Am Chem Soc 131:9494–9495
40. Labuta J, Ishihara S, Shundo A, Arai S, Takeoka S, Ariga K, Hill JP (2011) Chirality sensing by nonchiral porphines. Chem Eur J 17:3558–3561
41. Labuta J, Ishihara S, Šikorský T, Futera Z, Shundo A, Hanyková L, Burda JV, Ariga K, Hill JP (2013) NMR spectroscopic detection of chirality and enantiopurity in referenced systems without formation of diastereomers. Nat Commun 4:2188
42. Ishihara S, Labuta J, Futera Z, Mori S, Sato H, Ariga K, Hill JP (2018) NMR spectroscopic determination of enantiomeric excess using small prochiral molecules. J Phys Chem B 122:5114–5120
43. Labuta J, Ishihara S, Payne DT, Takimoto K, Sato H, Hanyková L, Ariga K, Hill JP (2021) Estimation of enantiomeric excess based on rapid host-guest exchange. Chemosensors 9:259
44. Hill JP, Karr PA, Uy RAZ, Subbaiyan NK, Futera Z, Ariga K, Ishihara S, Labuta J, D'Souza F (2022) Analyte interactions with oxoporphyrinogen derivatives: computational aspects. Curr Org Chem 26:580–595
45. Hill JP, Labuta J (2022) Oxoporphyrinogens: novel dyes based on the fusion of calix[4]pyrrole, quinonoids and porphyrins. In: Wakayama Y, Ariga K (eds) System-materials nanoarchitectonics. Springer Japan, Tokyo, pp 127–147

46. Norvaiša K, O'Brien JE, Osadchuk I, Twamley B, Borovkov V, Senge MO (2022) Importance of molecular symmetry for enantiomeric excess recognition by NMR. Chem Commun 58:5423–5426
47. Takimoto K, Ishihara S, Labuta J, Březina V, Payne DT, Hill JP, Ariga K, Sumita M, Mori S, Sato H (2020) Enantiomeric excess dependent splitting of NMR signal through dynamic chiral inversion and coligand exchange in a coordination complex. J Phys Chem Lett 8164–8169
48. Hynek J, Payne DT, Chahal MK, Sciortino F, Matsushita Y, Shrestha LK, Ariga K, Labuta J, Yamauchi Y, Hill JP (2021) Enhancement of singlet oxygen generation based on incorporation of oxoporphyrinogen (OxP) into microporous solids. Mater Today Chem 21:100534
49. Hynek J, Chahal MK, Payne DT, Liyanage A, D'Souza F, Hill JP (2023) Design rules for oxoporphyrinogen ('OxP') as a versatile chromophore for efficient singlet oxygen generation. J Porphyr Phthalocyanines 27:1108–1118
50. Hynek J, Payne DT, Shrestha LK, Chahal MK, Ma R, Dong J, Ariga K, Yamauchi Y, Hill JP (2024) Mild selective photochemical oxidation of an organic sulfide using OxP-polyimide porous polymers as singlet oxygen generators. Sci Technol Adv Mater 25:2322458
51. Březina V (May 2014) Spectroscopic and theoretical study of supramolecular complexes of symmetric porphyrins with chiral guests. MA thesis, Univerzita Karlova, Matematicko-fyzikální fakulta

Chapter 2
Quantitative Description of Supramolecular Binding

Supramolecular structures encompass molecular assemblies without formation of covalent bonds. The assemblies are held together by hydrogen bonding or by electrostatic forces. These weak binding forces enable incessant formation and dissociation of the supramolecular complex, implying that the supramolecular systems are dynamic even after they achieve equilibrium. Formation and dissociation of supramolecular complexes follow the general laws of kinetics and equilibrium for chemical reactions, which are explained in this chapter. From the theoretical point of view, the matter is described in detail in textbooks of Atkins [1] or McQuarrie and Simon [2]. From the practical point of view, the textbook of Connors [3] is the basic source.

2.1 Description of Chemical Reactions

2.1.1 Equilibrium Constant

A general chemical reaction converting reactants $\mathbf{R}_j$ to products $\mathbf{P}_j$ is described by the following scheme

Scheme 2.1 General scheme of a chemical reaction

$$|\nu_{\mathrm{R}1}|\mathbf{R}_1 + |\nu_{\mathrm{R}2}|\mathbf{R}_2 + \cdots \xrightleftharpoons{K_{\mathrm{eq}}} \nu_{\mathrm{P}1}\mathbf{P}_1 + \nu_{\mathrm{P}2}\mathbf{P}_2 + \cdots$$

V. Březina, *Supramolecular Complexes of Oxoporphyrinogens with Organic Molecules*, Springer Theses, https://doi.org/10.1007/978-3-031-68646-7_2

where ν_j are stoichiometric numbers,[1] $\nu_j < 0$ for reactants (educts) and $\nu_j > 0$ for products [1]. The quantity K_{eq} is called *equilibrium constant* (also denoted as binding or stability constant) [3].

Chemical reactions usually take place in contact with thermal reservoir at constant pressure. Therefore, description using Gibbs free energy potential $\mathrm{d}G = -S\mathrm{d}T + V\mathrm{d}p + \mu_{\mathbf{R}1}\mathrm{d}n_{\mathbf{R}1} + \mu_{\mathbf{R}2}\mathrm{d}n_{\mathbf{R}2} + \mu_{\mathbf{P}2}\mathrm{d}n_{\mathbf{P}2} + \mu_{\mathbf{P}1}\mathrm{d}n_{\mathbf{P}1} \ldots$ is convenient since at constant temperature and constant pressure $\mathrm{d}T = \mathrm{d}p = 0$. Thus, the Gibbs free energy reduces to

$$(\mathrm{d}G)_{p,T} = \mu_{\mathbf{R}1}\mathrm{d}n_{\mathbf{R}1} + \mu_{\mathbf{R}2}\mathrm{d}n_{\mathbf{R}2} + \mu_{\mathbf{P}2}\mathrm{d}n_{\mathbf{P}2} + \mu_{\mathbf{P}1}\mathrm{d}n_{\mathbf{P}1} \ldots, \tag{2.1}$$

where n_j is the amount of component j (in mol) and μ_j is the chemical potential of component j defined as $\mu_j = \frac{\partial G}{\partial n_j}$ (in $\mathrm{J.mol^{-1}}$).

Number of molecules n_j in the system deviate from their initial values n^{i}_j while approaching their equilibrium values. Thus, $n_j = n^{\mathrm{i}}_j + \nu_j\xi$, where ξ is the extent of reaction (in moles). Hence, the Gibbs energy change in Eq. (2.1) can be written as $(\mathrm{d}G)_{p,T} = \sum_j \mu_j \mathrm{d}n_j = \sum_j \nu_j \mu_j \mathrm{d}\xi$. Moreover, the differential can be expressed as $(\mathrm{d}G)_{p,T} = \left(\frac{\partial G}{\partial \xi}\right)_{p,T} \mathrm{d}\xi$, therefore by comparison

$$\left(\frac{\partial G}{\partial \xi}\right)_{p,T} = \sum_j \nu_j \mu_j . \tag{2.2}$$

In textbooks [2], the right-hand side of Eq. (2.2) is denoted as $\Delta_{\mathrm{r}}G$ (in $\mathrm{J.mol^{-1}}$), in particular,

$$\Delta_{\mathrm{r}}G = \sum_j \nu_j \mu_j . \tag{2.3}$$

Note that $\Delta_{\mathrm{r}}G$ is defined as a difference of chemical potentials (multiplied by appropriate stoichiometric numbers) and equals to the derivative of the Gibbs energy.

Principle of minimum Gibbs potential states that any unconstrained internal parameter (in this case the reaction extent ξ) minimizes the Gibbs potential in equilibrium [4], yielding the condition for equilibrium

$$\left(\frac{\partial G}{\partial \xi}\right)_{p,T} = 0 \Rightarrow \Delta_{\mathrm{r}}G = \sum_j \nu_j \mu_j = 0 . \tag{2.4}$$

This is a general condition valid for liquids as well as gases and solids. If $\Delta_{\mathrm{r}}G < 0$, the reaction spontaneously proceeds in forward direction and if $\Delta_{\mathrm{r}}G > 0$, the reaction spontaneously proceeds in backward direction until equilibrium is

[1] Stoichiometric numbers https://goldbook.iupac.org/terms/view/S06025.

achieved at $\Delta_r G = 0$. Hence, the crucial property of the system of interest is the exact form of chemical potential as function of concentration of all present chemical species.

The chemical potential of jth component of multi-component ideal gas is $\mu_j = \mu_j^\circ(p, T) + RT \ln x_j$, where x_j is the molar fraction of jth component, R is the molar gas constant and T is temperature. This also holds for the ideal liquid solution, where the liquids do not interact [5]. For non-ideal liquid solution, the chemical potential can be written in similar manner as

$$\mu_j = \mu_j^\circ(p, T) + RT \ln a_j \,, \tag{2.5}$$

where a_j is the *activity* of jth component defined so that Eq. (2.5) is exactly valid. Values of the *standard chemical potential* μ_j° are by definition related to a particular standard state in which $a_j = 1$. The choice of standard state includes values of pressure, temperature and concentrations of the experimental solution. There are multiple options for the choice of standard state (see p. 30 in [3] or Sect. 5F in [1]). The activity can be directly related to concentration by

$$a_j = \gamma_j^c [j] \,, \tag{2.6}$$

where $[j]$ is the concentration of jth component (in mol^{-1}). The (molar concentration) activity coefficients γ_j^c are a correction for non-ideality of the solution. For more details, see the next section.

Using the equation for chemical potential, Eq. (2.5), the equilibrium condition in Eq. (2.4) is rewritten as

$$\ln \frac{a_{\mathbf{P1}}^{\nu_{\mathrm{P1}}} a_{\mathbf{P2}}^{\nu_{\mathrm{P2}}} \cdots}{a_{\mathbf{R1}}^{\nu_{\mathrm{R1}}} a_{\mathbf{R2}}^{\nu_{\mathrm{R2}}} \cdots} = -\frac{\sum_j \nu_j \mu_j^\circ}{RT} \,. \tag{2.7}$$

Although the activities depend on concentration, the right-hand side and consequently the left-hand side depend only on pressure and temperature. Therefore, we define the concentration-independent *equilibrium constant* as

$$K_{\mathrm{eq}} = \exp\left(-\frac{\Delta_r G^\circ}{RT}\right) \,, \tag{2.8}$$

where $\Delta_r G^\circ = \sum_j \nu_j \mu_j^\circ$ is the *standard Gibbs energy of reaction* (in J.mol^{-1}). Hence, the standard Gibbs energy of reaction can be calculated from the equilibrium constant using

$$\Delta_r G^\circ = -RT \ln K_{\mathrm{eq}} \,. \tag{2.9}$$

The equilibrium constant binds together activities of all components in equilibrium,

$$K_{\mathrm{eq}} = \frac{a_{\mathbf{P1}}^{\nu_{\mathrm{P1}}} a_{\mathbf{P2}}^{\nu_{\mathrm{P2}}} \cdots}{a_{\mathbf{R1}}^{\nu_{\mathrm{R1}}} a_{\mathbf{R2}}^{\nu_{\mathrm{R2}}} \cdots} . \tag{2.10}$$

For dilute solutions ($\gamma_j^c = 1\ \mathrm{mol}^{-1}$) the equilibrium constant is equal to

$$K_{\mathrm{eq}} = \frac{[\mathbf{P1}]^{\nu_{\mathrm{P1}}} [\mathbf{P2}]^{\nu_{\mathrm{P2}}} \cdots}{[\mathbf{R1}]^{\nu_{\mathrm{R1}}} [\mathbf{R2}]^{\nu_{\mathrm{R2}}} \cdots} , \tag{2.11}$$

in this case K_{eq} is not a dimensionless quantity.

To conclude, value of $\Delta_{\mathrm{r}}G$ decides if the reaction is in equilibrium or proceeds in forward/backward direction. On the other hand, $\Delta_{\mathrm{r}}G^\circ$ determines the value of the equilibrium constant K_{eq} and consequently the equilibrium concentration of all chemical species. Because $\Delta_{\mathrm{r}}G^\circ = \Delta_{\mathrm{r}}H^\circ - T\Delta_{\mathrm{r}}S^\circ$, it is possible to determine the energetic contribution $\Delta_{\mathrm{r}}H^\circ$ (in J.mol^{-1}) and entropic contribution $\Delta_{\mathrm{r}}S^\circ$ (in J.K^{-1}.mol^{-1}) from temperature dependence of K_{eq}. If $\Delta_{\mathrm{r}}H^\circ$ and $\Delta_{\mathrm{r}}S^\circ$ are independent of temperature, plot of $\ln K_{\mathrm{eq}}$ versus $1/T$ derived from Eq. (2.9) yields straight line with the slope $-\Delta_{\mathrm{r}}H^\circ/R$ and intercept $\Delta_{\mathrm{r}}S^\circ/R$ according to

$$\ln K_{\mathrm{eq}} = -\frac{\Delta_{\mathrm{r}}H^\circ}{R}\frac{1}{T} + \frac{\Delta_{\mathrm{r}}S^\circ}{R} . \tag{2.12}$$

An equilibrium constant is defined not only for the overall reaction scheme according to Eq. (2.10) but also for every elementary reaction. Each elementary reaction has its own independent equilibrium constant and corresponding standard Gibbs energy of reaction.

2.1.2 Meaning of Activity Coefficients

Activity coefficients can be defined in multiple ways, depending on the use of a particular concentration scale (see p. 33 in [3]). Besides the molar concentration scale ($[j] = n_j/V$ [mol/l]) implying $a_j = \gamma_j^c[j]$, also molar fraction scale ($x_j = n_j/\sum n_j$ [mol/mol]) implying $a_j = \gamma_j^x x_j$ can be used. The latter approach is useful for interpretation of activity coefficients. The Gibbs energy of multicomponent solution $G = \sum_j n_j \mu_j$ can be reformulated as

$$\frac{G(T, p, n_j)}{n_{\mathrm{t}}} = \underbrace{\sum_j x_j \mu_j^\circ(T, p, n_j)}_{\text{pure components}} + \underbrace{RT\sum_j x_j \ln x_j}_{\text{mixing entropy}} + \underbrace{RT\sum_j x_j \ln \gamma_j^x}_{\text{non-ideality}} , \tag{2.13}$$

Fig. 2.1 Activity coefficients for benzene-acetonitrile solution at 298 K obtained from vapor pressure measurements at vapor-liquid equilibrium. Values adopted from [7]

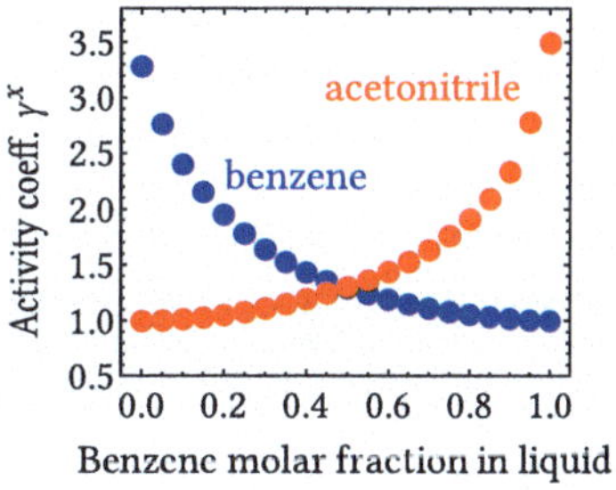

where $n_t = \sum n_j$ is the total concentration. The first two terms in Eq. (2.13) correspond to an ideal solution including mixing entropy but not mixing enthalpy (non-interacting particles). The last term stands for enthalpic and entropic contributions from non-ideality, which explains the physical meaning of activity coefficients. If $\gamma_j^x = 1$, the solution behaves as an ideal solution. This expression for Gibbs energy including activity coefficients is important not only for chemical reactions but also for other thermodynamic applications involving chemical potential, e.g. vapor pressure in vapor-liquid equilibrium or construction of phase diagrams [6].

Values of activity coefficients and exact meaning of μ_j° depend on the choice of reference state (defined at $\gamma_j^x = 1$) and standard state (at $a_j = 1$) (see p. 28 in [3]). If the reference state is defined at pure components, then the limit $x_j \to 1$ implies $\gamma_j^x \to 1$, and μ_j° is equal to molar Gibbs energy of molecules of the pure component j. On the other hand, if the reference state is at infinitely diluted components (solutes), then $x_j \to 0$ implies $\gamma_j^x \to 1$ (or $[j] \to 0$ implies $\gamma_j^c \to 1$). The latter case is applied in the definition of the equilibrium constant in Eq. (2.11). Illustration of actual measured activity coefficients γ_j^x in two-component solution is given in Fig. 2.1. Activity coefficients of solution mixtures can be obtained experimentally by vapor-liquid equilibria measurement (a standard technique) or other methods, such as IR spectroscopy [7–9].

The activity coefficients depend on temperature, pressure and $N_{\mathrm{spc}} - 1$ concentration/molar fraction values, where N_{spc} is the number of present chemical species. Gibbs-Duhem relation [4, 5] at constant temperature and pressure, $\sum n_j \mathrm{d}\mu_j = 0$, binds together chemical potentials of a multi-component mixture. After the substitution $\mu_j = \mu_j^\circ + RT \ln \gamma_j^x x_j$, the activity coefficients are related by (see p. 52 in [5])

$$\sum x_j \mathrm{d} \ln \gamma_j^x = 0\,. \tag{2.14}$$

For example, for two-component system with γ^x as a function of x_2 in the form $\gamma_1^x = \gamma_1^x(x_2)$, $\gamma_2^x = \gamma_2^x(x_2)$ it follows that (see p. 53 in [5])

$$\frac{\mathrm{d} \ln \gamma_2^x}{\mathrm{d}x_2} = -\frac{x_1}{x_2} \frac{\mathrm{d} \ln \gamma_1^x}{\mathrm{d}x_2}\,. \tag{2.15}$$

This condition must be obeyed for construction of models for activity coefficients. Simple Margules model employs polynomials in x_j, but for practical purposes, more realistic models are used, e.g. NRTL (non-random two-liquid), UNIQUAC (universal quasichemical) or UNIFAC (UNIQUAC Functional-group Activity Coefficients) models [5, 6, 10]. Dortmund Data Bank is a database of thermodynamic properties, where the activity coefficients of many chemical mixtures can be obtained [11].

2.1.3 The Rate Law

The central relation governing rates in chemical kinetics, *law of mass action*, was discovered in the second half of 19th century by contributions of Guldberg, Waage, van't Hoff and Horstmann [12, 13]. This law describes elementary reactions (not overall reactions) as collisions between the reacting molecules. The most common types of reactions are unimolecular reactions (e.g. decay of a complex or a conformational change) and bimolecular reactions (e.g. complex formation). Probability of the collisions is proportional to concentration of both reactants. If we consider a bimolecular reaction $\mathbf{R}_1 + \mathbf{R}_2 \rightleftharpoons \mathbf{P}_1 + \mathbf{P}_2$, then formally

$$\begin{aligned} \text{forward reaction rate} &= \kappa_{\text{fwd}}[\mathbf{R}_1][\mathbf{R}_2]\,, \\ \text{backward reaction rate} &= \kappa_{\text{rev}}[\mathbf{P}_1][\mathbf{P}_2]\,, \end{aligned} \tag{2.16}$$

where κ_{fwd} and κ_{rev} are the *reaction rate coefficients* for forward and backward (reverse) chemical reaction, respectively.[2] These coefficients depend on temperature and pressure but not on concentrations of the chemical species in the system. Difference of forward and backward reaction rates defines the *reaction rate* v [14] as

$$v = \kappa_{\text{fwd}}[\mathbf{R}_1][\mathbf{R}_2] - \kappa_{\text{rev}}[\mathbf{P}_1][\mathbf{P}_2]\,. \tag{2.17}$$

Time change of a particular chemical species concentration can be obtained from reaction rates of corresponding elementary reactions, for example in the case of one elementary reaction, $\frac{\mathrm{d}}{\mathrm{d}t}[\mathbf{R}_1] = -v$ or $\frac{\mathrm{d}}{\mathrm{d}t}[\mathbf{P}_2] = v$. Then, construction of the governing differential equations is straightforward,

[2] Although reaction rate coefficients $\kappa_\bullet$ are usually denoted in the literature as '$k_\bullet$', here, we apply the Greek letter 'κ' to avoid confusion with transition rate coefficients defined in Chap. 3.

$$\begin{aligned}
\frac{d[\mathbf{R}_1]}{dt} &= -\kappa_{\text{fwd}}[\mathbf{R}_1][\mathbf{R}_2] + \kappa_{\text{rev}}[\mathbf{P}_1][\mathbf{P}_2]\,, \\
\frac{d[\mathbf{R}_2]}{dt} &= -\kappa_{\text{fwd}}[\mathbf{R}_1][\mathbf{R}_2] + \kappa_{\text{rev}}[\mathbf{P}_1][\mathbf{P}_2]\,, \\
\frac{d[\mathbf{P}_1]}{dt} &= -\kappa_{\text{rev}}[\mathbf{P}_1][\mathbf{P}_2] + \kappa_{\text{fwd}}[\mathbf{R}_1][\mathbf{R}_2]\,, \\
\frac{d[\mathbf{P}_2]}{dt} &= -\kappa_{\text{rev}}[\mathbf{P}_1][\mathbf{P}_2] + \kappa_{\text{fwd}}[\mathbf{R}_1][\mathbf{R}_2]\,.
\end{aligned} \tag{2.18}$$

It is a set of first order nonlinear differential equations, which can be solved numerically. Notably, the equilibrium constant in dilute limit can be derived from the chemical kinetics because in equilibrium both reaction rates are equal, i.e. $v = 0$ (or equivalently, time derivatives in Eq. (2.18) are equal to zero), yielding

$$K_{\text{eq}} = \frac{\kappa_{\text{fwd}}}{\kappa_{\text{rev}}} = \frac{[\mathbf{P}_1][\mathbf{P}_2]}{[\mathbf{R}_1][\mathbf{R}_2]}\,. \tag{2.19}$$

All governing reactions must conserve the total mass, hence summing all subequations in Eq. (2.18) must yield $\frac{d}{dt}\left([\mathbf{R}_1] + [\mathbf{R}_2] + [\mathbf{P}_1] + [\mathbf{P}_2]\right) = 0$, i.e. all right-hand side terms of the subequations must sum to zero. The rate law is applicable to describe chemical kinetics in wide range of systems. For example, in enzyme-catalyzed reactions the chemical kinetics usually assumes short initial pre-steady-state mode (v dramatically changes in time) followed by approximate steady-state (v approximately constant) [15, 16].

2.1.4 Eyring Equation

Temperature dependence of reaction rate coefficients is obtained from the transition state theory of chemical reactions [1, 2, 17, 18]. Energetic scheme for bimolecular reaction is given in the following scheme (Scheme 2.2).

The reaction coordinate is an abstract 1D representation of geometric parameters of the two reacting molecules along the reaction pathway. The forward reaction

Scheme 2.2 Transition state in an elementary bimolecular reaction

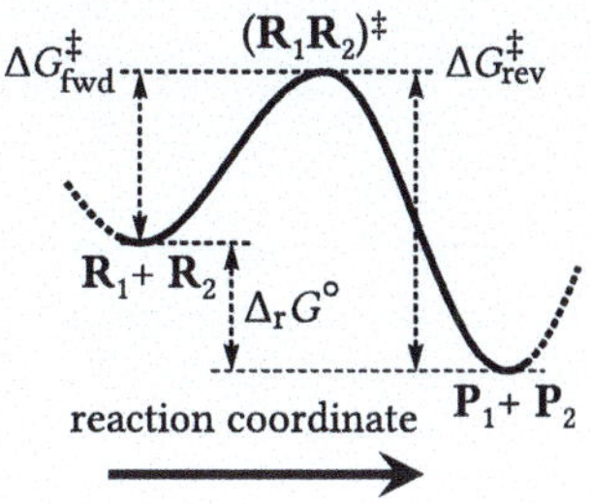

progresses when reactants $\mathbf{R}_1$ and $\mathbf{R}_2$ collide with energy big enough to overcome the barrier $\Delta G^{\ddagger}_{\mathrm{fwd}}$ and reach the transition state $(\mathbf{R}_1\mathbf{R}_2)^{\ddagger}$ and vice versa for the backward reaction. The transition state is a saddle point in the potential energy surface (depending on all atomic coordinates in the system), while states $\mathbf{R}_1 + \mathbf{R}_2$ or $\mathbf{P}_1 + \mathbf{P}_2$ are minima.

The reaction rate coefficients κ_{fwd} and κ_{rev} are directly related to barrier heights $\Delta G^{\ddagger}_{\mathrm{fwd}}$ and $\Delta G^{\ddagger}_{\mathrm{rev}}$, respectively according to the Eyring equation [19]

$$\kappa_j = \frac{\eta k_{\mathrm{Boltz}} T}{h} \exp\left(-\frac{\Delta G^{\ddagger}_j}{RT}\right), \tag{2.20}$$

where $\Delta G^{\ddagger}_j$ is the Gibbs energy barrier (in J.mol^{-1}) corresponding to κ_j, η is the transition probability, h is the Planck constant and R is the molar gas constant. The transition probability η describes how likely the system moves to the other minimum after it reaches the transition state, it is usually set to unity because it cannot be measured experimentally. The Eyring equation provides temperature dependence of reaction rate coefficients.

Relationship of the equilibrium constant and reaction rate coefficients $K_{\mathrm{eq}} = \kappa_{\mathrm{fwd}}/\kappa_{\mathrm{rev}}$ (for dilute solutions, cf. Eq. (2.19)) provides the relationship between the barrier heights and the standard reaction Gibbs energy. Using the definition of equilibrium constant (Eq. (2.8)) and the Eyring equation (2.20) for κ_{fwd} and κ_{rev}, it follows that

$$\exp\left(-\frac{\Delta_{\mathrm{r}} G^{\circ}}{RT}\right) = K_{\mathrm{eq}} = \frac{\kappa_{\mathrm{fwd}}}{\kappa_{\mathrm{rev}}} = \exp\left(-\frac{\Delta G^{\ddagger}_{\mathrm{fwd}} - \Delta G^{\ddagger}_{\mathrm{rev}}}{RT}\right). \tag{2.21}$$

Comparison of the first and last expressions yields

$$\Delta_{\mathrm{r}} G^{\circ} = \Delta G^{\ddagger}_{\mathrm{fwd}} - \Delta G^{\ddagger}_{\mathrm{rev}} \tag{2.22}$$

implying that $\Delta_{\mathrm{r}} G^{\circ}$ is the difference between minima in the Gibbs energy profile corresponding to $\mathbf{R}_1 + \mathbf{R}_2$ and $\mathbf{P}_1 + \mathbf{P}_2$, see Scheme 2.2.

Variable temperature measurements are a standard means to obtain the value of energy barrier of a molecular process. Similarly to the standard reaction Gibbs energy, the energy barrier can be decomposed to its enthalpic and entropic contributions,

$$\Delta G^{\ddagger}_j = \Delta H^{\ddagger}_j - T \Delta S^{\ddagger}_j. \tag{2.23}$$

Using this decomposition in the Eyring equation, the measured dependence of $\kappa_j(T)$ can be visualized in Eyring plot of $\ln(\kappa_j/T)$ versus $1/T$. Assuming $\Delta H^{\ddagger}_j$ and $\Delta S^{\ddagger}_j$ are constant along the reaction coordinate, the Eyring plot is a straight line because

$$\ln \frac{\kappa_j}{T} = -\frac{\Delta H^{\ddagger}_j}{R} \frac{1}{T} + \ln \frac{k_{\mathrm{Boltz}}}{h} + \frac{\Delta S^{\ddagger}_j}{R} \tag{2.24}$$

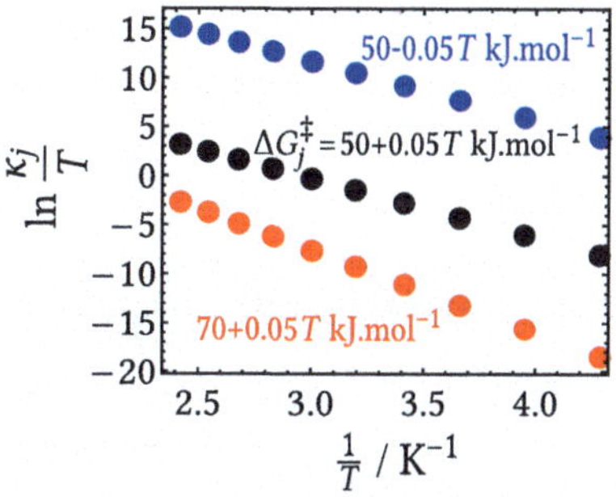

Fig. 2.2 Illustration of Eyring plot for various values of $\Delta G_j^\ddagger = \Delta H_j^\ddagger - T\Delta S_j^\ddagger$. Temperatures range from −40 to 140 °C with 20 °C step

as illustrated in Fig. 2.2. Here, we assume $\eta = 1$.

Values of $\Delta H_j^\ddagger$ and $\Delta S_j^\ddagger$ can be obtained from slope and intercept, respectively, from the Eyring plot according to Eq. (2.24), see details in Appendix C.6. Possible deviations from linearity in the Eyring plot can be caused by the presence of multiple processes with different energy barriers, which sum to the observed apparent reaction rate coefficient [15]. Another possible cause is dependence of $\Delta H_j^\ddagger$ and $\Delta S_j^\ddagger$ on temperature, which can be modeled by additional terms with molar heat capacity change $\Delta C_p^\ddagger$ in the Eyring equation [20, 21].

2.2 Host-Guest Binding

In many cases of supramolecular binding, the interacting molecules are denoted as host (usually a large molecule) and guest (usually a small molecule), nevertheless, the choice is rather arbitrary. Common binding mechanisms are, for example, hydrogen bonding, ion pairing or formation of coordination complexes (metal-ligand interaction) [22]. The host molecule often contains hydrogen bond donor sites or Lewis basic donor atoms (with lone electron pair). On the other hand, the guest molecule often contains hydrogen bond acceptor sites or Lewis acidic acceptor atoms [23]. Examples of host molecules include proteins, cage molecules, cyclic molecules with central cavity (e.g. crown ethers or cyclodextrin) or zeolites. Guest molecules include, for example, metal cations, halide anions, hormones, pheromones and other small organic molecules with functional group(s) [23].

During the construction of host-guest binding models in the following sections, infinite dilution will be assumed and consequently, all activity coefficients are set to unity (i.e. $\gamma_j = 1$). As a result, these models cannot describe the host-guest binding in the whole concentration range. The formalism of host-guest binding can be readily used for description of protonation with an acid in nonpolar medium, where the cation and anion originating from the acid are held closely together due to electrostatic forces. In polar medium (e.g. in water), the cation and anion strongly interact with the medium thus moving freely in the solution [24]. In addition, ionic species in polar solvent strongly behave non-ideally and the approximation $\gamma_j \approx 1$ is not valid even for dilute solutions.

2.2.1 H:G 1:1 Binding Model

In this work, host molecules are consistently denoted as **H** and guest molecules as **G**. Although host-guest complexes $\mathbf{H}_m\mathbf{G}_n$ with arbitrary stoichiometry can generally exist, in this work we assume the presence of only one host in the complex. The most elementary interaction is formation of **HG** complex with 1:1 **H**:**G** stoichiometry described by the following scheme (Fig. 2.3).

Binding strength of the complex is characterized by the equilibrium constant $K_{\mathbf{HG}}$ defined as $K_{\mathbf{HG}} = \exp\left(-\frac{\Delta_r G^\circ}{RT}\right)$, where $\Delta_r G^\circ = \mu^\circ_{\mathbf{HG}} - \mu^\circ_{\mathbf{H}} - \mu^\circ_{\mathbf{G}}$ (see Eq. (2.4)). According to Eq. (2.11), the equilibrium constant is related to concentrations of the chemical species. Together with the balance of total mass, the complete set of equations is obtained,

$$K_{\mathbf{HG}} = \frac{[\mathbf{HG}]}{[\mathbf{H}][\mathbf{G}]}, \tag{2.25a}$$

$$[\mathbf{H}]_t = [\mathbf{H}] + [\mathbf{HG}], \tag{2.25b}$$

$$[\mathbf{G}]_t = [\mathbf{G}] + [\mathbf{HG}], \tag{2.25c}$$

where $[\bullet]_t$ denotes total concentration. Then the concentrations can be calculated analytically as

$$[\mathbf{H}] = \frac{K_{\mathbf{HG}}([\mathbf{H}]_t - [\mathbf{G}]_t) - 1 + \sqrt{[K_{\mathbf{HG}}([\mathbf{H}]_t - [\mathbf{G}]_t) - 1]^2 + 4K_{\mathbf{HG}}[\mathbf{H}]_t}}{2K_{\mathbf{HG}}}, \tag{2.26a}$$

$$[\mathbf{HG}] = \frac{K_{\mathbf{HG}}([\mathbf{H}]_t + [\mathbf{G}]_t) + 1 - \sqrt{[K_{\mathbf{HG}}([\mathbf{H}]_t - [\mathbf{G}]_t) - 1]^2 + 4K_{\mathbf{HG}}[\mathbf{H}]_t}}{2K_{\mathbf{HG}}}, \tag{2.26b}$$

$$[\mathbf{G}] = \frac{K_{\mathbf{HG}}([\mathbf{G}]_t - [\mathbf{H}]_t) - 1 + \sqrt{[K_{\mathbf{HG}}([\mathbf{H}]_t - [\mathbf{G}]_t) - 1]^2 + 4K_{\mathbf{HG}}[\mathbf{H}]_t}}{2K_{\mathbf{HG}}}. \tag{2.26c}$$

Scheme 2.3 Host-guest binding scheme (**H**:**G** 1:1 stoichiometry)

$$\mathbf{H} + \mathbf{G} \overset{K_{\mathbf{HG}}}{\rightleftharpoons} \mathbf{HG}$$

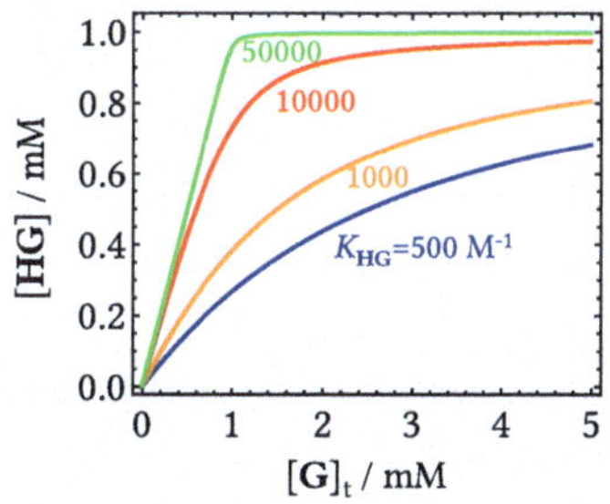

Fig. 2.3 Concentration of complexed host as function of $[\mathbf{G}]_t$ in host-guest 1:1 binding model for different equilibrium constants. Total host concentration $[\mathbf{H}]_t = 1$ mM is constant

The above equations enable calculation of concentrations of all species when the total concentrations and $K_{\mathbf{HG}}$ value are known. On the other hand, measurement of concentration of the chemical species can be used for determination of $K_{\mathbf{HG}}$. The effect of different $K_{\mathbf{HG}}$ values is illustrated in Fig. 2.3. Implementation of Eq. (2.26) in *Mathematica* is given in Appendix A.1.

The host-guest binding model does not account for interactions with the medium **M** (i.e. solvent), which can be approximately described as $\mathbf{HM} + \mathbf{GM} \rightleftharpoons \mathbf{HG} + \mathbf{MM}$. Hence, the host-guest binding model is relevant only in situations, where the contributions of medium-medium and medium-solute (i.e. medium-host and medium-guest) interaction energies to the standard reaction Gibbs energy are negligible. Consequently, host-guest models cannot be applied in the case of weak binding corresponding to values of $K_{\mathbf{HG}} \lessapprox 1$ (see p. 6 and 13 in [3]). For example, hydrophobic association in polar solvents is governed mainly by strong medium-medium interactions.

2.2.2 *Aspects of Multiple Binding Sites*

Host molecules can possess more than one binding site for the guest. Let us consider two binding sites denoted as a and b, which enables formation of four chemical species for the host, namely **H**, $\mathbf{HG}_a$, $\mathbf{HG}_b$ and $\mathbf{HG}_2$ as shown in Scheme 2.4a. The species $\mathbf{HG}_a$ and $\mathbf{HG}_b$ are denoted according to the occupied binding site.

Each binding step in Scheme 2.4a is described by a *microscopic binding constant* defined as

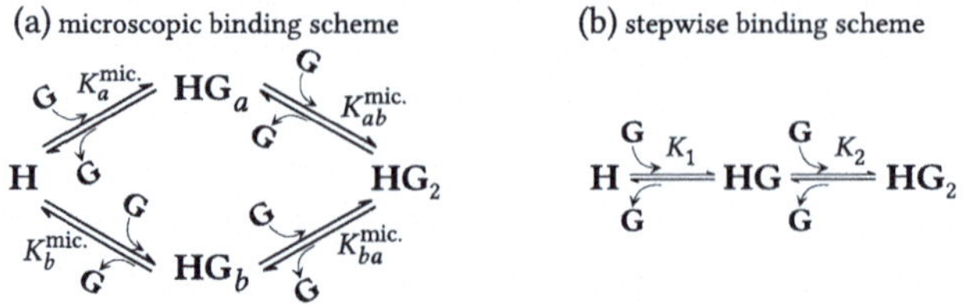

Scheme 2.4 Host-guest 1:2 binding model. (**a**) microscopic binding scheme described by microscopic binding constants, (**b**) stepwise binding scheme described by stepwise binding constants

Table 2.1 Number of microscopic and stepwise binding constants for a host-guest system, where the host molecule contains n binding sites

	Total No. microscopic	No. independent microscopic	No. stepwise
$n = 2$	4	3	2
$n = 3$	12	7	3
$n = 4$	32	15	4
$n = 5$	80	31	5
$n = 6$	192	63	6
General n	$n2^{n-1}$	$2^n - 1$	n

$$K_a^{\text{mic.}} = \frac{[\mathbf{HG}_a]}{[\mathbf{H}][\mathbf{G}]}, \tag{2.27a}$$

$$K_{ab}^{\text{mic.}} = \frac{[\mathbf{HG}_2]}{[\mathbf{HG}_a][\mathbf{G}]}, \tag{2.27b}$$

$$K_b^{\text{mic.}} = \frac{[\mathbf{HG}_b]}{[\mathbf{H}][\mathbf{G}]}, \tag{2.27c}$$

$$K_{ba}^{\text{mic.}} = \frac{[\mathbf{HG}_2]}{[\mathbf{HG}_b][\mathbf{G}]}. \tag{2.27d}$$

The total number of microscopic binding constants increases rapidly with the number of binding sites n, but not all of them are independent [25], see Table 2.1.

It is often impossible to distinguish between binding at site a or site b experimentally. Therefore, it is convenient to introduce *stepwise binding constants* as illustrated in Scheme 2.4b. The binding constant K_1 accounts for binding at either of the two sites, and K_2 accounts for binding of a second guest molecule at either $\mathbf{HG}_a$ or $\mathbf{HG}_b$, formally

$$K_1 = \frac{[\mathbf{HG}_a] + [\mathbf{HG}_b]}{[\mathbf{H}][\mathbf{G}]} = \frac{[\mathbf{HG}]}{[\mathbf{H}][\mathbf{G}]}, \tag{2.28a}$$

$$K_2 = \frac{[\mathbf{HG}_2]}{([\mathbf{HG}_a] + [\mathbf{HG}_b])\,[\mathbf{G}]} = \frac{[\mathbf{HG}_2]}{[\mathbf{HG}][\mathbf{G}]}, \tag{2.28b}$$

where $[\mathbf{HG}] = [\mathbf{HG}_a] + [\mathbf{HG}_b]$. In a general binding scheme, a stepwise binding constant K_j describes binding of j-th guest to the complex $\mathbf{HG}_{j-1}$ according to the definition

$$K_j = \frac{[\mathbf{HG}_j]}{[\mathbf{HG}_{j-1}][\mathbf{G}]} \,. \tag{2.29}$$

The stepwise binding constants are uniquely determined from the (independent) microscopic binding constants but not vice versa because the number of the latter is always higher then the former, see Table 2.1. Thus, microscopic binding constants cannot be calculated from the stepwise binding constants and are mostly experimentally inaccessible except for symmetric systems, see Sect. 2.2.4 (although they have been determined in an asymmetric two-site system from measurements with many different guests [26]). Stepwise binding constants are sufficient for construction of binding models describing experimental data. However, microscopic binding constants describe the interaction between binding sites during cooperative binding as explained later in Sect. 2.2.4.

For an overall reaction $\mathbf{H} + \mathrm{n}\mathbf{G} \rightleftharpoons \mathbf{HG}_\mathrm{n}$, the *overall binding constant* is defined as

$$\beta_j = \frac{[\mathbf{HG}_\mathrm{j}]}{[\mathbf{H}][\mathbf{G}]^j} = \prod_{j=1}^{n} K_j \,, \tag{2.30}$$

where $j = 1 \ldots n$ (cf. p. 47 in [3]). The overall binding constant is a product of stepwise binding constants. In some cases, one or more complexes $\mathbf{HG}_j$ are present at undetectable concentrations in the sample solution (e.g. for highly cooperative binding steps, see Sect. 2.2.4). Then, the corresponding stepwise binding constants cannot be determined, but the binding can still be described using the product of corresponding stepwise binding constants.

It is impossible to find an analytical solution for concentrations of all species in an overall reaction $\mathbf{H} + \mathrm{n}\mathbf{G} \rightleftharpoons \mathbf{HG}_\mathrm{n}$ for arbitrary n. Thus, following the approach given in the paper of Hargrove [27], the concentrations of host-related species (i.e. $[\mathbf{H}]$ and $[\mathbf{HG}_j]$) and $[\mathbf{G}]_\mathrm{t}$ are expressed as functions of $[\mathbf{H}]_\mathrm{t}$ and $[\mathbf{G}]$. Subsequently, the free guest concentration is calculated numerically from a polynomial equation in the form $f([\mathbf{G}];\, K_j,\, [\mathbf{H}]_\mathrm{t},\, [\mathbf{G}]_\mathrm{t}) = 0$. Here, we give an example of host-guest binding with 1:4 stoichiometry. The equations for stepwise binding constants and total mass balance are

$$K_j = \frac{[\mathbf{HG}_j]}{[\mathbf{HG}_{j-1}][\mathbf{G}]}, \quad \text{for } j = 1, \ldots, 4 \,, \tag{2.31a}$$

$$[\mathbf{H}]_\mathrm{t} = [\mathbf{H}] + [\mathbf{HG}] + [\mathbf{HG}_2] + [\mathbf{HG}_3] + [\mathbf{HG}_4] \,, \tag{2.31b}$$

$$[\mathbf{G}]_\mathrm{t} = [\mathbf{G}] + [\mathbf{HG}] + 2[\mathbf{HG}_2] + 3[\mathbf{HG}_3] + 4[\mathbf{HG}_4] \,. \tag{2.31c}$$

The concentrations of host-related species have a simple functional dependence on the free guest concentration as calculated from Eq. (2.31),

$$[\mathbf{H}] = \frac{[\mathbf{H}]_t}{1 + K_1[\mathbf{G}] + K_1K_2[\mathbf{G}]^2 + K_1K_2K_3[\mathbf{G}]^3 + K_1K_2K_3K_4[\mathbf{G}]^4}, \quad (2.32a)$$

$$[\mathbf{HG}] = \frac{K_1[\mathbf{G}][\mathbf{H}]_t}{1 + K_1[\mathbf{G}] + K_1K_2[\mathbf{G}]^2 + K_1K_2K_3[\mathbf{G}]^3 + K_1K_2K_3K_4[\mathbf{G}]^4}, \quad (2.32b)$$

$$[\mathbf{HG}_2] = \frac{K_1K_2[\mathbf{G}]^2[\mathbf{H}]_t}{1 + K_1[\mathbf{G}] + K_1K_2[\mathbf{G}]^2 + K_1K_2K_3[\mathbf{G}]^3 + K_1K_2K_3K_4[\mathbf{G}]^4}, \quad (2.32c)$$

$$[\mathbf{HG}_3] = \frac{K_1K_2K_3[\mathbf{G}]^3[\mathbf{H}]_t}{1 + K_1[\mathbf{G}] + K_1K_2[\mathbf{G}]^2 + K_1K_2K_3[\mathbf{G}]^3 + K_1K_2K_3K_4[\mathbf{G}]^4}, \quad (2.32d)$$

$$[\mathbf{HG}_4] = \frac{K_1K_2K_3K_4[\mathbf{G}]^4[\mathbf{H}]_t}{1 + K_1[\mathbf{G}] + K_1K_2[\mathbf{G}]^2 + K_1K_2K_3[\mathbf{G}]^3 + K_1K_2K_3K_4[\mathbf{G}]^4}. \quad (2.32e)$$

In the next step, Eq. (2.32a–e) are substituted into Eq. (2.31c) and rearranged, yielding the aforementioned equation $f([\mathbf{G}]; K_j, [\mathbf{H}]_t, [\mathbf{G}]_t) = 0$ in the form

$$K_1K_2K_3K_4[\mathbf{G}]^5 + K_1K_2K_3(4K_4[\mathbf{H}]_t - K_4[\mathbf{G}]_t + 1)[\mathbf{G}]^4 + K_1K_2(3K_3[\mathbf{H}]_t - K_3[\mathbf{G}]_t + 1)[\mathbf{G}]^3$$
$$+ K_1(2K_2[\mathbf{H}]_t - K_2[\mathbf{G}]_t + 1)[\mathbf{G}]^2 + (K_1[\mathbf{H}]_t - K_1[\mathbf{G}]_t + 1)[\mathbf{G}] - [\mathbf{G}]_t = 0. \quad (2.33)$$

Concentration $[\mathbf{G}]$ is then calculated numerically from this polynomial equation in the interval $[\mathbf{G}] \in [0, [\mathbf{G}]_t]$, concentrations of other species are obtained from Eq. (2.32). Implementation and derivation of the binding models for up to four binding sites in *Mathematica* is given in Appendix A.2–A.4.

2.2.3 Competitive Host-Ligand Binding

Competitive binding is a situation where two different ligands bind to a single binding site at the host molecule (and the binding of one ligand prevents binding of another ligand). Assuming the host molecule possesses one binding site (1:1 binding stoichiometry) and denoting the competitive ligands as **G** (guest) and **W** (for example, water), the binding can be described by two binding constants $K_{\mathbf{HG}}$ and $K_{\mathbf{HW}}$ according to Scheme 2.5 [15, 28].

The corresponding equations for binding constants and mass balance are

Scheme 2.5 Competitive 1:1 host-ligand binding scheme with ligands **G** and **W**

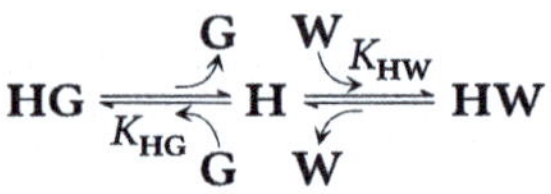

$$K_{\mathrm{HG}} = \frac{[\mathbf{HG}]}{[\mathbf{H}][\mathbf{G}]}, \tag{2.34a}$$

$$K_{\mathrm{HW}} = \frac{[\mathbf{HW}]}{[\mathbf{H}][\mathbf{W}]}, \tag{2.34b}$$

$$[\mathbf{H}]_{\mathrm{t}} = [\mathbf{H}] + [\mathbf{HG}] + [\mathbf{HW}], \tag{2.34c}$$

$$[\mathbf{G}]_{\mathrm{t}} = [\mathbf{G}] + [\mathbf{HG}], \tag{2.34d}$$

$$[\mathbf{W}]_{\mathrm{t}} = [\mathbf{W}] + [\mathbf{HW}]. \tag{2.34e}$$

Numeric solution, analogous to Eqs. (2.32) and (2.33), requires solving a system of two second-order polynomial equations in $[\mathbf{G}]$ and $[\mathbf{W}]$, see Appendix A.5.1. Here, we show another solution, which leads to numeric treatment of third-order polynomial equation in $[\mathbf{H}]$. First, the following equations

$$[\mathbf{HG}] = \frac{K_{\mathrm{HG}}[\mathbf{H}][\mathbf{G}]_{\mathrm{t}}}{1 + K_{\mathrm{HG}}[\mathbf{H}]}, \tag{2.35a}$$

$$[\mathbf{HW}] = \frac{K_{\mathrm{HW}}[\mathbf{H}][\mathbf{W}]_{\mathrm{t}}}{1 + K_{\mathrm{HW}}[\mathbf{H}]}, \tag{2.35b}$$

$$[\mathbf{G}] = \frac{[\mathbf{G}]_{\mathrm{t}}}{1 + K_{\mathrm{HG}}[\mathbf{H}]}, \tag{2.35c}$$

$$[\mathbf{W}] = \frac{[\mathbf{W}]_{\mathrm{t}}}{1 + K_{\mathrm{HW}}[\mathbf{H}]} \tag{2.35d}$$

are obtained by combination of Eqs. (2.34a, b) and (2.34d, e). Then, substitution of Eq. (2.35a,b) into Eq. (2.34c) yields the aforementioned third-order polynomial equation in $[\mathbf{H}]$,

$$[\mathbf{H}]^3 K_{\mathrm{HG}} K_{\mathrm{HW}} + [\mathbf{H}]^2 (K_{\mathrm{HG}} + K_{\mathrm{HW}} + [\mathbf{G}]_{\mathrm{t}} K_{\mathrm{HG}} K_{\mathrm{HW}} - [\mathbf{H}]_{\mathrm{t}} K_{\mathrm{HG}} K_{\mathrm{HW}} + K_{\mathrm{HG}} K_{\mathrm{HW}} [\mathbf{W}]_{\mathrm{t}}) \\ + [\mathbf{H}](1 + [\mathbf{G}]_{\mathrm{t}} K_{\mathrm{HG}} - [\mathbf{H}]_{\mathrm{t}} K_{\mathrm{HG}} - [\mathbf{H}]_{\mathrm{t}} K_{\mathrm{HW}} + K_{\mathrm{HW}} [\mathbf{W}]_{\mathrm{t}}) - [\mathbf{H}]_{\mathrm{t}} = 0. \tag{2.36}$$

This equation is solved numerically in the interval $[\mathbf{H}] \in [0, [\mathbf{H}]_{\mathrm{t}}]$. The resulting value of $[\mathbf{H}]$ is then substituted into Eq. (2.35), yielding the concentrations of other chemical species. Implementation of this solution and derivation of the corresponding equations in *Mathematica* is given in Appendix A.5.2.

2.2.4 Cooperativity

Cooperative binding relates to host molecules with multiple binding sites. It arises, when the presence of a bound guest increases or decreases affinity to bind another guest (cf. [29] and p. 78 in [3]). Let us illustrate this phenomenon with the host-guest binding model, where the host molecule possesses two binding sites *a* and *b*.

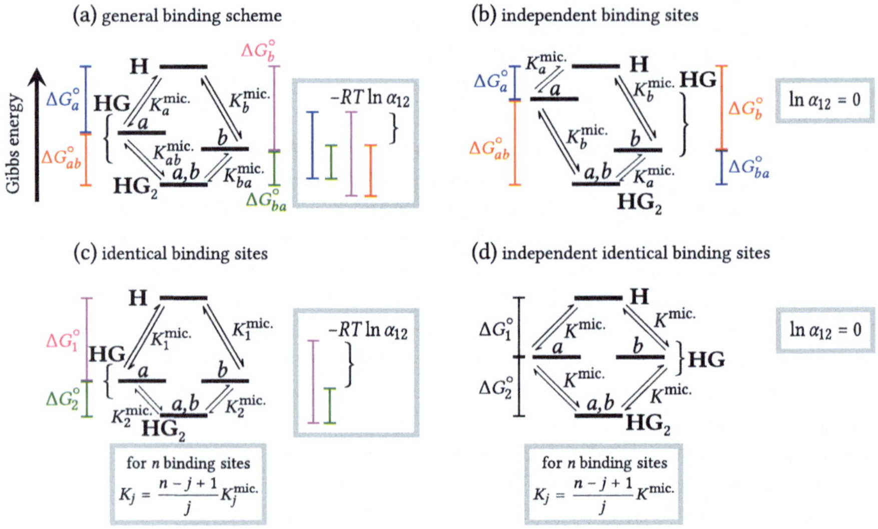

Scheme 2.6 Gibbs energy profile for host-guest 1:2 binding model. (**a**) General case, (**b**) independent (i.e. non-interacting) binding sites, (**c**) identical binding sites, (**d**) identical and independent binding sites. In **c** and **d**, $\Delta G_1^\circ = \Delta G_a^\circ = \Delta G_b^\circ$ and $\Delta G_2^\circ = \Delta G_{ab}^\circ = \Delta G_{ba}^\circ$

Here, binding is described by four microscopic binding constants (Scheme 2.4a). For example, cooperative binding at the site a arises when binding steps $\mathbf{H} \rightarrow \mathbf{HG}_a$ (in the absence of another bound guest) and $\mathbf{HG}_b \rightarrow \mathbf{HG}_2$ (in the presence of another bound guest) have different affinities (i.e. $K_a^{\text{mic.}} \neq K_{ba}^{\text{mic.}}$). From the point of view of Gibbs energy profile, $\Delta G_a^\circ \neq \Delta G_{ba}^\circ$ (subscript 'r' in Δ_r omitted in the following text) as shown in Scheme 2.6a. On the other hand, independence of binding sites implies non-cooperative binding (for binding sites located far apart, e.g. at distant parts of a large molecule) and requires $\Delta G_a^\circ = \Delta G_{ba}^\circ$ and $\Delta G_b^\circ = \Delta G_{ab}^\circ$, see Scheme 2.6b.

The four microscopic binding constants are not independent, which can be inferred from the Gibbs energy profile. The energy difference between the $\mathbf{HG}_2$ complex and the free host and guest species should be equal for both branches $a \rightarrow ab$ and $b \rightarrow ba$, i.e. $\mu_{\mathbf{HG}_2}^\circ - \mu_{\mathbf{H}}^\circ - 2\mu_{\mathbf{G}}^\circ = \Delta G_a^\circ + \Delta G_{ab}^\circ = \Delta G_b^\circ + \Delta G_{ba}^\circ$ (see Eq. (2.4)). Hence, using the formula $\Delta G_j^\circ = -RT \ln K_j$ (Eq. (2.9)) implies $-RT \ln \left(K_a^{\text{mic.}} K_{ab}^{\text{mic.}}\right) = -RT \ln \left(K_b^{\text{mic.}} K_{ba}^{\text{mic.}}\right)$, and thus $K_a^{\text{mic.}} K_{ab}^{\text{mic.}} = K_b^{\text{mic.}} K_{ba}^{\text{mic.}}$. Subsequently, an interaction parameter α_{12}, describing interaction between both binding sites (cf. p. 23 in [3]), can be defined as

$$\alpha_{12} = \frac{K_{ba}^{\text{mic.}}}{K_a^{\text{mic.}}} = \frac{K_{ab}^{\text{mic.}}}{K_b^{\text{mic.}}}\,. \tag{2.37}$$

Thus, α_{12} describes cooperativity of binding at both a and b sites, see Scheme 2.6a. It is also a measure of standard reaction Gibbs energy difference since

$\Delta G^\circ_{ab} - \Delta G^\circ_b = \Delta G^\circ_{ba} - \Delta G^\circ_a = -RT \ln \alpha_{12}$. Interacting sites imply $\alpha_{12} \neq 1$. If $\alpha_{12} > 1$, then binding at one site increases binding affinity for the other site (i.e. $K^{\text{mic.}}_{ba} > K^{\text{mic.}}_a$ and $K^{\text{mic.}}_{ab} > K^{\text{mic.}}_b$). If $\alpha_{12} < 1$, the binding affinity decreases upon binding of single guest molecule (i.e. $K^{\text{mic.}}_{ba} < K^{\text{mic.}}_a$ and $K^{\text{mic.}}_{ab} < K^{\text{mic.}}_b$). Hence, the cooperativity assumes three qualitatively different modes according to the value of the interaction parameter,

$$\begin{aligned}
\alpha_{12} &< 1 \text{ (negative cooperativity)}\,,\\
\alpha_{12} &= 1 \text{ (no cooperativity)}\,,\\
\alpha_{12} &> 1 \text{ (positive cooperativity)}\,.
\end{aligned}$$

The case $\alpha_{12} = 1$ must be taken cautiously when assessing independence of the binding sites because a fortuitous compensation of different effects in interacting sites can be manifested as $\alpha_{12} \approx 1$ (cf. [26] and p. 86 in [3]).

The general relationship between stepwise and microscopic binding constants derived in [25] is

$$K_j = \frac{\sum^n_{l_1<l_2<\cdots<l_j} K^{\text{mic.}}_{l_1} K^{\text{mic.}}_{l_1 l_2} \ldots K^{\text{mic.}}_{l_1 l_2 \ldots l_j}}{\sum^n_{l_1<l_2<\cdots<l_{j-1}} K^{\text{mic.}}_{l_1} K^{\text{mic.}}_{l_1 l_2} \ldots K^{\text{mic.}}_{l_1 l_2 \ldots l_{j-1}}}\,, \tag{2.38}$$

where n is the number of binding sites and $l_j = a, b, c, \ldots n$. The binding site labels are ordered as $a < b < c < \ldots n$, thus, for example, $K^{\text{mic.}}_{acb}$ is not present in the formula as it is dependent on other constants. For $n = 2$, the relationships are $K_1 = K^{\text{mic.}}_a + K^{\text{mic.}}_b$ and $K_2 = \frac{K^{\text{mic.}}_a K^{\text{mic.}}_{ab}}{K^{\text{mic.}}_a + K^{\text{mic.}}_b}$. Note that the dependent $K^{\text{mic.}}_{ba}$ does not appear in these formulae as expected. There are three important special cases for a host with n binding sites:

Case 1—Independent binding sites as shown in Scheme 2.6b. Here, each binding site has its own binding constant $K^{\text{mic.}}_{l_j}$ independent of occupancy of other binding sites, i.e. $K^{\text{mic.}}_{l_1 l_2 \ldots l_j} = K^{\text{mic.}}_{l_j}$ where $l_j = a, b, c, \ldots n$. Substitution into Eq. (2.38) yields

$$K_j = \frac{\sum^n_{l_1<l_2<\cdots<l_j} K^{\text{mic.}}_{l_1} K^{\text{mic.}}_{l_2} \ldots K^{\text{mic.}}_{l_j}}{\sum^n_{l_1<l_2<\cdots<l_{j-1}} K^{\text{mic.}}_{l_1} K^{\text{mic.}}_{l_2} \ldots K^{\text{mic.}}_{l_{j-1}}}\,, \tag{2.39}$$

where $l_j = a, b, c, \ldots n$. For $n = 2$, the relationships are $K_1 = K^{\text{mic.}}_a + K^{\text{mic.}}_b$ and $K_2 = \frac{K^{\text{mic.}}_a K^{\text{mic.}}_b}{K^{\text{mic.}}_a + K^{\text{mic.}}_b}$. Systems with independent binding sites are always non-cooperative.

Case 2—Identical binding sites as shown in Scheme 2.6c. Here, symmetry of the host molecule implies equal energy levels for a particular complex $\mathbf{HG}_j$. Therefore, the microscopic binding constants only depend on the number of bound guests, not on the particular site, hence $K^{\text{mic.}}_{l_1 l_2 \ldots l_j} = K^{\text{mic.}}_j$, where $j = 1, 2, \ldots, n$. Their relationship with stepwise binding constants can be obtained either by substitution into Eq. (2.38) or by counting the occupied and unoccupied binding sites as follows

Table 2.2 Relationship between stepwise and microscopic binding constants for a host-guest system, where the host molecule contains n identical binding sites

	stepwise and microscopic binding constants					
$n = 2$	$K_1 = 2K_1^{\text{mic.}}$	$K_2 = \frac{1}{2}K_2^{\text{mic.}}$				
$n = 3$	$K_1 = 3K_1^{\text{mic.}}$	$K_2 = K_2^{\text{mic.}}$	$K_3 = \frac{1}{3}K_3^{\text{mic.}}$			
$n = 4$	$K_1 = 4K_1^{\text{mic.}}$	$K_2 = \frac{3}{2}K_2^{\text{mic.}}$	$K_3 = \frac{2}{3}K_3^{\text{mic.}}$	$K_4 = \frac{1}{4}K_4^{\text{mic.}}$		
$n = 5$	$K_1 = 5K_1^{\text{mic.}}$	$K_2 = 2K_2^{\text{mic.}}$	$K_3 = K_3^{\text{mic.}}$	$K_4 = \frac{1}{2}K_4^{\text{mic.}}$	$K_5 = \frac{1}{5}K_5^{\text{mic.}}$	
$n = 6$	$K_1 = 6K_1^{\text{mic.}}$	$K_2 = \frac{5}{2}K_2^{\text{mic.}}$	$K_3 = \frac{4}{3}K_3^{\text{mic.}}$	$K_4 = \frac{3}{4}K_4^{\text{mic.}}$	$K_5 = \frac{2}{5}K_5^{\text{mic.}}$	$K_6 = \frac{1}{6}K_6^{\text{mic.}}$

$$K_j = \frac{\text{No. unoccupied sites on } \mathbf{HG}_{j-1}}{\text{No. occupied sites on } \mathbf{HG}_j} K_j^{\text{mic.}} = \frac{n-(j-1)}{j} K_j^{\text{mic.}} = \frac{n-j+1}{j} K_j^{\text{mic.}} . \quad (2.40)$$

Overview of these relationships for different n is given in Table 2.2.

Systems with identical binding sites might be either cooperative or non-cooperative. Noteworthy, the microscopic binding constants can be directly calculated from the stepwise constants.

Case 3—Independent identical binding sites as shown in Scheme 2.6d. Here, all binding steps are fully described by only one microscopic binding constant $K^{\text{mic.}}$ (cf. p. 52 in [3]). The simplification $K_j^{\text{mic.}} = K^{\text{mic.}}$ in Eq. (2.40) yields

$$K_j = \frac{n-j+1}{j} K^{\text{mic.}} . \quad (2.41)$$

For $n = 2$, the relationships are $K_1 = 2K^{\text{mic.}}$ and $K_2 = \frac{1}{2}K^{\text{mic.}}$. These systems are always non-cooperative and the microscopic constant can be determined from the stepwise constants.

The interaction parameter has been defined for the case $n = 2$ as ratio of microscopic binding constants, which can be generalized to arbitrary n provided the binding sites are identical (see Case 2 above). Thus, for binding step $\mathbf{HG}_{j-1} \rightarrow \mathbf{HG}_j$ it holds

$$\alpha_{j-1,j} = \frac{K_j^{\text{mic.}}}{K_{j-1}^{\text{mic.}}} . \quad (2.42)$$

Using this definition, $-RT \ln \alpha_{j-1,j}$ can be interpreted as a difference of standard Gibbs energies of reaction (i.e. $-RT \ln \alpha_{j-1,j} = \Delta G_j^\circ - \Delta G_{j-1}^\circ$, where ΔG_j° corresponds to binding step $\mathbf{HG}_{j-1} \rightarrow \mathbf{HG}_j$), see Scheme 2.6c. Using Eqs. (2.40) and (2.42), the interaction parameters can be calculated from the stepwise binding constants as (cf. Eq. 2) in [30])

Table 2.3 Interaction parameters for host-guest binding, where the host molecule contains n identical binding sites

	Interaction parameter				
$n = 2$	$\alpha_{12} = \frac{4K_2}{K_1}$				
$n = 3$	$\alpha_{12} = \frac{3K_2}{K_1}$	$\alpha_{23} = \frac{3K_3}{K_2}$			
$n = 4$	$\alpha_{12} = \frac{8K_2}{3K_1}$	$\alpha_{23} = \frac{9K_3}{4K_2}$	$\alpha_{34} = \frac{8K_4}{3K_3}$		
$n = 5$	$\alpha_{12} = \frac{5K_2}{2K_1}$	$\alpha_{23} = \frac{2K_3}{K_2}$	$\alpha_{34} = \frac{2K_4}{K_3}$	$\alpha_{45} = \frac{5K_5}{2K_4}$	
$n = 6$	$\alpha_{12} = \frac{12K_2}{5K_1}$	$\alpha_{23} = \frac{15K_3}{8K_2}$	$\alpha_{34} = \frac{16K_4}{9K_3}$	$\alpha_{45} = \frac{15K_5}{8K_4}$	$\alpha_{56} = \frac{12K_6}{5K_5}$

$$\alpha_{j-1,j} = \frac{(n-j+2)j}{(n-j+1)(j-1)} \frac{K_j}{K_{j-1}} . \tag{2.43}$$

Expressions for interaction parameters for different n are listed in Table 2.3. When the binding sites are not identical, binding step $\mathbf{HG}_{j-1} \rightarrow \mathbf{HG}_j$ cannot be described by one interaction parameter only. However, in these cases the microscopic binding constants are usually experimentally inaccessible.

The type of cooperativity described above is called ‘allosteric cooperativity’, which can be either ‘homotropic’ or ‘heterotropic’. The former refers to cooperative binding of one type of guests as in our case, the latter refers to cooperative binding of different types of guests to the same host molecule. There are also other types of cooperative binding for single guest molecules, which attach to the host at multiple sites at once or for hydrogen bonding in DNA double helix (cf. p. 175 and 215 in [31]). A well-known example of allosteric cooperative binding is oxygen binding to haemoglobin at its four binding sites. This system is described by various models in the literature, one of them is based on the formalism described here, which is referred to as Adair-Klotz model [32].

References

1. Atkins PW, Paula JD, Keeler J (2018) Atkins’ physical chemistry. Oxford University Press
2. McQuarrie DA, Simon JD (1997) Physical chemistry: a molecular approach. University Science Books, Aug
3. Connors KA (1987) Binding constants: the measurement of molecular complex stability. Wiley
4. Callen HB (1985) Thermodynamics and an introduction to thermostatistics. Wiley
5. Lue L (2009) Chemical thermodynamics. Ventus Publishing, Place of publication not identified
6. Gmehling J, Kolbe B, Kleiber M, Rarey J (2012) Chemical thermodynamics for process simulation. Wiley
7. Srivastava R, Smith BD (1986) Total pressure vapor-liquid equilibrium data for benzene + acetonitrile, diethylamine + ethylacetate, and propylamine + diethylamine binary systems. J Chem Eng Data 31:94–99

8. Hiak T, Kurihara K, Kojima K (1994) Vapor-liquid equilibria for acetone + chloroform + methanol and constituent binary systems at 101.3 kPa. J Chem Eng Data 39:714–719
9. Aveyard R, Briscoe BJ, Chapman J (1973) Activity coefficients and association of n-alkanols in n-octane. J Chem Soc, Faraday Trans 1: Phys Chem Condens Phases 69:1772–1778
10. Renon H, Prausnitz JM (1968) Local compositions in thermodynamic excess functions for liquid mixtures. AIChE J 14:135–144
11. Dortmund Data Bank, http://www.ddbst.com/
12. Mysels KJ (1956) Textbook errors: VII. The laws of reaction rates and of equilibrium. J Chem Educ 33:178
13. Guggenheim EA (1956) Textbook errors: IX. More about the laws of reaction rates and of equilibrium. J Chem Educ 33:544
14. Laidler KJ (1981) Symbolism and terminology in chemical kinetics. Pure Appl Chem 53:753–771
15. Březina V, Hanyková L, Velychkivska N, Hill JP, Labuta J (2022) NMR lineshape analysis using analytical solutions of multi-state chemical exchange with applications to kinetics of host-guest systems. Sci Rep 12:17369
16. Mulquiney PJ, Kuchel PW (2003) Modelling metabolism with Mathematica: detailed examples including erythrocyte metabolism. CRC Press, Boca Raton, Fla
17. Connors KA (1990) Chemical kinetics: the study of reaction rates in solution. Wiley, Aug
18. Laidler KJ, King MC (1983) Development of transition-state theory. J Phys Chem 87:2657–2664
19. Glasstone S, Laidler KJ, Eyring H (1941) The theory of rate processes: the kinetics of chemical reactions, viscosity, diffusion and electrochemical phenomena. McGraw-Hill Book Company, Incorporated
20. Arcus VL, Prentice EJ, Hobbs JK, Mulholland AJ, Van der Kamp MW, Pudney CR, Parker EJ, Schipper LA (2016) On the temperature dependence of enzyme-catalyzed rates. Biochemistry 55:1681–1688
21. van der Kamp MW, Prentice EJ, Kraakman KL, Connolly M, Mulholland AJ, Arcus VL (2018) Dynamical origins of heat capacity changes in enzyme-catalysed reactions. Nat Commun 9:1177
22. Kyba EP, Helgeson RC, Madan K, Gokel GW, Tarnowski TL, Moore SS, Cram DJ (1977) Host-guest complexation. 1. Concept and illustration. J Am Chem Soc 99:2564–2571
23. Steed JW, Atwood JL (2009) Supramolecular chemistry, 2nd edn. Wiley, Chichester, UK
24. Edsall JT, Wyman J, Hoffman JG (1958) Thermodynamics, electrostatics, and the biological significance of the properties of matter. Phys Today 11:50–52
25. Klotz I, Hunston D (1975) Protein interactions with small molecules. Relationships between stoichiometric binding constants, site binding constants, and empirical binding parameters. J Biol Chem 250:3001–3009
26. Connors KA, Pendergast DD (1984) Microscopic binding constants in cyclodextrin systems: complexation of α-cyclodextrin with sym-1,4-disubstituted benzenes. J Am Chem Soc 106:7607–7614
27. Hargrove AE, Zhong Z, Sessler JL, Anslyn EV (2010) Algorithms for the determination of binding constants and enantiomeric excess in complex host: guest equilibria using optical measurements. New J Chem 34:348–354
28. Labuta J, Ishihara S, Šikorský T, Futera Z, Shundo A, Hanyková L, Burda JV, Ariga K, Hill JP (2013) NMR spectroscopic detection of chirality and enantiopurity in referenced systems without formation of diastereomers. Nat Commun 4:2188
29. Hunter CA, Anderson HL (2009) What is Cooperativity? Angewandte Chemie International Edition 48:7488–7499
30. Ercolani G (2003) Assessment of Cooperativity in Self-Assembly. J Am Chem Soc 125:16097–16103
31. Anslyn EV, Dougherty DA (2006) Modern physical organic chemistry. University Science Books
32. Stefan MI, Edelstein SJ, Le Novère N (2009) Computing phenomenologic Adair-Klotz constants from microscopic MWC parameters. BMC Syst Biol 3:68

Chapter 3
Nuclear Magnetic Resonance Spectroscopy

Nuclear magnetic resonance (NMR) is a phenomenon when electromagnetic waves in the MHz range are absorbed in atomic nuclei immersed in external magnetic field and then emitted back at a particular resonant frequency. NMR spectroscopy exploits the NMR related phenomena to obtain useful information about matter at atomic level. In synthetic chemistry, it is applied for structure determination of newly synthesized substances as well as for sample purity checks. In chemical physics, binding and dynamic properties of various systems can be investigated. There are many textbooks dealing with theory of NMR, e.g. Levitt [1], Cavanagh et al. [2], Macomber [3] or Slichter [4]. In this chapter, short introduction to NMR spectroscopy is given as well as detailed treatment of chemical exchange in NMR as described in our paper [5], parts of the text are taken from that publication.

3.1 Basic Phenomena in NMR

3.1.1 Nuclei in Magnetic Field

Atomic nuclei with nonzero spin I also possess the magnetic moment μ, whose magnitude is described by element-specific gyromagnetic ratio γ (ratio of magnetic dipole moment and angular momentum),

$$\boldsymbol{\mu} = \gamma \hbar \mathbf{I} \,. \tag{3.1}$$

In the presence of external magnetic field, energy levels of nuclear spins split into $2I + 1$ levels. This phenomenon is called the *Zeeman effect*. In the following text,

V. Březina, *Supramolecular Complexes of Oxoporphyrinogens with Organic Molecules*, Springer Theses, https://doi.org/10.1007/978-3-031-68646-7_3

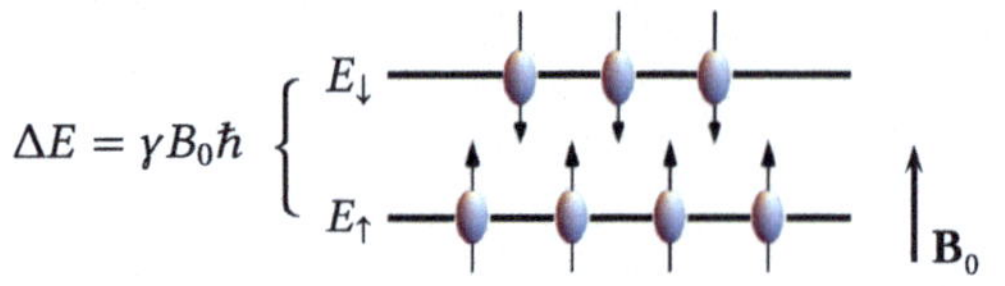

Fig. 3.1 Illustration of Zeeman splitting for spin-1/2 nucleus with $\gamma > 0$ (spin and magnetic moments are oriented in the same direction). Populations of energy levels are not to scale

let us consider only nuclei with spin $I = 1/2$ resulting in Zeeman splitting into two energy levels. Energy of a magnetic dipole in a magnetic field $\mathbf{B}_0$ is

$$E_{\text{mag}} = -\boldsymbol{\mu}.\mathbf{B}_0\,, \tag{3.2}$$

thus, the Zeeman splitting for $I = 1/2$ induces two energy levels, $E_\uparrow$ for spin-up $|\uparrow\rangle$ and $E_\downarrow$ for spin-down $|\downarrow\rangle$. Hence, the energy difference is

$$\Delta E = E_\uparrow - E_\downarrow = -\gamma B_0 \hbar = \omega_0 \hbar \tag{3.3}$$

as illustrated in Fig. 3.1. The quantity

$$\omega_0 = -\gamma B_0 \tag{3.4}$$

is called the *Larmor frequency*.

For $\gamma > 0$, the energy level $E_\uparrow$ is lower than $E_\downarrow$ (i.e. $\Delta E < 0$), hence, in equilibrium, state $|\uparrow\rangle$ is more populated than state $|\downarrow\rangle$ according to Boltzmann distribution,

$$\frac{p_\uparrow}{p_\downarrow} = \exp\left(-\frac{\Delta E}{k_{\text{Boltz}} T}\right)\,. \tag{3.5}$$

At room temperature, the population difference $p_\uparrow - p_\downarrow$ for hydrogen nucleus ($\gamma =$ 42.58 MHz/T) is very small, about 10^{-5}. The observed NMR signal is weak since it is proportional to the population difference.

In reality, most nuclear spins are not in their eigenstates $|\uparrow\rangle$ or $|\downarrow\rangle$ but in superposition of these two states (see Sect. 11.1 in [1]). Thus, proper description of the whole ensemble of nuclear spins requires the use of spin density matrix $\hat{\rho}$ with full quantum mechanical approach. However, a classical macroscopic model can be applied for description of the basics of NMR measurement, while some phenomena in NMR require quantum mechanical approach. From macroscopic point of view, the ensemble of nuclear spins can be described by *magnetization*[1] **M**, i.e. density of magnetic moments in a volume V,

$$\mathbf{M} = \frac{1}{V}\sum \boldsymbol{\mu}\,. \tag{3.6}$$

Let us assume that the external magnetic field points in the z-direction, i.e. $\mathbf{B}_0 = B_0\mathbf{e}_z$. Then equilibrium implies that the magnetization is also aligned with z-axis (in order to minimize the energy $E_{\mathrm{mag}} = -\mathbf{M}.\mathbf{B}_0$). When initially tilted away from the z-axis, the magnetization starts to precess around the z-axis with the Larmor frequency ω_0 and relaxes back to equilibrium.

3.1.2 Chemical Shift

Local magnetic field B_{loc} at the position of reference nucleus is the sum of the external field B_0 and additional field B_{ind} induced by the presence of electrons in the surroundings. The induced field is proportional to the external field, $B_{\mathrm{ind}} = -\sigma B_0$, where σ is the shielding constant. In general, the shielding constant is a tensor, which is averaged to a scalar due to rapid motion in liquid samples. Hence, $B_{\mathrm{loc}} = B_0 + B_{\mathrm{ind}}$ and the nuclear magnetic spins precess with modified Larmor frequency

$$\omega_0 = -\gamma B_{\mathrm{loc}} = -\gamma(1-\sigma)B_0\,. \tag{3.7}$$

This influence of surrounding electrons is called *chemical shift* (see Sect. 3.7 in [1] or 3.1 in [7]). It enables differentiation of symmetrically nonequivalent nuclear spins in NMR measurement. Thus, NMR provides a useful tool to identify chemical substances, as every molecule has its unique NMR fingerprint.

It is common practice to measure the chemical shift relative to a chemical standard present in the sample with Larmor frequency ω_{std} and shielding constant σ_{std}. Conversion from angular frequency ω to chemical shift δ is defined as [1, 2, 7]

$$\delta = \frac{\omega - \omega_{\mathrm{std}}}{2\pi\nu_0} \times 10^6 = (\sigma - \sigma_{\mathrm{std}}) \times 10^6\,, \tag{3.8}$$

where ν_0 is the spectrometer frequency corresponding to $\sigma = 0$ (i.e. $2\pi\nu_0 = -\gamma B_0$ from Eq. (3.7)). The unit of chemical shift δ is 'ppm' (parts per million) since the frequency difference $(\omega - \omega_{\mathrm{std}})/(2\pi)$ is in orders of Hz or kHz while ν_0 is in orders of MHz. Chemical shift only depends on shielding constants and does not depend on

[1] Nuclear magnetization of water in external field $B_0 = 11.7$ T is $M(H_2O) = 2c(H_2O)\,N_A\,\mu\,(p_\uparrow - p_\downarrow) = 9.6 \times 10^{-4}$ A.m^{-1}, where $c(H_2O) = 55.5$ M is the molar water concentration, N_A the Avogadro constant, $\mu = 2.2 \times 10^{-27}$ A.m^2 the magnetic moment of hydrogen nucleus and $p_\uparrow - p_\downarrow = \frac{\hbar\gamma B_0}{2k_B T} = 6.4 \times 10^{-6}$ the population difference at room temperature. This value is much smaller than the magnetization of a typical bar magnet $M(\text{magnet}) = 3.7 \times 10^5$ A.m^{-1} [6].

the external field B_0 because it is calculated as a relative frequency difference. Hence, measured chemical shifts of a particular molecule are identical on spectrometers with different B_0. Note that due to historical reasons, the chemical shift axis in ppm is plotted in increasing order from right to left. Equation (3.8) provides a simple formula for the conversion from ppm to rad.s^{-1} valid for frequency difference,

$$\Delta\omega[\text{rad.s}^{-1}] = 2\pi \times \nu_0\,[\text{MHz}] \times \Delta\delta\,[\text{ppm}]\,. \qquad (3.9)$$

Tetramethylsilan (TMS) is commonly used in nonpolar solvents as a NMR standard because it is inert to common chemicals and its nuclei have rather stable Larmor frequency value (temperature change by 100 °C leads to change of δ_{TMS} by 0.1 ppm [8]). It has twelve hydrogen atoms with identical chemical shift which provides strong NMR signal even at low TMS concentrations.

3.1.3 Relaxation

There are several types of interaction of the nuclear spins with each other and with their surroundings. These interactions are mostly averaged out by molecular motion in liquid or gaseous samples and thus not directly detectable in NMR measurements. Yet they are the cause of relaxation of the magnetization **M** from any starting position to the equilibrium $\mathbf{M} \parallel \mathbf{B}_0$. There are two qualitatively different relaxation mechanisms (see Sects. 2.6 and 2.7 in [1]).

The first mechanism, spin-lattice relaxation, is described by the relaxation time T_1. Here, energy is transferred from the observed spins to the surroundings (called lattice) to achieve thermodynamic equilibrium with minimum energy $E_{\text{mag}} = -\mathbf{M}.\mathbf{B}_0$ [3]. This type of relaxation involves only the M_z component of the magnetization, hence, it is also called *longitudinal relaxation*.

The other relaxation mechanism with the relaxation time T_2 is called spin-spin relaxation. It is the result of direct dipole-dipole interaction between nuclear magnetic moments. Their interaction causes fluctuations in their local magnetic field B_{loc} and consequently their Larmor frequency, which leads to gradual dephasing during their precession. In this case, no energy is dissipated, but entropy of the system increases.

Longitudinal relaxation rate $R_1 = 1/T_1$ and transverse relaxation rate $R_2 = 1/T_2$ are also used instead of relaxation times. For small molecules with low viscosity, such as pure water, at room temperature, typical values of T_1 and T_2 are about 3–4 s. However, in solutions of macromolecules or proteins the relaxation times are an order of magnitude or more shorter. In addition, T_2 is always smaller than T_1.

3.1.4 J-Coupling

An interaction, which is not averaged due to fast molecular motion, is indirect spin-spin coupling mediated by electrons in molecular bonds. It is also called J-coupling and its strength is described by the coupling constant J (in Hz). J-coupling is manifested in NMR by splitting of a single NMR resonance at Larmor frequency ω_0 into two or more resonances (see Sect. 3.8 in [1]). For example, if the observed nuclear spin I_1 is J-coupled to another nuclear spin $I_2 = 1/2$, then two resonances for I_1 are observed. In particular, observed angular frequency of I_1 is $\omega_0 + \pi J$ if I_2 is in state $|\uparrow\rangle$, and $\omega_0 - \pi J$ if I_2 is in state $|\downarrow\rangle$. Generally, if spin $I_2 > 1/2$ or the spin I_1 is coupled to several other nuclear spins, the observed Larmor frequency splits into more than two frequencies. J-coupling is a quantum phenomenon and it cannot be described by the classical model (see Sect. 17.8 in [1]).

3.1.5 Bloch Equations

Classical derivation of magnetization dynamics and consequently the spectral lineshapes for NMR is based on Bloch equations [9, 10]. Here, single nuclear spin can be represented by the classical angular momentum $\mathbf{L}$ of some rotating object related to the magnetic moment as $\boldsymbol{\mu} = \gamma \mathbf{L}$. Time change of $\mathbf{L}$ can be induced by torque $\boldsymbol{\tau}$, i.e. $\frac{\mathrm{d}\mathbf{L}}{\mathrm{d}t} = \boldsymbol{\tau}$. The presence of external field causes torque on the magnetic moment, $\boldsymbol{\tau} = \boldsymbol{\mu} \times \mathbf{B}$ [11]. Combination of the expressions for torque yields $\frac{\mathrm{d}\boldsymbol{\mu}}{\mathrm{d}t} = \gamma \boldsymbol{\mu} \times \mathbf{B}$, which is valid for a single magnetic moment. Assuming homogeneous magnetic field in the whole sample, the same expression holds for the magnetization (i.e. volume density of magnetic moments). However, apart from the torque, both longitudinal (T_1) and transverse (T_2) relaxation mechanisms have to be taken into account, hence the resulting *Bloch equations* describing the evolution of magnetization are

$$\begin{aligned} \frac{\mathrm{d}M_x}{\mathrm{d}t} &= \gamma(\mathbf{M} \times \mathbf{B})_x - \frac{M_x}{T_2}, \\ \frac{\mathrm{d}M_y}{\mathrm{d}t} &= \gamma(\mathbf{M} \times \mathbf{B})_y - \frac{M_y}{T_2}, \\ \frac{\mathrm{d}M_z}{\mathrm{d}t} &= \gamma(\mathbf{M} \times \mathbf{B})_z + \frac{M_0 - M_z}{T_1}. \end{aligned} \tag{3.10}$$

Note that the torque $\gamma \mathbf{M} \times \mathbf{B}$ is perpendicular to both the magnetization $\mathbf{M}$ and the magnetic field $\mathbf{B}$, which causes precession of $\mathbf{M}$ around $\mathbf{B}$.

3.1.6 Effect of Radiofrequency Field

The macroscopic magnetization can be manipulated by application of radiofrequency (RF) electromagnetic field of amplitude B_{RF} perpendicular to the external field $\mathbf{B}_0$, whose frequency ω_{RF} is close to the Larmor frequency of the nuclei. Because of this resonance, even small B_{RF} has substantial effect on the spin states. It can be shown (see Sect. 8.4.2 in [1]) that during the RF irradiation, the magnetization undergoes nutation movement, which tilts the magnetization away from the equilibrium position ($\mathbf{M} \parallel \mathbf{B}_0$). Solution of the Bloch equations (Eq. (3.10)) shows that the nutation frequency is

$$\omega_{\mathrm{nut}} = |\frac{1}{2}\gamma B_{RF}|\,. \tag{3.11}$$

For example, if the RF field is applied during a pulse of duration τ_{p}, the magnetization rotates by the angle $\beta_{\mathrm{p}} = \omega_{\mathrm{RF}}\tau_{\mathrm{p}}$. In NMR spectroscopy, this is used to manipulate the direction of magnetization using short pulses. For example, in order to tilt $\mathbf{M}$ into the x-y plane, the pulse amplitude B_{RF} and τ_{p} are chosen such that $\omega_{\mathrm{RF}}\tau_{\mathrm{p}} = \frac{\pi}{2}$. This type of pulse is called '$(\frac{\pi}{2})$ pulse' (see Sect. 10.8 in [1]), its typical duration is $\tau_{\mathrm{p}} \approx 10\ \mu$s.

3.2 Measurement of NMR

3.2.1 NMR Spectrometer

NMR is measured with a NMR spectrometer as shown in Fig. 3.2a. The sample is located in a probe, which is inside a superconducting magnet (cooled with liquid helium and liquid nitrogen) generating the external magnetic field $\mathbf{B}_0$, see Fig. 3.2b. RF field is generated by transverse coil wound perpendicular to the sample tube, see Fig. 3.2c. The coil serves as transmitter, inducing the NMR signal, and receiver, detecting the NMR signal. NMR spectroscopy requires high magnetic fields in order to obtain good spectral resolution, for example, magnetic field $B_0 = 11.7$ T induces Larmor frequency of 500 MHz for ^{1}H nuclei. Note that frequency of the RF field produced by NMR (in orders of 100 MHz) is too high for the current analog-digital converters. Hence, a reference wave with frequency ω_{ref} is subtracted from the NMR signal in so-called 'quadrature detection' (see Sects. 4.5.2 and A.5 in [1]; [13]), and the NMR signal with frequency ω_0 is digitized with the offset frequency $\Omega_0 = \omega_0 - \omega_{\mathrm{ref}}$. Morover, signal in complex form is obtained from quadrature detection. Importantly, ω_{ref} is equal to the frequency of the RF pulse ω_{RF} (see Sect. 3.1.6) since they are produced by the same RF wave synthesizer (see Sect. 4.6 and Fig. 4.14 in [1]). This fact is conveniently used for theoretical assessment, when a frame of reference rotating with frequency ω_{ref} is introduced (see Sect. 10.6 in [1]).

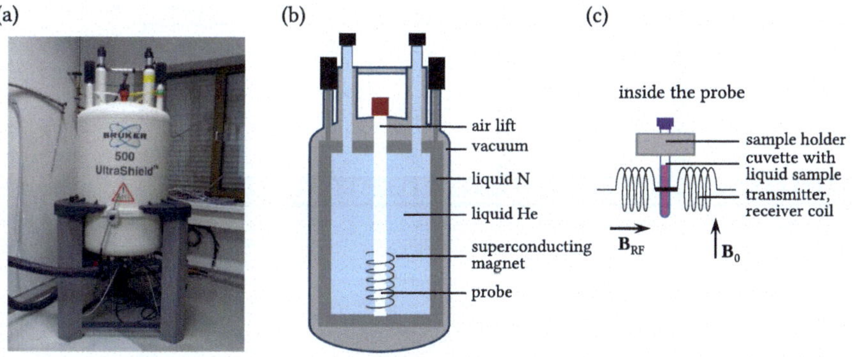

Fig. 3.2 NMR spectrometer. **a** Photo of Bruker 500 MHz commercial spectrometer (500 MHz is the Larmor frequency for ^{1}H nuclei), **b** scheme of the spectrometer parts and **c** detailed view on the sample in the probe. The spectrometer scheme was created according to [12]

3.2.2 Pulsed NMR Experiment

The production and detection of NMR signal is illustrated in Fig. 3.3. The procedure starts from equilibrium (**1**), with the magnetization along the z-axis. Then, the magnetization $\mathbf{M}$ is tilted to the x-y plane perpendicular to $\mathbf{B}_0$ after a $(\frac{\pi}{2})$ RF pulse from the transmitter coil (**2**). Thereafter, during so-called free induction decay (FID) period, it undergoes free precession with Larmor frequency ω_0 (**3**) until it relaxes back to equilibrium (**4**) due to the longitudinal and transverse relaxation mechanisms. Note that $|\mathbf{M}(t)|$ changes during the evolution due to the presence of relaxation.[2] Also, according to the definition $\omega_0 = -\gamma B_0$, nuclei with $\gamma > 0$ precess clockwise and nuclei with $\gamma < 0$ precess counterclockwise. During the relaxation period (**3**), the NMR signal is produced in the sample and detected as induced electric current in the receiver coil, which is proportional to the precessing magnetization.

Since the magnetization precesses around the z-axis, only the transverse magnetization component (transverse to the external magnetic field $\mathbf{B}_0$) is detected. The NMR signal is acquired with quadrature detection (see the previous section) producing complex signal in the form of $M_x(t) + iM_y(t)$, where the original Larmor frequency ω_0 is detected as an offset frequency $\Omega_0 = \omega_0 - \omega_{\text{ref}}$ (where $\omega_{\text{ref}} = \omega_{\text{RF}}$). Usually, frequency domain is preferred over the time domain to analyze NMR signal, for which Fourier transform (FT) is used.[3]

To measure the weak NMR signal (mainly due to low $p_\uparrow - p_\downarrow$), high sample concentrations about 0.1–1 mM are typically used to obtain hydrogen spectrum.

---

[2] Bloch equations (Eq. (3.10)) have the form $\frac{\mathrm{d}}{\mathrm{d}t}\mathbf{M} = \gamma\mathbf{M} \times \mathbf{B} + \mathbf{f}(T_1, T_2)$. Therefore, the magnitude of magnetization $|\mathbf{M}|$ changes in time due to the relaxation terms $\mathbf{f}(T_1, T_2)$ since $\frac{\mathrm{d}}{\mathrm{d}t}(\mathbf{M}.\mathbf{M}) = 2\mathbf{M}.\frac{\mathrm{d}}{\mathrm{d}t}\mathbf{M} = 2\mathbf{M}.\{\gamma\mathbf{M} \times \mathbf{B} + \mathbf{f}(T_1, T_2)\} = 2\mathbf{M}.\mathbf{f}(T_1, T_2) \neq 0$.

[3] Pulsed FT NMR spectroscopy is the state-of-the-art technique and superseded the original continuous-wave (CW) NMR spectroscopy [14].

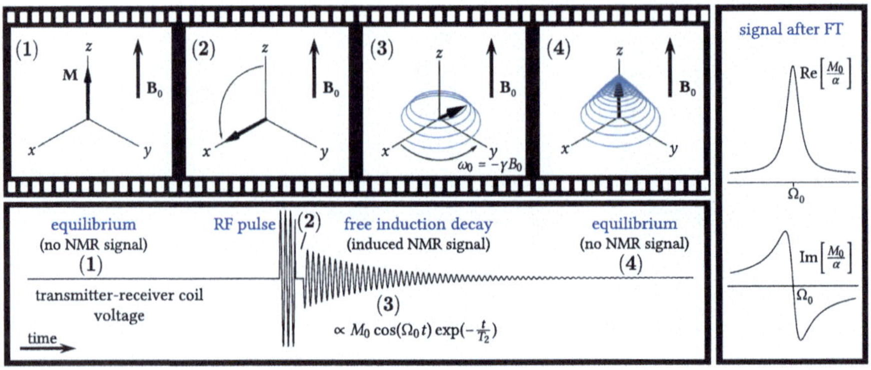

Fig. 3.3 Illustration of pulsed NMR experiment. Snapshots (**1**–**4**) in the upper left panel correspond to points (**1**–**4**) in the bottom left panel showing the voltage in transmitter-receiver coil. The right panel shows Fourier transform of the NMR signal

Also, data acquisition described in Fig. 3.3 is repeated many times with subsequent data averaging, typically the number of scans $N_{\text{scan}} < 100$ for hydrogen nuclei and $N_{\text{scan}} = 2000$–20000 for carbon nuclei.

3.2.3 *Lorentzian Spectral Lineshape*

The NMR signal is detected during the free induction decay (FID) period ((3) in Fig. 3.3), hence it is highly relevant for calculation of the observed spectral lineshapes. During FID, total magnetic field is equal to the constant external field $\mathbf{B} = (0, 0, B_0)$, thus defining the complex transverse magnetization

$$M_{xy} = M_x + iM_y \tag{3.12}$$

Bloch equations (Eq. (3.10)) reduce to

(laboratory frame)

$$\frac{\mathrm{d}M_{xy}}{\mathrm{d}t} = i\omega_0 M_{xy} - R_2 M_{xy}\,, \tag{3.13a}$$

$$\frac{\mathrm{d}M_z}{\mathrm{d}t} = R_1(M_0 - M_z)\,. \tag{3.13b}$$

The complex transverse magnetization M_{xy} is proportional to so-called -1 quantum coherence (nondiagonal element of density matrix $\hat{\rho}$) in quantum description of NMR, which is directly proportional to the acquired NMR signal during FID.

Theory of NMR spectroscopy is often constructed in the frame of reference rotating with ω_{ref} (see Sect. 3.2.1). In this frame, magnetization due to spins with Larmor frequency ω_0 rotates with $\Omega_0 = \omega_0 - \omega_{\mathrm{ref}}$. Quadrature detection yields digitized NMR signal with this offset frequency Ω_0, therefore, the acquired signal is described by Bloch equations in rotating frame,

(rotating frame)

$$\frac{\mathrm{d}M_{xy}}{\mathrm{d}t} = i\Omega_0 M_{xy} - R_2 M_{xy}\,, \tag{3.14a}$$

$$\frac{\mathrm{d}M_z}{\mathrm{d}t} = R_1(M_0 - M_z)\,. \tag{3.14b}$$

In the following text, we will consider spin dynamics only in the rotating frame of reference. Without the loss of generality, let us assume the initial position of magnetization after the $(\frac{\pi}{2})$ pulse along the x-axis, i.e. $M_{xy}(0) = M_x(0) = M_0$ (and $M_y = M_z = 0$), where M_0 is the initial equilibrium magnetization. Using this initial condition, the solution of Eq. (3.14) during FID is

$$M_{xy}(t) = M_0 \exp\left(i\Omega_0 t - R_2 t\right)\,, \tag{3.15a}$$

$$M_z(t) = M_0\left[1 - \exp\left(-R_1 t\right)\right]\,. \tag{3.15b}$$

Note that $M_{xy} \to 0$ and $M_z \to M_0$ as $t \to \infty$.

In order to calculate the spectral lineshape in frequency domain $S(\Omega)$, we apply Fourier transform (FT)[4] on the above solutions for $M_{xy}(t)$. As the signal is collected from $t = 0$, the corresponding magnetization is multiplied by the Heaviside step function $\theta(t)$ or equivalently the integration proceeds in time interval $(0, +\infty)$ instead of $(-\infty, +\infty)$,

$$S(\Omega) = \mathrm{FT}[M_{xy}\theta(t)] = \int_{-\infty}^{+\infty} M_{xy}\theta(t)\exp(-i\Omega t)\mathrm{d}t = \int_{0}^{+\infty} M_{xy}\exp(-i\Omega t)\mathrm{d}t\,. \tag{3.16}$$

With the use of Eq. (3.15a), the solution of Bloch equations for transverse magnetization during FID, i.e. the spectral lineshape, is

$$S(\Omega) = \frac{M_0}{R_2 + i(\Omega - \Omega_0)} = \frac{M_0}{\alpha}\,. \tag{3.17}$$

For compact notation, we have introduced a quantity

$$\alpha = R_2 + i(\Omega - \Omega_0)\,. \tag{3.18}$$

[4] The convention for FT used in this work is $\hat{f}(\Omega) = \int_{-\infty}^{+\infty} f(t)e^{-i\Omega t}\mathrm{d}t$. The corresponding inverse FT is then $f(t) = \frac{1}{2\pi}\int_{-\infty}^{+\infty} \hat{f}(\Omega)e^{i\Omega t}\mathrm{d}\Omega$.

Real and imaginary parts of $S(\Omega)$ are called 'absorption' and 'dispersion' lineshapes, respectively,[5]

$$\mathrm{Re}[S(\Omega)] = \mathrm{Re}\left[\frac{M_0}{\alpha}\right] = \frac{M_0 R_2}{R_2^2 + (\Omega - \Omega_0)^2}, \tag{3.19a}$$

$$\mathrm{Im}[S(\Omega)] = \mathrm{Im}\left[\frac{M_0}{\alpha}\right] = \frac{M_0(\Omega - \Omega_0)}{R_2^2 + (\Omega - \Omega_0)^2}. \tag{3.19b}$$

These lineshapes are shown in Fig. 3.3 (right panel). Full width at half maximum of the absorption lineshape is proportional to R_2, i.e. transverse relaxation causes so-called 'homogeneous broadening' of the NMR lineshapes. However, further 'inhomogeneous broadening' is caused by inhomogeneity of the external magnetic field B_0. The sign of the dispersion lineshape depends on the convention in FT. More details about the Lorentzian lineshape are given in Appendix B.1.1.

3.3 Chemical Exchange

NMR spectroscopy is sensitive to molecular processes, which change chemical environments, i.e. states $A, B, \ldots$, of a particular reference nuclear spin. Such processes, when they are reversible, are called *chemical exchange* in NMR terminology and lead to non-Lorentzian spectral lineshapes. We denote these transitions between states as *spin kinetics*. Chemical exchange is characterized by transition rate coefficients k_{ij} (in units of s^{-1}), where $i, j = A, B, \ldots; i \neq j$. These coefficients are often denoted as 'rate constants', although, due to their possible dependence on concentration (for example, in a bimolecular reaction, see discussion in Sect. 3.4.1), we apply the term 'transition rate coefficients'. The simplest and most common case is a two-state chemical exchange between A and B spin states (i.e. between states with different chemical environments and thus different offset Larmor frequencies Ω_A and Ω_B) described by two transition rate coefficients k_{AB}, k_{BA} as illustrated in Scheme 3.1.

Basic principles of chemical exchange are illustrated in Fig. 3.4 with the example of symmetric two-state exchange. In that case, transition rate coefficients k are identical in both directions (i.e. $k_{AB} = k_{BA} = k$), see Fig. 3.4a. Lifetime τ in both states is exponentially distributed [15] and $k = 1/\langle\tau\rangle$ plays a role of mean frequency of transitions (inverse of the mean lifetime), see Fig. 3.4b. Figure 3.4c shows that the reference nuclear spin precesses with offset frequency Ω_A while in state A and with Ω_B while in state B (decay of the FID signal due to T_2 relaxation is not vis-

Scheme 3.1 Spin kinetics of two-state exchange

$$A \underset{k_{BA}}{\overset{k_{AB}}{\rightleftharpoons}} B$$

[5] The terms 'absorption' and 'dispersion' have no physical meaning and are purely historical [1].

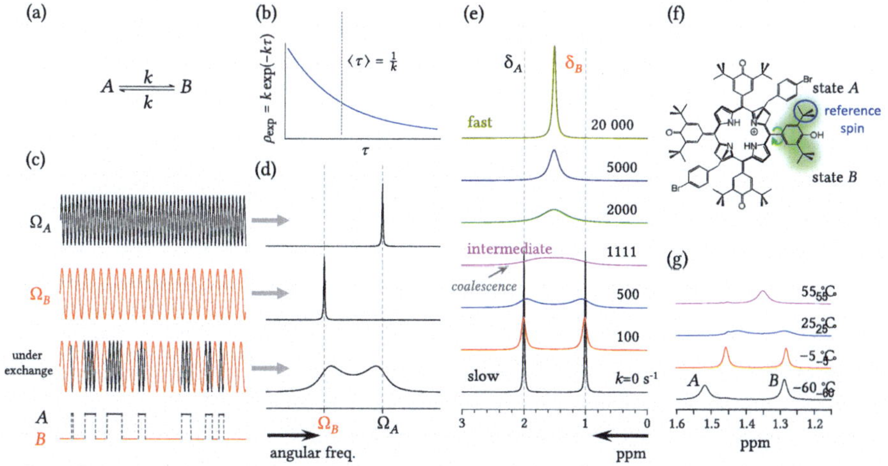

Fig. 3.4 Illustration of symmetric two-state exchange with $R_2^A = R_2^B$. **a** Transition in both directions is characterized by the same transition rate coefficient k. **b** Lifetime between transitions has exponential distribution ρ_{exp}, k is the inverse of mean lifetime $\langle \tau \rangle$. **c** FID signals without and with transitions between states A and B with different Larmor frequencies, **d** the corresponding spectra. **e** Exchange spectra in slow, intermediate (including the coalescence point at $k = 1111\ \text{s}^{-1}$) and fast regimes. **f** Symmetric two-state exchange in a real system, rotation of a bulky side group provides two different environments for the reference spin (denoted by blue circle). **g** Hidden old layer in the image is visible. Transition rate coefficient increases with temperature. Note that the Larmor frequencies Ω_A and Ω_B change with temperature

ible during such short time period). Chemical exchange causes "mixing" of these Larmor frequencies resulting to nontrivial alteration of the resulting NMR spectrum, see Fig. 3.4d (bottom spectrum). There are three different regimes depending on the magnitude of k compared to the Larmor frequency difference $\Delta\Omega_{AB} = |\Omega_A - \Omega_B|$ as shown in Fig. 3.4e:

1. **Slow exchange regime** occurs when $k \ll \Delta\Omega_{AB}$. The spectral lineshape is very close to two Lorentzians placed at Ω_A and Ω_B. Mixing of the two Larmor frequencies is too slow and does not induce any significant spectral change.
2. **Intermediate exchange regime** occurs when $k \approx \Delta\Omega_{AB}$. The spectrum is broadened and significantly differs from the Lorentzian lineshape. The two peaks merge at a so-called 'coalescence point' k_c (when k increases, the minimum between two peaks disappears exactly at k_c).
3. **Fast exchange regime** occurs when $k \gg \Delta\Omega_{AB}$. Fast mixing of Larmor frequencies averages the signal into a single Lorentzian-like peak located at $p_A\Omega_A + p_B\Omega_B$.

Note that the intermediate regime is sometimes called 'slow intermediate' or 'fast intermediate regime' depending from which side the coalescence point is approached [1]. Furthermore, the area below the spectral lineshape (i.e. peak intensity) is not altered when k changes, see Appendix B.

Transition rate coefficient at the coalescence point k_c for symmetric two-state exchange can be calculated from the analytical lineshape [1, 5]. Assuming $R_2^A = R_2^B = R_2$, k_c can be obtained by the following approximative formula

$$k_c \approx \frac{\Delta\Omega_{AB}}{2\sqrt{2}}\left[1 - 1.138\frac{R_2}{\Delta\Omega_{AB}}\right]. \tag{3.20}$$

This equation can be used for rapid estimation of k_c from a spectrum measured exactly at the coalescence point (no need for lineshape fitting) [16, 17]. Furthermore, the concept of coalescence can also be generalized for asymmetric two-state exchange (i.e. $k_{AB} \neq k_{BA}$). In contrast to symmetric two-state exchange, where the coalescence transition rate coefficient is a single value k_c, asymmetric exchange exhibits infinitely many pairs of transition rate coefficients $\{k_{AB,c}, k_{BA,c}\}$ at which the spectrum has coalescence lineshapes (provided constant values of Ω_A, Ω_B, and R_2). This generalization as well as the approximative formula for non-zero R_2 in Eq. (3.20) were published in our paper [5].

An actual example of a molecular process inducing symmetric two-state exchange is shown in Fig. 3.4f,g. Rotation of a bulky side group (Fig. 3.4f) causes jumps of a *tert*-butyl reference spin (all nine hydrogen nuclei denoted by blue circle have identical Larmor frequency due to the motional averaging – fast rotation around CH_3–C bond and *tert*-butyl–C bond) between state A in the vicinity of bromobenzyl and state B further from the bromobenzyl. The corresponding NMR spectra (Fig. 3.4g) show increase of k with temperature. Note that the offset Larmor frequencies Ω_A and Ω_B change with temperature.

NMR lineshape fitting is a straightforward method to obtain transition rate coefficients and often requires only the acquisition of simple 1D NMR spectra [18–20] as described in Sect. 3.2.2. For more complicated molecules (e.g. proteins), 2D experiments are required. For example, HSQC spectra (see Sect. 7.1 in [2]) can be analyzed, where either a 1D crosssection [21–23] or the full 2D spectrum [24–27] lineshape is fitted. NMR lineshape fitting is suitable for the analysis of exchange processes with transition rate coefficients approximately in the range of 10–10^5 s^{-1}.

Methods other than lineshape fitting exist for the determination of transition rate coefficients. However, those methods usually require elaborate NMR sequences and extended experimental time. They are also aimed at the analysis of different ranges of transition rate coefficients (both higher and lower). Reviews of these methods (e.g. ZZ-exchange, EXSY, R_2 relaxation dispersion) are available from Bain [28], Kleckner [29] or Furukawa [30].

3.3.1 Kinetics of Spin States

Chemical exchange of a reference spin (i.e. the spin of interest within the structure of molecule) can be modeled with discrete states $A, B, C, \ldots$ assuming that the transition time between any two states is short compared to the average lifetime in any state. Relative populations of states are denoted as $p_A, p_B, p_C, \ldots$, where $\sum p_j = 1$. For illustration, let us consider two-state spin kinetics, where $[A]$ and $[B]$ are concentrations of spins in states A and B, respectively. Then the time change of $[A]$ is

$$\frac{\mathrm{d}[A]}{\mathrm{d}t} = -k_{AB}[A] + k_{BA}[B]\,. \tag{3.21}$$

The first term on the right-hand side accounts for decrease of $[A]$ due to the $A \to B$ transitions, the second term accounts for its increase due to the $B \to A$ transitions. The equation for time change of $[B]$ is constructed analogously. Using the definitions $p_A = \frac{[A]}{[A]+[B]}$ and $p_B = \frac{[B]}{[A]+[B]}$, the kinetic equations can be formulated as (assuming constant total concentration $[A] + [B]$)

$$\frac{\mathrm{d}}{\mathrm{d}t}\begin{pmatrix} p_A \\ p_B \end{pmatrix} = \mathbf{K}\begin{pmatrix} p_A \\ p_B \end{pmatrix}, \tag{3.22}$$

where

$$\mathbf{K} = \begin{pmatrix} -k_{AB} & k_{BA} \\ k_{AB} & -k_{BA} \end{pmatrix} \tag{3.23}$$

is the kinetic matrix for two states. Note that the sum of left-hand sides of Eq. (3.22) yields zero since $\frac{\mathrm{d}}{\mathrm{d}t}(p_A + p_B) = \frac{\mathrm{d}}{\mathrm{d}t}1 = 0$. Hence, the sum of right-hand sides of Eq. (3.22) also yields zero, which implies that all elements of jth column of the kinetic matrix $\mathbf{K}$ must sum to zero for arbitrary j, i.e. $\sum_i \mathbf{K}_{ij} = 0$. This finding holds for any number of states and can be used to test if $\mathbf{K}$ is correctly constructed.

For a general number of states N, let us define the vector of populations $\mathbf{p} = (p_A, p_B, \ldots, p_N)^\mathsf{T}$, then the time dependence of populations is governed by

$$\frac{\mathrm{d}\mathbf{p}}{\mathrm{d}t} = \mathbf{K}\mathbf{p}\,. \tag{3.24}$$

Note that these first-order differential equations are linear, unlike the governing equations for chemical kinetics (Eq. (2.18)).

We are interested in the equilibrium state, which requires the time derivative at the left-hand side of Eq. (3.24) to be zero and converts this differential equation into an algebraic equation

$$\mathbf{K}\,\mathbf{p} = \mathbf{0}\,. \tag{3.25}$$

Solution of this equation can be readily obtained by a diagrammatic method introduced in King and Altman [31] or Hill [32].

Apart from equilibrium, this solution also describes the steady-state. To understand the difference between equilibrium and steady-state, let us define the *population flux* from state i to state j as

$$J_{ij} = p_i k_{ij} - p_j k_{ji}\,. \tag{3.26}$$

This definition implies $J_{ij} = -J_{ji}$.

The condition for *equilibrium* is that population flux between each state i and j is equal to zero (i.e. $J_{ij} = 0$). In contrast, in *steady-state* mode non-zero fluxes (constant in time) are present, but the net flux $\mathcal{J}_j$ to any state j is zero, so that the populations do not change over time,

$$\mathcal{J}_j = \sum_{\substack{i=A,B,\ldots \\ i\neq j}} J_{ij} = 0\,, \tag{3.27}$$

for all $j = A, B, \ldots$. Note that $\sum_{j=A,B,\ldots} \mathcal{J}_j = 0$ even outside equilibrium or steady-state because the spin in states $A, B, \ldots$ is an isolated system. The condition for zero population flux at equilibrium can be reformulated using populations and transition rate coefficients setting $J_{ij} = 0$ in Eq. (3.26). Hence, at equilibrium (but not at steady-state),

$$\frac{p_j^{\mathrm{eq}}}{p_i^{\mathrm{eq}}} = \frac{k_{ij}}{k_{ji}}\,. \tag{3.28}$$

As follows from these relationships, the equilibrium conditions reduce the number of independent transition rate coefficients compared to any out-of-equilibrium state. This must be considered during the analysis of actual experimental data. Note that a steady-state spin kinetics can be achieved only in schemes containing a closed cycle [32]. Closed cycle exists in the scheme if it is possible to return to the same state through a different process (i.e. through a different transition state).

3.3.2 *Kinetics of Two-State Exchange*

As already mentioned above, two-state spin kinetics is characterized by the kinetic matrix $\mathbf{K} = \begin{pmatrix} -k_{AB} & k_{BA} \\ k_{AB} & -k_{BA} \end{pmatrix}$. Equilibrium populations calculated from $\mathbf{Kp} = 0$ are given in Eq. (3.29) in Table 3.1a. Since closed cycles are not contained in the simple two-state exchange (Table 3.1a), a steady-state mode is not feasible. It also follows from Eqs. (3.23) and (3.25) that $J_{AB} = J_{BA} = 0$, which is equivalent to the equilibrium condition in Eq. (3.28). However, even in two-state kinetics, a closed cycle can arise if the transition between states is accomplished by more than one process.

To illustrate this case, see the scheme of two-state kinetics with two reversible processes in Table 3.1b. There are two independent reversible processes I (k^{I}_{AB} and k^{I}_{BA}) and II (k^{II}_{AB} and k^{II}_{BA}) both of which cause interconversion of A and B states. Since a closed cycle is present, in theory, there can be non-zero population flux in clockwise ($J^{\mathrm{I}}_{AB} = J^{\mathrm{II}}_{BA} > 0$), or counterclockwise ($J^{\mathrm{I}}_{AB} = J^{\mathrm{II}}_{BA} < 0$) directions. However, the NMR spectral lineshape for a two-process two-state exchange (Table 3.1b) is in principle indistinguishable from one-process two-state exchange (Table 3.1a). Consequently, steady-state and equilibrium modes in the two-process case also cannot be distinguished (both can be mapped on an equilibrium one-process spectral lineshape). There are many examples of two-state exchange with more than one process, some of which enable steady-state, including the conversion of a substrate to a product in reversible enzyme-catalyzed reactions according to the Michaelis-Menten scheme [33], conformational variation in dimesityl systems, where both mesityl rings can flip about the connecting single bonds [17, 34, 35] or other compounds with correlated internal motions [36, 37]. These examples are described in detail in our paper [5].

Table 3.1 Three-state spin kinetics and its special cases

Type	Steady-state populations	Equilibrium populations
(a) General		
$A \underset{k_{BA}}{\overset{k_{AB}}{\rightleftharpoons}} B$	No closed cycle, steady-state impossible	$p_A^{\mathrm{eq}} = \frac{k_{BA}}{k_{AB}+k_{BA}}$ (3.29a) $p_B^{\mathrm{eq}} = \frac{k_{AB}}{k_{AB}+k_{BA}}$ (3.29b)
(b) Two processes		
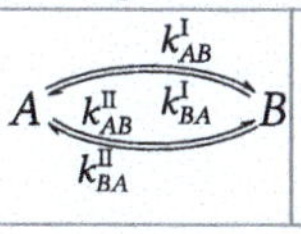	$p_A^{\mathrm{ss}} = \frac{k^{\mathrm{I}}_{BA}+k^{\mathrm{II}}_{BA}}{k^{\mathrm{I}}_{AB}+k^{\mathrm{II}}_{AB}+k^{\mathrm{I}}_{BA}+k^{\mathrm{II}}_{BA}}$ (3.30) $p_B^{\mathrm{ss}} = \frac{k^{\mathrm{I}}_{AB}+k^{\mathrm{II}}_{AB}}{k^{\mathrm{I}}_{AB}+k^{\mathrm{II}}_{AB}+k^{\mathrm{I}}_{BA}+k^{\mathrm{II}}_{BA}}$ (3.31)	$p_A^{\mathrm{eq}} = \frac{k^{\mathrm{I}}_{BA}}{k^{\mathrm{I}}_{AB}+k^{\mathrm{I}}_{BA}} = \frac{k^{\mathrm{II}}_{BA}}{k^{\mathrm{II}}_{AB}+k^{\mathrm{II}}_{BA}}$ (3.32a) $p_B^{\mathrm{eq}} = \frac{k^{\mathrm{I}}_{AB}}{k^{\mathrm{I}}_{AB}+k^{\mathrm{I}}_{BA}} = \frac{k^{\mathrm{II}}_{AB}}{k^{\mathrm{II}}_{AB}+k^{\mathrm{II}}_{BA}}$ (3.32b)

3.3.3 Kinetics of Three-State Exchange

Transitions among three states are described by six transition rate coefficients according to the scheme in Table 3.2a with the corresponding kinetic matrix

$$\mathbf{K} = \begin{pmatrix} -k_{AB} - k_{AC} & k_{BA} & k_{CA} \\ k_{AB} & -k_{BA} - k_{BC} & k_{CB} \\ k_{AC} & k_{BC} & -k_{CA} - k_{CB} \end{pmatrix} . \tag{3.33}$$

The equilibrium/steady-state populations can be determined using the equation $\mathbf{Kp} = \mathbf{0}$ with populations $\mathbf{p} = (p_A, p_B, p_C)^\mathsf{T}$, which can also be reformulated using population fluxes (Eq. (3.26)) as

$$J_{BA} + J_{CA} = 0 , \tag{3.34a}$$

$$J_{AB} + J_{CB} = 0 , \tag{3.34b}$$

$$J_{AC} + J_{BC} = 0 . \tag{3.34c}$$

The expressions for populations of the general three-state kinetics and two special cases are summarized in Table 3.2. In contrast to simple two-state kinetics (Table 3.1a), general three-state kinetics (Table 3.2a) contains a cycle and thus enabling the existence of steady-state. However, imposing symmetry (Table 3.2b) or complexity reduction (Table 3.2c) to the general three-state scheme leads only to equilibrium solution and any steady-state solution is absent.

In contrast to simple two-state kinetics (Table 3.1a), general three-state kinetics (Table 3.2a) contains a cycle and thus enabling the existence of steady-state. The expressions for steady-state and equilibrium populations of the general three-state kinetics and two special cases are summarized in Table 3.2. For the special cases, it can be seen that imposing symmetry (Table 3.2b) or complexity reduction (Table 3.2c) to the general three-state scheme leads only to solutions in equilibrium and a steady-state solution is absent (for more details, see text below).

In the case of general three-state kinetics, the populations for steady-state p_j^{ss} are given in Eq. (3.29) in Table 3.2a.

At equilibrium, this solution further simplifies when it is combined with the equilibrium conditions ($\frac{p_j^{\mathrm{eq}}}{p_i^{\mathrm{eq}}} = \frac{k_{ij}}{k_{ji}}$, Eq. (3.28)). Since the transition rate coefficients are not mutually independent at equilibrium, we obtain

$$1 = \frac{p_C^{\mathrm{eq}}}{p_B^{\mathrm{eq}}} \frac{p_B^{\mathrm{eq}}}{p_A^{\mathrm{eq}}} \frac{p_A^{\mathrm{eq}}}{p_C^{\mathrm{eq}}} = \frac{k_{BC}}{k_{CB}} \frac{k_{AB}}{k_{BA}} \frac{k_{CA}}{k_{AC}} . \tag{3.39}$$

If we choose, for example, k_{BA} as the dependent rate coefficient, then the equilibrium populations can be obtained according to Eq. (3.36abc) in Table 3.2a. The dependent transition rate coefficient k_{BA} is calculated according to Eq. (3.36d). See Appendix B.3 for derivation of the populations in *Mathematica*. It is possible to select any other

Table 3.2 Three-state spin kinetics and its special cases

Type	Steady-state populations	Equilibrium populations
(a) General		
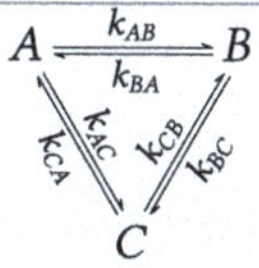	$p_j^{ss} = \frac{\pi_j}{\pi_A+\pi_B+\pi_C}$ (3.35) $j = A, B, C$ $\pi_A = k_{BA}k_{CA} + k_{BC}k_{CA} + k_{BA}k_{CB}$ $\pi_B = k_{AB}k_{CA} + k_{AB}k_{CB} + k_{AC}k_{CB}$ $\pi_C = k_{AC}k_{BA} + k_{AB}k_{BC} + k_{AC}k_{BC}$	$p_A^{eq} = \frac{k_{BC}k_{CA}}{k_{AC}k_{BC}+k_{AC}k_{CB}+k_{BC}k_{CA}}$ (3.36a) $p_B^{eq} = \frac{k_{AC}k_{CB}}{k_{AC}k_{BC}+k_{AC}k_{CB}+k_{BC}k_{CA}}$ (3.36b) $p_C^{eq} = \frac{k_{AC}k_{BC}}{k_{AC}k_{BC}+k_{AC}k_{CB}+k_{BC}k_{CA}}$ (3.36c) $k_{AB} = \frac{k_{AC}k_{BA}k_{CB}}{k_{BC}k_{CA}}$ (3.36d)
(b) Half-symmetric		
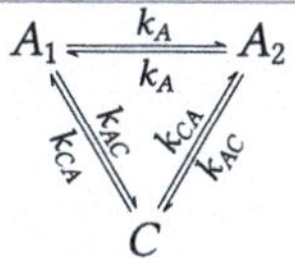	Symmetry prevents steady-state	$p_{A_1}^{eq} = p_{A_2}^{eq} = \frac{k_{CA}}{2k_{CA}+k_{AC}}$ (3.37a) $p_C^{eq} = \frac{k_{AC}}{2k_{CA}+k_{AC}}$ (3.37b)
(c) Consecutive		
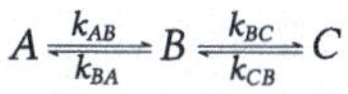	No closed cycle, steady-state impossible	$p_A^{eq} = \frac{k_{BA}k_{CB}}{k_{BA}k_{CB}+k_{AB}k_{BC}+k_{AB}k_{CB}}$ (3.38a) $p_B^{eq} = \frac{k_{AB}k_{CB}}{k_{BA}k_{CB}+k_{AB}k_{BC}+k_{AB}k_{CB}}$ (3.38b) $p_C^{eq} = \frac{k_{AB}k_{BC}}{k_{BA}k_{CB}+k_{AB}k_{BC}+k_{AB}k_{CB}}$ (3.38c)

transition rate coefficient as dependent. It follows from these relationships that the equilibrium condition reduces the number of independent transition rate coefficients during lineshape fitting. Prior to use of the three-state lineshape formula derived later in Sect. 3.3.7, the populations and transition rate coefficients for steady-state should be related using Eq. (2.29a) and for equilibrium using Eq. (3.36).

The first special case of three-state kinetics is denoted *half-symmetric* and is shown in Table 3.2b. In this case, the states A and B are denoted as A_1 and A_2, respectively, due to the symmetry in the corresponding chemical species. Hence, the states A_1 and A_2 are equally populated, and the system is described by three transition rate coefficients: k_{AC}, k_{CA} and k_A. Populations given by Eq. (3.37) already imply equilibrium because they obey the equilibrium conditions in Eq. (3.28). Steady-state with non-zero net fluxes is not possible for this kinetic scheme. The second special case is a *consecutive* kinetic scheme, which is shown in Table 3.2c. As this scheme lacks any closed cycle, only equilibrium is possible with populations expressed in Eq. (3.38).

3.3.4 Bloch-McConnell Equations

Exchange spin dynamics is described by *Bloch-McConnell equations* [2, 38], which originate from Bloch equations (Eq. (3.10)) and possess additional terms accounting for transfer of magnetization through the exchange process. For two-state chemical exchange the Bloch-McConnell equations are as follows

(laboratory frame)

$$\frac{\mathrm{d}M_x^A}{\mathrm{d}t} = \gamma(\mathbf{M}^A \times \mathbf{B}^A)_x - R_2^A M_x^A - k_{AB} M_x^A + k_{BA} M_x^B \,, \tag{3.40a}$$

$$\frac{\mathrm{d}M_y^A}{\mathrm{d}t} = \gamma(\mathbf{M}^A \times \mathbf{B}^A)_y - R_2^A M_y^A - k_{AB} M_y^A + k_{BA} M_y^B \,, \tag{3.40b}$$

$$\frac{\mathrm{d}M_z^A}{\mathrm{d}t} = \gamma(\mathbf{M}^A \times \mathbf{B}^A)_z + R_1^A (M_0^A - M_z^A) - k_{AB} M_z^A + k_{BA} M_z^B \,, \tag{3.40c}$$

$$\frac{\mathrm{d}M_x^B}{\mathrm{d}t} = \gamma(\mathbf{M}^B \times \mathbf{B}^B)_x - R_2^B M_x^B - k_{BA} M_x^B + k_{AB} M_x^A \,, \tag{3.40d}$$

$$\frac{\mathrm{d}M_y^B}{\mathrm{d}t} = \gamma(\mathbf{M}^B \times \mathbf{B}^B)_y - R_2^B M_y^B - k_{BA} M_y^B + k_{AB} M_y^A \,, \tag{3.40e}$$

$$\frac{\mathrm{d}M_z^B}{\mathrm{d}t} = \gamma(\mathbf{M}^B \times \mathbf{B}^B)_z + R_1^B (M_0^B - M_z^B) - k_{BA} M_z^B + k_{AB} M_z^A \,. \tag{3.40f}$$

Let us use these equations to describe FID in constant external field $\mathbf{B}_0$. Using the definition for complex transverse magnetization $M_{xy}^j = M_x^j + i M_y^j$ of nuclei in states $j = A, B$, we obtain in the rotating frame of reference (see Sect. 3.2.1)

(rotating frame)

$$\frac{\mathrm{d}M_{xy}^A}{\mathrm{d}t} = i\Omega_A M_{xy}^A - R_2^A M_{xy}^A - k_{AB} M_{xy}^A + k_{BA} M_{xy}^B \,, \tag{3.41a}$$

$$\frac{\mathrm{d}M_{xy}^B}{\mathrm{d}t} = i\Omega_B M_{xy}^B - R_2^B M_{xy}^B - k_{BA} M_{xy}^B + k_{AB} M_{xy}^A \,. \tag{3.41b}$$

Equation (3.41) can be conveniently expressed in matrix form for arbitrary number of states N. Let us define a vector of complex transverse magnetizations $\mathbf{M}_{xy} = (M_{xy}^A, M_{xy}^B, \ldots, M_{xy}^N)^\mathrm{T}$. Then the McConnell equations of general N-state exchange (of non-J-coupled spins) for FID signal are

(rotating frame)

$$\frac{\mathrm{d}}{\mathrm{d}t}\mathbf{M}_{xy}(t) = (i\mathbf{L} - \mathbf{R_2} + \mathbf{K})\,\mathbf{M}_{xy}(t) \,, \tag{3.42}$$

where the diagonal[6] matrix $\mathbf{L} = \mathrm{diag}(\Omega_A, \Omega_B, \ldots, \Omega_N)$ describes evolution due to the external magnetic field, the diagonal matrix $\mathbf{R_2} = \mathrm{diag}(R_2^A, R_2^B, \ldots, R_2^N)$ accounts for transverse relaxation, and the kinetic matrix $\mathbf{K}$ characterizes flux of magnetization from one state to another due to the chemical exchange as discussed in detail in previous sections. Bloch-McConnell equations can be solved analytically or numerically in the frequency domain, which will be explained in the next section.

[6] 'diag(•)' represents a diagonal matrix with diagonal elements listed in the parentheses.

Note that this classical description is not applicable in the presence of J-coupling, which requires quantum mechanical approach [39, 40].

3.3.5 Solution of Bloch-McConnell Equations

In literature reports [1, 2, 18, 28] the differential Bloch-McConnell equations (Eq. (3.42)) are usually solved using matrix exponential involving matrix diagonalization. Subsequently, the time-dependent vector of complex transverse magnetizations $\mathbf{M}_{xy}(t)$ is converted to the frequency domain using FT. This method is suitable for symmetric two-state exchange, however, it is already inconvenient for the case of an asymmetric two-state exchange. Therefore, other studies [41–43] use Bloch-McConnell equations in the frequency domain for numerical simulations of exchange spectra. FT converts the differential Bloch-McConnell equations into algebraic equations, which can be solved either numerically or analytically using standard methods.

The following derivation holds for an arbitrary number of states and arbitrary kinetic schemes. First, FT is applied to Eq. (3.42). Note that the signal from the spectrometer is not actually infinite, but it is equal to zero for $t < 0$. Therefore, the functions $f^j(t) = \theta(t)M^j_{xy}(t)$ should be transformed instead of $M^j_{xy}(t)$, $j = A, B, \ldots, N$. The function $\theta(t)$ is the Heaviside step function, equal to zero for $t < 0$ and unity otherwise. The left-hand side of Eq. (3.42) multiplied by the Heaviside function can be rewritten using the Dirac delta function $\delta(t)$,

$$\theta(t)\frac{\mathrm{d}M^j_{xy}(t)}{\mathrm{d}t} = \frac{\mathrm{d}}{\mathrm{d}t}\left[\theta(t)M^j_{xy}(t)\right] - M^j_{xy}(t)\frac{\mathrm{d}\theta(t)}{\mathrm{d}t} = \frac{\mathrm{d}f(t)}{\mathrm{d}t} - M^j_{xy}(t)\delta(t)\,. \quad (3.43)$$

FT of the first term on the right-hand side is[7]

$$\widehat{\frac{\mathrm{d}f(t)}{\mathrm{d}t}} = i\Omega\hat{f}(\Omega)\,, \quad (3.44)$$

[7] The convention for FT used in this work is $\hat{f}(\Omega) = \int_{-\infty}^{+\infty} f(t)e^{-i\Omega t}\mathrm{d}t$. The corresponding inverse FT is then $f(t) = \frac{1}{2\pi}\int_{-\infty}^{+\infty} \hat{f}(\Omega)e^{i\Omega t}\mathrm{d}\Omega$.

and FT of the second term is

$$\int_{-\infty}^{+\infty} M_{xy}^{j}(t)\delta(t)e^{-i\Omega t}\mathrm{d}t = M_{xy}^{j}(t=0)\,. \tag{3.45}$$

Without loss of generality, let us assume that the initial transverse magnetization is real and positive, $M_{xy}^{j}(t=0) = M_0^j$. Using Eqs. (3.43), (3.44) and (3.45), FT of Eq. (3.42) yields

$$[i(\Omega\mathbf{1} - \mathbf{L}) + \mathbf{R} - \mathbf{K}]\begin{pmatrix}\hat{f}_A(\Omega)\\ \hat{f}_B(\Omega)\\ \vdots\\ \hat{f}_N(\Omega)\end{pmatrix} = \begin{pmatrix}M_0^A\\ M_0^B\\ \vdots\\ M_0^N\end{pmatrix}, \tag{3.46}$$

where $\mathbf{1}$ is the unit matrix, $\mathbf{L} = \mathrm{diag}(\Omega_A, \Omega_B, \ldots, \Omega_N)$, $\mathbf{R} = \mathrm{diag}(R_2^A, R_2^B, \ldots, R_2^N)$ and $\mathbf{K}$ is the appropriate kinetic matrix. Using the definition $\alpha_j = R_j + i(\Omega - \Omega_j)$ we can rewrite Eq. (3.46) as

$$[\mathbf{A} - \mathbf{K}]\begin{pmatrix}\hat{f}_A(\Omega)\\ \hat{f}_B(\Omega)\\ \vdots\\ \hat{f}_N(\Omega)\end{pmatrix} = M_0\begin{pmatrix}p_A\\ p_B\\ \vdots\\ p_N\end{pmatrix}, \tag{3.47}$$

where $\mathbf{A} = \mathrm{diag}(\alpha_A, \alpha_B, \ldots, \alpha_N)$ and $M_0^j = M_0 p_j$.

Solution of this set of linear equations for $\hat{f}_j(\Omega)$ is straightforward. Resulting complex NMR signal is detected as the sum of all components,

$$S(\Omega) = \hat{f}_A(\Omega) + \hat{f}_B(\Omega) + \cdots + \hat{f}_N(\Omega)\,. \tag{3.48}$$

Solution for two- and three-state exchange is given below. Implementation of this analytical procedure in *Mathematica* to obtain four- and five-state exchange spectral lineshapes is shown in Appendix B.2. The derivation of the analytical solution also provides a scheme for the numerical simulation of the spectrum [41, 42]. Multiplying Eq. (3.47) by $[\mathbf{A} - \mathbf{K}]^{-1}$ from left yields $\hat{f}_j(\Omega)$ values for a fixed Ω. Hence, for N states, inversion of an $N \times N$ matrix is required for each frequency value. Numerical solution in *MATLAB* is shown in our paper [5].

So-called steady-state approximation (in the time domain) for continuous-wave NMR spectroscopy [44–46] provides an equation fully analogous to Eq. (3.47) without a requirement to apply FT. However, the angular frequency Ω in those derivations stands for the frequency of applied continuous radiofrequency field, not for the frequency of the detected signal (as in pulse NMR approach).

3.3.6 Two-State Exchange Spectral Lineshape

The solution of the system of Bloch-McConnell equations (Eq. (3.42)) for two-state exchange is a standard part of literature reports in the symmetric [1, 2, 28] and also asymmetric case [18, 47–49]. The formula in complex form reads

$$S_{\text{two-state exch.}}(\Omega) = M_0 \frac{p_A\alpha_B + p_B\alpha_A + k_{AB} + k_{BA}}{\alpha_A\alpha_B + k_{AB}\alpha_B + k_{BA}\alpha_A}, \tag{3.49a}$$

where

$$\alpha_j = R_2^j + i(\Omega - \Omega_j), \tag{3.49b}$$

for $j = A, B$. Total transverse magnetization is denoted as $M_0 = M_0^A + M_0^B$. At equilibrium, the populations and transition rate coefficients in Eq. (3.49) are connected by the algebraic Eq. (3.29).

The expression α_j^{-1} describes the complex Lorentzian lineshape with Larmor frequency Ω_j and transverse relaxation R_2^j, see Sect. 3.2.3. This is useful for calculation of two limiting cases explaining the spectral behavior in Fig. 3.4e: (**i**) The infinitely slow exchange limit gives

$$\lim_{k_{AB},k_{BA}\to 0} S_{\text{two-state exch.}}(\Omega) = M_0 \left(\frac{p_A}{\alpha_A} + \frac{p_B}{\alpha_B}\right), \tag{3.50}$$

which is, as expected, sum of two Lorentzian peaks. (**ii**) The infinitely fast exchange limit should be done in equilibrium regime, i.e. keeping the ratio of transition rate coefficients constant and equal to the ratio of populations, which yields

$$\lim_{k_{AB},k_{BA}\to \infty} S_{\text{two-state exch.}}(\Omega) \left[\frac{k_{AB}}{k_{BA}} = \frac{p_B}{p_A}\right] = \frac{M_0}{p_A\alpha_A + p_B\alpha_B}. \tag{3.51}$$

This expression is a single Lorentzian peak located at averaged frequency $p_A\Omega_A + p_B\Omega_B$ with averaged relaxation rate $p_A R_2^A + p_B R_2^B$. Lineshapes with larger number of states behave analogously in slow and fast limit.

For lineshape fitting, it is convenient to express the real (absorption) part of the spectrum from Eq. (3.49). The result and its implementation in *Mathematica* can be found in Appendix B.1.2. The analytical lineshape can also be used to calculate the position of coalescence point in Eq. (3.20), detailed treatment is given in our paper [5].

3.3.7 Three-State Exchange Spectral Lineshape

The analytical solution of the system of Bloch-McConnell equations (Eq. (3.42)) for three-state exchange was, to the best of our knowledge, first published in our paper [5] (although special cases were presented during the early years of NMR spectroscopy [47, 50]). It should be noted that Kovrigin [41] analyzed numerically several properties of three-state (and four-state) chemical exchange. There also exist NMR spectral simulation programs (e.g. *NmrLineGuru* [42] and *TITAN* [24]), which enable the numerical fitting of multi-state exchange lineshapes.

The spectral lineshape for three-state chemical exchange, derived using the procedure in Sect. 3.3.5, is as follows

$$S_{\text{three-state exch.}}(\Omega) = M_0 \frac{\mathcal{P}}{\mathcal{Q}}, \tag{3.52}$$

where

$$\begin{aligned}
M_0 &= M_0^A + M_0^B + M_0^C , \\
\mathcal{P} &= p_A[\alpha_B\alpha_C + \alpha_B(k_{CA} + k_{CB} + k_{AC}) + \alpha_C(k_{BA} + k_{BC} + k_{AB})] \\
&\quad + p_B[\alpha_A\alpha_C + \alpha_A(k_{CA} + k_{CB} + k_{BC}) + \alpha_C(k_{AB} + k_{AC} + k_{BA})] \\
&\quad + p_C[\alpha_A\alpha_B + \alpha_A(k_{BA} + k_{BC} + k_{CB}) + \alpha_B(k_{AB} + k_{AC} + k_{CA})] \\
&\quad + \pi_A + \pi_B + \pi_C , \\
\mathcal{Q} &= \alpha_A\alpha_B\alpha_C + \alpha_A\alpha_B(k_{CA} + k_{CB}) + \alpha_A\alpha_C(k_{BA} + k_{BC}) + \alpha_B\alpha_C(k_{AB} + k_{AC}) \\
&\quad + \alpha_A\pi_A + \alpha_B\pi_B + \alpha_C\pi_C , \\
\pi_A &= k_{BA}k_{CA} + k_{BC}k_{CA} + k_{BA}k_{CB} , \\
\pi_B &= k_{AB}k_{CA} + k_{AB}k_{CB} + k_{AC}k_{CB} , \\
\pi_C &= k_{AC}k_{BA} + k_{AB}k_{BC} + k_{AC}k_{BC} .
\end{aligned}$$

Quantities π_j are proportional to the steady-state population flux as can be deduced from Eq. (3.29).

Details about lineshape fitting in different modes (i.e. equilibrium, steady-state and out-of-steady-state mode with time-dependent populations) of the spin kinetics and discussion about the interdependency of the fitted parameters during lineshape analysis can be found in our paper [5]. For lineshape fitting, it is convenient to express the real (absorption) part of the spectrum from Eq. (3.52). The result and its implementation in *Mathematica* can be found in Appendix B.1.3. It is also worth mentioning that a steady-state in the general three-state exchange scheme cannot be distinguished from an equilibrium by analyzing a single NMR spectrum since the equilibrium lineshape (Eq. (3.52) and (3.36)) can also be fitted to the steady-state lineshape (Eq. (3.52) and (3.29), although they are not mathematically identical [5]. Analytical three-state exchange spectral lineshape has been used for fitting of actual NMR data in [5, 51].

3.3.8 Exchange Schemes Containing a Fast Process

Let us consider a consecutive exchange with slow or intermediate exchange between states A and B and fast exchange between B and C, see Scheme 3.2a.

The two fast-exchanging peaks merge, so that only two peaks can be observed. This situation can be modeled as limit of $k_{BC}, k_{CB} \to \infty$ in Eq. (3.52), where the ratio $k_{BC}/k_{CB} = p_C/p_B$ is kept constant. The resulting equation is

$$S_{\text{three-state exch.limit}}(\Omega) = M_0 \frac{\mathcal{P}}{\mathcal{Q}}, \tag{3.53}$$

where

$$M_0 = M_0^A + M_0^B + M_0^C,$$
$$\mathcal{P} = p_A \frac{p_B \alpha_B + p_C \alpha_C}{p_B + p_C} + (p_B + p_C)\alpha_A + \frac{k_{BA} p_B}{p_B + p_C} + k_{AB},$$
$$\mathcal{Q} = \alpha_A \frac{p_B \alpha_B + p_C \alpha_C}{p_B + p_C} + \frac{k_{BA} p_B}{p_B + p_C} \alpha_A + k_{AB} \frac{p_B \alpha_B + p_C \alpha_C}{p_B + p_C}.$$

In comparison with Eq. (3.49), it becomes obvious that three-state exchange containing a fast process can be modeled as a two-state exchange between states A and B' with transition rate coefficients k_{AB} and $k_{B'A}$, where $k_{B'A}$ has the following relationship to the transition rate coefficient k_{BA} with physical meaning

$$k_{B'A} = k_{BA} \frac{p_B}{p_B + p_C}. \tag{3.54}$$

The offset resonant frequency of the state B' is the population weighted average of the frequencies Ω_B and Ω_C, in particular $\Omega_{B'} = (p_B \Omega_B + p_C \Omega_C)/(p_B + p_C)$ (by analogy with fast asymmetric two-state exchange). Similarly, it holds that $R_2^{B'} = (p_B R_2^B + p_C R_2^C)/(p_B + p_C)$ for the relaxation rate. We can view the B' replacement

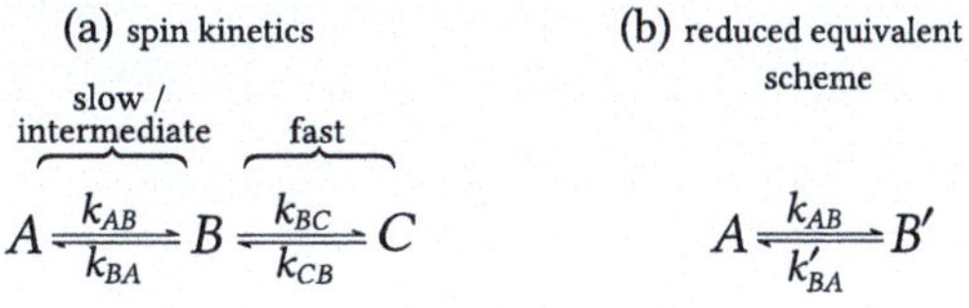

Scheme 3.2 Consecutive three-state exchange containing a fast exchange process. (**a**) Spin kinetics scheme and (**b**) the corresponding reduced equivalent scheme

state as being merged B and C since $M_{B'} = M_B + M_C$ (and consequently $p_{B'} = p_B + p_C$). The full kinetic scheme is simplified to a *reduced equivalent scheme* as seen in Scheme 3.2b. To maintain the transition rate from B' to A (one-way population flux, i.e. $k_{B'A} p_{B'}$) the same as the transition rate from B to A ($k_{BA} p_B$) in the original scheme, the equality $k_{B'A} p_{B'} = k_{BA} p_B$ must hold. Because $p_{B'}$ is larger than p_B, the modified constant $k_{B'A}$ should be smaller than the original constant k_{BA}. Note that only the transition rate coefficient *out* of the replacement state B' is modified.

Modified transition rate coefficient $k_{B'A}$ can also be obtained directly from the differential equations $\frac{\mathrm{d}\mathbf{p}}{\mathrm{d}t} = \mathbf{Kp}$ without the use of analytical lineshapes. In this case, the equations are as follows

$$\frac{\mathrm{d}p_A}{\mathrm{d}t} = -k_{AB} p_A + k_{BA} p_B \,, \tag{3.55a}$$

$$\frac{\mathrm{d}p_B}{\mathrm{d}t} = -(k_{BA} + k_{BC}) p_B + k_{AB} p_A + k_{CB} p_C \,, \tag{3.55b}$$

$$\frac{\mathrm{d}p_C}{\mathrm{d}t} = -k_{CB} p_C + k_{BC} p_B \,. \tag{3.55c}$$

After summing Eq. (3.55b) and Eq. (3.55c), expanding $p_B = \frac{p_B}{p_B + p_C}(p_B + p_C)$ and substituting $p_B + p_C = p_{B'}$ we obtain

$$\frac{\mathrm{d}p_A}{\mathrm{d}t} = -k_{AB} p_A + k_{BA} \frac{p_B}{p_B + p_C} p_{B'} \,, \tag{3.56a}$$

$$\frac{\mathrm{d}p_{B'}}{\mathrm{d}t} = -k_{BA} \frac{p_B}{p_B + p_C} p_{B'} + k_{AB} p_A \,. \tag{3.56b}$$

Prefactors of $p_{B'}$ in the above equations yield the correct form of $k_{B'A}$ (cf. Eq. (3.54)).

Reduced equivalent schemes have been used for analysis of actual NMR data with fast chemical exchange in [5, 52].

3.4 NMR of Host-Guest Systems

3.4.1 Chemical Exchange in Host-Guest Complexes

In this section, we provide the relationship between *reaction rate coefficients* κ_{ij} (introduced in Sect. 2.1.3), which describe the kinetics of host-guest binding, and *transition rate coefficients* k_{ij} which describe transitions of a particular spin between corresponding states. Reaction rate coefficients are independent of concentration and their temperature dependence is governed by the Eyring equation (Eq. (2.20)). On the other hand, transition rate coefficients may or may not be dependent on concentration. The relationship between transition and reaction rate coefficients also automatically

Scheme 3.3 Two-state exchange with 1:1 host-guest binding. Schemes for **a** chemical kinetics and **b** corresponding spin kinetics of a nucleus located at the host molecule

(a) chemical kinetics

$$\mathbf{H} + \mathbf{G} \underset{\kappa_{BA}}{\overset{\kappa_{AB}}{\rightleftharpoons}} \mathbf{HG}$$

(b) spin kinetics

$$A \underset{k_{BA}}{\overset{k_{AB}}{\rightleftharpoons}} B$$

reveals the concentration dependence of transition rate coefficients, cf. Sect. 5.6.1 in [2].

Let us consider the 1:1 binding of a host (**H**) with a guest (**G**) in Scheme 3.3a with an equilibrium constant

$$K_{\mathbf{HG}} = \frac{[\mathbf{HG}]}{[\mathbf{H}][\mathbf{G}]} = \frac{\kappa_{AB}}{\kappa_{BA}} . \tag{3.57}$$

Guldberg-Waage's law of mass action (Sect. 2.1.3) assumes that the forward and backward rates of a chemical reaction are proportional to the concentrations of reacting molecules,

$$\text{forward reaction rate} = \kappa_{AB}[\mathbf{G}][\mathbf{H}] , \tag{3.58a}$$

$$\text{backward reaction rate} = \kappa_{BA}[\mathbf{HG}] . \tag{3.58b}$$

The free host **H** can be assigned to state A and the complex **HG** to state B ($[A] = [\mathbf{H}]$ and $[B] = [\mathbf{HG}]$). The variation of concentration with time of the spin state B associated with the **HG** complex, i.e. the transition rate, is given by Scheme 3.3b and Eq. (3.24) (multiplied by $[\mathbf{H}]_\mathrm{t}$ to transform populations to concentrations, subscript 't' denotes total concentration), yielding $\frac{\mathrm{d}[B]}{\mathrm{d}t} = k_{AB}[A] - k_{BA}[B]$. In the latter equation, the two right-hand side terms are identified as forward and backward transition rates, that is

$$\text{forward transition rate} = k_{AB}[A] , \tag{3.59a}$$

$$\text{backward transition rate} = k_{BA}[B] . \tag{3.59b}$$

A comparison of Eq. (3.58) with Eq. (3.59), using $[A] = [\mathbf{H}]$ and $[B] = [\mathbf{HG}]$, implies that k_{AB} depends on free guest concentration unlike k_{BA},

$$k_{AB} = \kappa_{AB}[\mathbf{G}] , \tag{3.60a}$$

$$k_{BA} = \kappa_{BA} . \tag{3.60b}$$

The free guest concentration $[\mathbf{G}]$ and consequently the value of k_{AB} increases with $[\mathbf{G}]_\mathrm{t}$ and decreases with $[\mathbf{H}]_\mathrm{t}$. For bimolecular elementary reactions the transition rate coefficients usually depend linearly on concentration, while for unimolecular reaction (i.e. decay of the complex) the transition rate coefficients are independent of concentration.

The above formulae (Eq. (3.60)) imply that simple 1:1 host-guest binding does not allow (in general) for symmetric two-state exchange because during the titration experiments (i.e. measurement of spectra at different guest concentrations when the guest is gradually added into the host solution), the p_B/p_A ratio always increases upon addition of guest molecules. Only two out of the three parameters κ_{AB}, κ_{BA} and $K_{\mathbf{HG}}$ are independent. The dependence of [**H**], [**HG**] and [**G**] on the total host and guest concentrations ([**H**]$_t$ and [**G**]$_t$) can be expressed in an analytical form for 1:1 host-guest binding, see Sect. 2.2.1.

Figure 3.5 illustrates a simulated titration of a host with a guest, which form a complex according to Scheme 3.3a. Simulated spectra shown in Fig. 3.5a contain two resonances in an intermediate exchange regime. The resonances belong to the host in states A (corresponding to **H**) and B (corresponding to **HG**). As the total amount of guest ([**G**]$_t$) increases from zero to one equivalent (from bottom to the top in Fig. 3.5a), the population of state B increases (Fig. 3.5b) as well as the concentration of free and bound guest ([**G**] and [**HG**] in Fig. 3.5c). Because the exchange is in an intermediate regime, the apparent peak maxima shift towards each other and eventually coalesce (Fig. 3.5d, details about coalescence in asymmetric two-state exchange can be found in our paper [5]). In accordance with Eq. (3.60), the transition rate coefficient k_{AB} depends on guest concentration, while k_{BA} does not. Note that although the function $k_{AB}([\mathbf{G}])$ is linear, the function $k_{AB}([\mathbf{G}]_t)$ is nonlinear and can be calculated using Eq. (3.60a) and Eq. (2.26c).

Another example of a three-state exchange in the competitive host-ligand binding model, including the lineshapes during a simulated titration experiment can be found in our paper [5].

3.4.2 NMR Titration Binding Isotherms

Titration of the host with the guest is a standard experiment to investigate host-guest supramolecular binding. At the beginning, the sample solution contains only the host molecule and the guest is gradually added from a stock solution (the stock solution may contain the same concentration of host as in the sample to prevent dilution). If the host and guest molecules do not interact, the measured spectrum is just sum of the pure **H** and pure **G** spectra. Their interaction is manifested by appearance of new peaks due to the **HG** complex (slow exchange), by shift of resonances (fast exchange) or by more complex change of the spectral lineshapes (intermediate exchange regime).

A *binding isotherm* describes dependence of a spectral feature (e.g. chemical shift or integrated peak intensity) on the total concentration of added guest in a titration at constant temperature. The binding isotherm can also be calculated from a binding model, which enables fitting of the experimental data and determination of binding constants. Figure 3.6 illustrates construction of binding isotherms using the example of 1:1 host-guest binding from the previous section. Slow exchange between

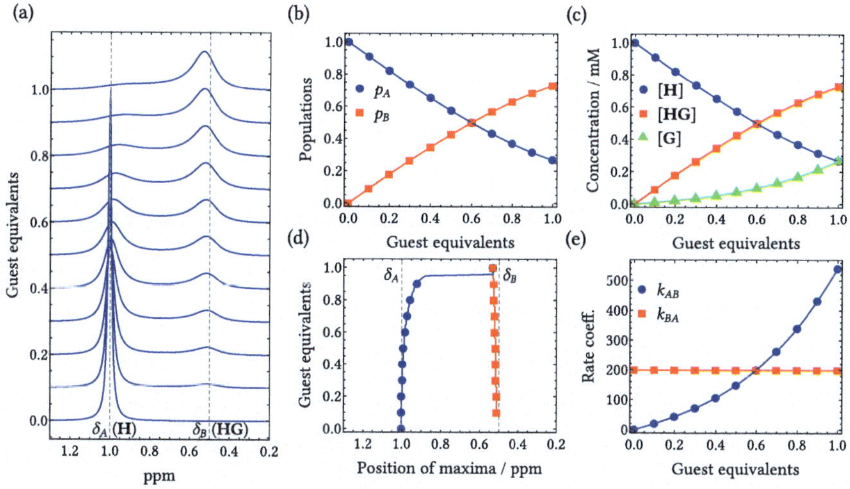

Fig. 3.5 Illustration of two-state exchange combined with 1:1 host-guest binding in a simulated titration of host with guest (at constant $[\mathbf{H}]_t$). **a** Simulated spectral lineshapes (y-scaling of spectra is adjusted for clarity). Concentration of guest corresponds to the value where the simulated spectrum meets the y-axis. The spectrum of the free host is cropped for clarity. Concentration dependence of **b** populations, **c** concentration of all species, **d** apparent positions of peak maxima and **e** transition rate coefficients. Discrete points in plots of various quantities correspond to the spectra shown on the left. Simulation parameters: $R_2^A = R_2^B = 30\ \mathrm{s}^{-1}$, $\kappa_{AB} = 2 \times 10^6\ \mathrm{s}^{-1}\mathrm{M}^{-2}$, $K_{\mathbf{HG}} = 1 \times 10^4\ \mathrm{M}^{-1}$, $[\mathbf{H}]_t = 1$ mM, spectrometer frequency $\nu_0 = 500.13$ MHz (which is used for conversion from ppm to rad.s^{-1}). Chemical shifts of pure A (corresponding to **H**) and pure B (**HG**) states are $\delta_A = 1$ ppm and $\delta_B = 0.5$ ppm, respectively. Equivalents of guest are defined as $[\mathbf{G}]_t/[\mathbf{H}]_t$. The spectral lineshapes were calculated using Eq. (3.49) and concentrations determined using Eq. (2.26)

free and complexed host enables integration of isolated peaks (numerically or using lineshape fitting) due to free and complexed host as shown in Fig. 3.6a. The binding isotherm corresponding to observed complexed fraction of host can be constructed as $p_{\mathbf{HG},\mathrm{obs}} = \frac{\mathrm{area}(A)}{\mathrm{area}(A)+\mathrm{area}(B)}$, see Fig. 3.6b. Complexed fraction can also be calculated from the host-guest binding model using Eq. (2.26b), $p_{\mathbf{HG},\mathrm{calc}} = \frac{[\mathbf{HG}]}{[\mathbf{H}]_t}$. Hence, $p_{\mathbf{HG},\mathrm{obs}}$ obtained from the experiment can be fitted with $p_{\mathbf{HG},\mathrm{calc}}$ with $K_{\mathbf{HG}}$ as the fitting parameter. To conclude, slow exchange binding isotherm can be constructed as

$$\text{complexed fraction} = \frac{\sum \text{area(peaks due to the complex)}}{\sum \text{area(all peaks)}} . \tag{3.61}$$

Fast exchange regime causes averaging of observed signals and only one peak is observed (cf. Eq. (3.51)), whose observed chemical shift δ_{obs} is the population-weighted average of chemical shifts δ_j of all states j, which are under mutual fast exchange. Mathematically

$$\delta_{\mathrm{obs}} = \sum_{j=A,B,\ldots} p_j \delta_j . \tag{3.62}$$

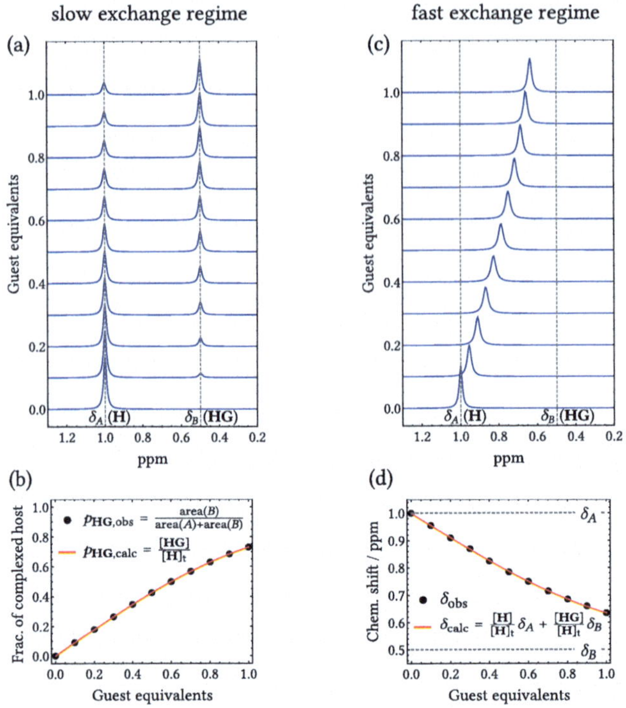

Fig. 3.6 Illustration of NMR titration isotherms for two-state exchange combined with 1:1 host-guest binding in simulated titrations of host with guest (at constant $[\mathbf{H}]_t$). **a** Simulated spectral lineshapes in slow exchange regime. **b** Fraction of complexed host and the slow exchange binding isotherm. **c** Simulated spectral lineshapes in fast exchange regime. **d** Position of the observed averaged peak (δ_{obs}) and the fast exchange binding isotherm. Concentration of guest corresponds to the value where the simulated spectrum meets the y-axis in **a**, **c**, y-scaling of the spectra is adjusted for clarity. Discrete points in **b**, **d** correspond to the spectra shown in **a**, **c**, respectively. Simulation parameters are $\kappa_{AB} = 2 \times 10^4\ \text{s}^{-1}\text{M}^{-2}$ in **a** and $\kappa_{AB} = 2 \times 10^8\ \text{s}^{-1}\text{M}^{-2}$ in **b**, the other parameters are identical as in Fig. 3.5. Equivalents of guest are defined as $[\mathbf{G}]_t/[\mathbf{H}]_t$. The spectral lineshapes were calculated using Eq. (3.49) and concentrations determined using Eq. (2.26)

Considering our example with two states shown in Fig. 3.6c, the binding isotherm is $\delta_{\text{obs}} = p_A\delta_A + p_B\delta_B$ (see page 194 in [53]). The observed chemical shift can be fitted with $\delta_{\text{calc}} = \frac{[\mathbf{H}]}{[\mathbf{H}]_t}\delta_A + \frac{[\mathbf{HG}]}{[\mathbf{H}]_t}\delta_B$ with $K_{\mathbf{HG}}$ as the fitting parameter, where $[\mathbf{H}]$ and $[\mathbf{HG}]$ are calculated using Eq. (2.26), see Fig. 3.6d.

If the NMR resonances are in intermediate regime as in Fig. 3.5a, then the spectral pattern must be fitted with the appropriate exchange lineshape and the aforementioned slow and fast binding isotherms are not applicable.

3.4.3 Determination of Enantiomeric Excess

Mirror image of a *chiral molecule* is non-superposable with the original (i.e. it lacks symmetry elements of mirror plane, inversion center or improper rotation axis [54]). The two mirror forms are called *enantiomers* and are often denoted as (*R*) and (*S*).[8] Synthesis of chiral molecules mostly produces a mixture of both enantiomers, which is characterized by *enantiomeric excess* e.e. defined as

$$\text{e.e.} = \frac{[R] - [S]}{[R] + [S]} \,. \tag{3.63}$$

Thus, e.e. = 1 and e.e. = −1 imply enantiopure (*R*)- and (*S*)-form, respectively; e.e. = 0 implies racemate (i.e. 1:1 mixture of (*R*)- and (*S*)-form).

NMR spectra of both enantiomers are identical and thus indistinguishable. Their distinction is enabled by the presence of additional chiral agents in the sample (alternatively, the solvent itself can serve as chiral agent [56]), which form covalent (chiral derivatizing agents) or non-covalent (chiral solvating agents, chiral lanthanide shift reagents) bonds with the chiral analyte [56–58]. Binding of the chiral agent with two enantiomeric forms of chiral analyte produces so-called *diastereomers*, i.e. stereoisomers which are not mutual mirror images[9] (possible types of diastereomers in the case of 1:1 binding are (*R*,*R*) and (*R*,*S*) or alternatively (*S*,*R*) and (*S*,*S*)). Diastereomers have different NMR spectra thus enabling determination of e.e. The use of chiral derivatizing agents is limited by kinetic resolution (i.e. different reaction rate for (*R*)- and (*S*)-enantiomers) or by racemization (e.g. by heat) during the analyte-agent reaction [59]. These disadvantages are not present, when non-covalently bound agents are used.

Let us consider non-covalent binding of a chiral solvating agent denoted as host **H** (enantiopure, can be either (*R*)- or (*S*)-form) with enantiomeric mixture of guest analytes (*R*)-**G** and (*S*)-**G** according to the scheme in Fig. 3.7a. Fast exchange between free molecules and their complex is required to produce sharp signals for peak integration. The reference spin of (*R*)-**G** undergoes exchange between states R_{free} and R_{bound} forming a replacement state R' (see Sect. 3.3.8), similarly, reference spin of (*S*)-**G** undergoes exchange between states S_{free} and S_{bound} forming a replacement state S', see Fig. 3.7b. According to Eq. (3.62) signals of states R' and S' are located at

$$\delta_{R'} = p_{R,\text{free}}\delta_{R,\text{free}} + p_{R,\text{bound}}\delta_{R,\text{bound}} \,, \tag{3.64a}$$

$$\delta_{S'} = p_{S,\text{free}}\delta_{S,\text{free}} + p_{S,\text{bound}}\delta_{S,\text{bound}} \,. \tag{3.64b}$$

[8] Descriptors (*R*)/(*S*) are used, when chirality is caused by a particular asymmetric center in the molecule, mostly an sp^3 carbon atom. Other descriptors exist, see page 304 in [55].

[9] Diastereomers https://goldbook.iupac.org/terms/view/D01679.

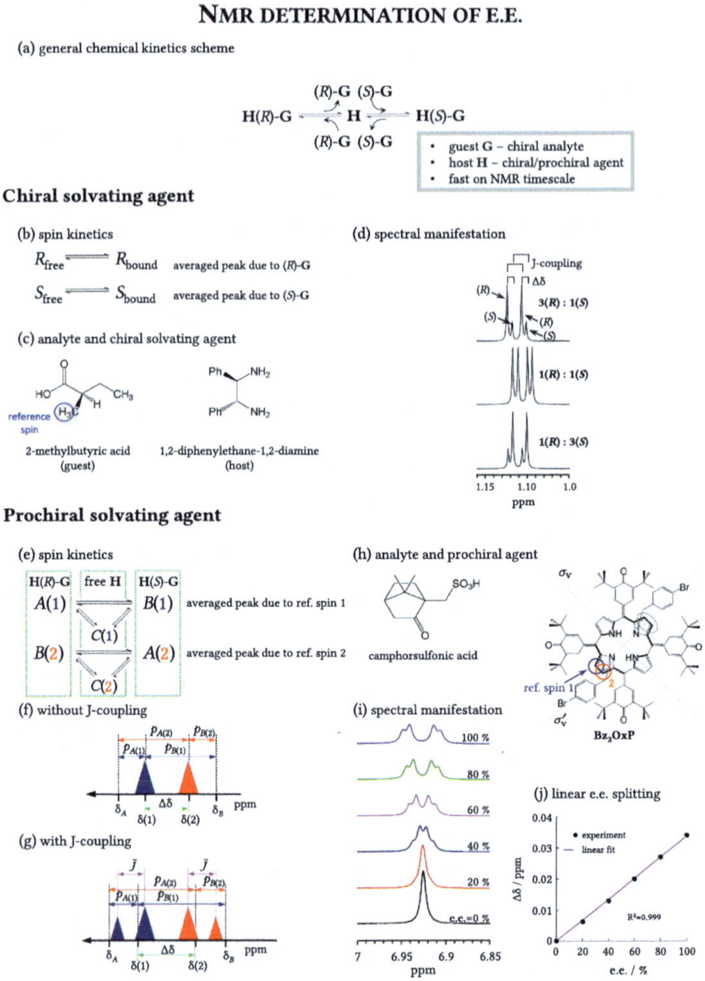

Fig. 3.7 Determination of enantiomeric excess (e.e.) using chiral or prochiral solvating agents. **a** Chemical kinetics scheme usable in both methods. The processes must be fast on NMR timescale. **Chiral solvating agent**: **b** Spin kinetics of two reference spins located at two different molecules (R)-**G** and (S)-**G**. **c–d** Example of a chiral analyte and a chiral solvating agent adopted from [58]. The reference methyl peak due to **G** (denoted by blue circle in **c**) forms a doublet due to J-coupling, and further splits by $\Delta\delta$ as the diastereomers are formed. Intensity ratio of peaks due to (R)-**G** and (S)-**G** is p_R/p_S. **Prochiral solvating agent**: **e** Spin kinetics of two reference spins located at the same host molecule in **HG** complex. **f** Determination of $\Delta\delta$ from spectrum without J-coupling and (**g**) with J-coupling. The former spectral shape consists of two Lorentzians, the latter can be found in [60] or in Appendix B.1.4. Splitting due to J-coupling $\tilde{J}$ is related to the coupling constant J by $\tilde{J}$ [ppm] $= J$ [Hz]$/\nu_0$ [MHz], where ν_0 is the spectrometer frequency). Schemes in **e**, **f** are valid for $p_{\mathbf{H}} = 0$. **h–j** Example of chiral analyte and prochiral solvating agent **Bz₂OxP** adopted from [61]. The enantiotopic reporter groups of prochiral **Bz₂OxP** are its β-protons denoted by color circles in **h** (reference spin 1 and 2). Another pair of identical reporter groups denoted by gray ellipse lies symmetrically relative to the mirror plane σ_v. Spectral splitting in **i** is linear with e.e. according to Eq. (3.67) as shown in **j**. (**e**, **j**) also have problem with visible old hidden layer

Despite fast exchange, two signals are present because they originate from different guest molecules. The splitting $\Delta\delta = \delta_{R'} - \delta_{S'}$ is equal to [57]

$$\Delta\delta = p_{R,\text{free}}\delta_{R,\text{free}} + p_{R,\text{bound}}\delta_{R,\text{bound}} - p_{S,\text{free}}\delta_{S,\text{free}} - p_{S,\text{bound}}\delta_{S,\text{bound}} . \quad (3.65)$$

Most importantly, intensity of the peaks is proportional to the concentration of each analyte enanantiomer, which enables determination of e.e. by peak integration. Different binding affinity of the agent **H** to (*R*)-**G** or (*S*)-**G** has no effect on resulting areas of the split peaks because of the undergoing exchange between free and bound forms of guest.

Determination of e.e. is illustrated with 2-methylbutyric acid analyte (guest) and 1,2-diphenylethane-1,2-diamine chiral solvating agent (host) [58], see Fig. 3.7c. Signal due to the reference guest methyl group (denoted by blue circle in Fig. 3.7c) forms a doublet due to J-coupling as shown in Fig. 3.7d. Additional splitting by $\Delta\delta$ is attributed to the formation of diastereomers. Then the integrated intensities of peaks due to (*R*)- and (*S*)-form reflect the enantiomeric excess. Note that $\Delta\delta$ must be large enough so that the two signals are separated, its value differs for each chiral solvating agent [56]. Furthermore, $\Delta\delta$ is the biggest if the chiral solvating agent is enantiopure. This method also works when 1:2 host-guest complexes are formed, however, value of $\Delta\delta$ differs for each stoichiometry [56].

A different method of e.e. determination was introduced recently. It is based on complexation of chiral analytes with so-called prochiral solvating agents [60, 62–65], in particular with oxoporphyrinogens introduced in Sect. 1.2. To understand this method, we will explain a few terms related to so-called 'topicity', which is illustrated in Fig. 3.8. Topicity [66] is a property of a pair of sites, in our case the sites are hydrogen atoms. Figure 3.8a illustrates *homotopic*[10] pair of sites, where substitution of the red hydrogen atom for chlorine atom leads to the same result as substitution of the green hydrogen atom. Sites are homotopic when they can interchange through proper rotation $\mathbf{C}_n$. Figure 3.8b illustrates *enantiotopic*[11] pair of sites, where substitution of the red or green hydrogen atoms results into two different enantiomers (replacement with a chiral group leads to diastereomers). Enantiotopic molecular sites are related through improper rotation[12] $\mathbf{S}_n$, but not through proper rotation $\mathbf{C}_n$. Note that sites located exactly in the mirror symmetry plane are not enantiotopic. Finally, Fig. 3.8c illustrates *diastereotopic*[13] pair of sites, where substitution of the red or green hydrogen atoms results into two different diastereomers. Diastereotopic sites are constitutionally equivalent (e.g. bound to the same C atom) but not related by any symmetry operation [66]. Enantiotopic or diastereotopic nuclei are also called 'heterotopic'. Pair of atomic nuclei located at homotopic or enantiotopic sites possess identical NMR chemical shift unlike diastereotopic sites. Topicity enables us

---

[10] Homotopic https://goldbook.iupac.org/terms/view/H02857.

[11] Enantiotopic https://goldbook.iupac.org/terms/view/E02083.

[12] Improper rotation includes inversion **i** ($\mathbf{S}_1 = \mathbf{C}_1\sigma_\text{h} = \mathbf{i}$) and mirror reflection σ_h ($\mathbf{S}_2 = \mathbf{C}_2\sigma_\text{h} = \sigma_\text{h}$) [54].

[13] Diastereotopic https://goldbook.iupac.org/terms/view/D01685.

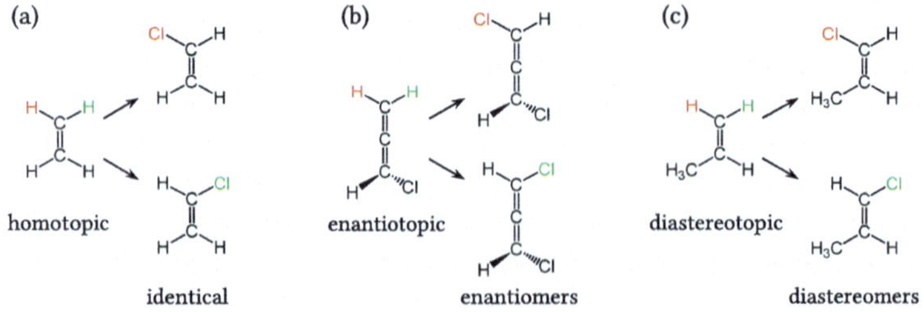

Fig. 3.8 Illustration of topicity. The colored pair of hydrogen atoms are **a** homotopic, when replacement of the red hydrogen results to identical structure as replacement of the green hydrogen, **b** enantiotopic, when the replacement leads to enantiomers and **c** diastereotopic, when replacement produces diastereomers

to define *prochiral* molecules as those having at least one pair of enantiotopic sites [66, 67]. Generally, prochirality is "the geometric property of an achiral object (or spatial arrangement of points or atoms) which is capable of becoming chiral in a single desymmetrization step".[14] The mirror symmetry plane between enantiotopic sites is denoted as *prochiral plane*.

When a prochiral solvating agent (host) forms a complex with a chiral analyte (guest) according to the scheme in Fig. 3.7a, the complexes **H**(*R*)-**G** and **H**(*S*)-**G** are mutual mirror images as follows from the definition of prochirality. Hence, there is chirality transfer from guest to host [60]. Supramolecular complexes are bound weakly, therefore the guest molecule undergoes fast movement at the host binding site forming so-called 'chiral field' [60]. Symmetry of the chiral field depends on symmetry of the host and guest molecules, see Chapter 7 for details. Pair of enantiotopic nuclear spins of host are chosen as reference spins and their NMR spectrum is observed. As shown in Fig. 3.7e, both reference spins are in state C in free host. In the **H**(*R*)-**G** complex, the reference spin 1 is in state A and the reference spin 2 is in state B. Since the reference spins are enantiotopic, complex **H**(*S*)-**G** induces state B the reference spin 1 and state A in the reference spin 2. In other words, the presence of prochiral plane in the host shapes the chiral field so that the exchange $(R) \leftrightarrow (S)$ of enantiomers in the complex causes swapping of states $A \rightarrow B$ and $B \rightarrow A$ of the two reference enantiotopic spins. Host-guest complex formation must be fast on the NMR timescale to permit measurement of e.e., thus, only one averaged peak can be observed for each reference spin. According to Eq. (3.62), chemical shifts of the observed averaged peaks are

$$\delta(1) = p_{\mathbf{HG}}(p_{A(1)}\delta_A + p_{B(1)}\delta_B) + p_{\mathbf{H}}\delta_C \,, \tag{3.66a}$$

$$\delta(2) = p_{\mathbf{HG}}(p_{B(2)}\delta_B + p_{A(2)}\delta_A) + p_{\mathbf{H}}\delta_C \,, \tag{3.66b}$$

where $p_{A(1)}$, $p_{B(1)}$, $p_{A(2)}$ and $p_{B(2)}$ are relative fractions of the reference spins 1 and 2 in the corresponding states A or B in the host-guest complex, it holds $p_{A(1)} + p_{B(1)} =$

[14] Prochirality https://goldbook.iupac.org/terms/view/P04859.

1 and $p_{A(2)} + p_{B(2)} = 1$. Furthermore, $p_{\mathbf{H}} + p_{\mathbf{HG}} = 1$ holds for populations of free ($p_{\mathbf{H}}$) and complexed ($p_{\mathbf{HG}}$) host. Let us denote relative fractions of (*R*)- and (*S*)-form of guest as p_R and p_S, respectively, where $p_R + p_S = 1$. Using Eq. (3.63) in the form of e.e. $= \frac{p_R - p_S}{p_R + p_S}$ and using the relationships $p_{A(1)} = p_R$, $p_{B(1)} = p_S$, $p_{A(2)} = p_S$ and $p_{B(2)} = p_R$, Eq. (3.66) imply that splitting $\Delta\delta = \delta(1) - \delta(2)$ between signals of reference spins 1 and 2 equals to [60]

$$\Delta\delta = \Delta\delta_{\text{max}} \times \text{e.e.} \times p_{\mathbf{HG}} , \tag{3.67}$$

where $\Delta\delta_{\text{max}} = \delta_A - \delta_B$. Hence, the observed splitting $\Delta\delta$ is directly proportional to e.e. Fig. 3.7f, g show how $\Delta\delta$ is determined without or with the presence of J-coupling. These schemes are valid for $_p\mathbf{HG} = 1$, otherwise the observed peak positions are both shifted towards δ_C, and $\Delta\delta$ would decrease according to Eq. (3.67). Note that both averaged peaks have the same intensity. Determination of $\Delta\delta$ for J-coupled nuclei in Fig. 3.7g requires lineshape fitting, the relevant formula for the spectral lineshape is given in Appendix B.1.4.

An actual example is e.e. determination of camphorsulfonic acid analyte using $\mathbf{Bz_2OxP}$ prochiral solvating agent [61], see Fig. 3.67h. The reporter spins (reference spins 1 and 2) belong to enantiotopic β-hydrogens denoted by blue and red circles (another pair of identical reporter groups denoted by gray ellipse lie symmetrically relative to the mirror plane σ_{v}, their signals overlap with the signal of the reference spins). These atoms are located symmetrically with respect to prochiral mirror plane σ_{v}' and are mutually J-coupled, hence the resulting spectral shapes shown in Fig. 3.67i follow the scheme in Fig. 3.67g. Spectral lineshape fitting revealed linear dependence of $\Delta\delta$ on e.e. shown in Fig. 3.67j as expected from Eq. (3.67). Note that $C_{2\text{v}}$ symmetry of $\mathbf{Bz_2OxP}$ causes that each nuclear spin of the molecule has its enantiotopic counterpart reflected through σ_{v} or σ_{v}' symmetry plane, which, at the same time, are not related by the C_2 rotation axis. Only the N atoms and the hydrogens of NH groups are an exception as they lie exactly on the symmetry plane.

To conclude, contrary to the use of chiral solvating agents, where e.e. is determined by integration, prochiral solvating agents determine e.e. from the extent of peak splitting $\Delta\delta$. Importantly, sign of e.e. cannot be determined from one measurement since e.e. and −e.e. produce identical peak splitting. The sign can be determined when, for example, small amount of enantiopure (*R*)-form of analyte is added into the sample, then $\Delta\delta$ increases upon addition if e.e. > 0 or decreases if e.e. < 0. Also, value of $\Delta\delta_{\text{max}}$ must be obtained independently using enantiopure analyte. Oxoporphyrinogens or TPP (see Sect. 1.2) are well-described prochiral solvating agents, furthermore, a sp^3 *meso*-carbon porphyrinogen [68] or a coordination complex of Zn with salen-like ligand [69], acting as prochiral solvating agents, have been published. Determination of e.e. using prochiral solvating agents is also possible in host-guest complexes with 1:2 stoichiometry [64]. Other methods of chiral discrimination in NMR use chiral liquid crystals [70, 71].

References

1. Levitt MH (2001) Spin dynamics: basics of nuclear magnetic resonance. Wiley
2. Cavanagh J, Fairbrother WJ, Palmer AG, Skelton NJ (1996) Protein NMR spectroscopy: principles and practice. Academic
3. Macomber RS (1998) A complete introduction to modern NMR spectroscopy. Wiley
4. Slichter CP (1996) Principles of magnetic resonance. Springer Science & Business Media
5. Březina V, Hanyková L, Velychkivska N, Hill JP, Labuta J (2022) NMR lineshape analysis using analytical solutions of multi-state chemical exchange with applications to kinetics of host-guest systems. Sci Rep 12:17369
6. Barman B, Petrou A (2019) Measuring the magnetization of a permanent magnet. Am J Phys 87:275–278
7. Günther H, Spectroscopy NMR (2013) Basic principles, concepts and applications in chemistry. Wiley
8. Hoffman RE (2003) Variations on the chemical shift of TMS. J Magn Reson 163:325–331
9. Bloch F (1946) Nuclear induction. Phys Rev 70:460–474
10. Bloch F, Hansen WW, Packard M (1946) The nuclear induction experiment. Phys Rev 70:474–485
11. Griffiths DJ (2013) Introduction to electrodynamics, 4th edn. Pearson, Boston
12. Antcliffe D, Gordon AC (2016) Metabonomics and intensive care. Crit Care 20:68
13. Bain AD, Duns G (1996) A unified approach to dynamic NMR based on a physical interpretation of the transition probability. Can J Chem 74:819–824
14. Belton PS (1996) Chapter 17 Continuous wave and rapid scan correlation NMR. In: Rutledge DN (ed) Data Handling in Science and Technology, vol 18. Signal treatment and signal analysis in NMR. Elsevier, pp 362–373
15. Szalay Z, Rohonczy J (2010) Monte Carlo simulation of NMR lineshapes in chemically exchanging spin systems. Prog Nucl Magn Reson Spectrosc 56:198–216
16. Anamimoghadam O, Jones LO, Cooper JA, Beldjoudi Y, Nguyen MT, Liu W, Krzyaniak MD, Pezzato C, Stern CL, Patel HA, Wasielewski MR, Schatz GC, Stoddart JF (2021) Discrete Open-Shell Tris(bipyridinium radical cationic) Inclusion Complexes in the Solid State. J Am Chem Soc 143:163–175
17. Biali SE, Nugiel DA, Rappoport Z (1989) Stable simple enols. Part 19. Steric effects and threshold rotational mechanisms in 1-substituted 2,2-dimesitylethenols. J Am Chem Soc 111:846–852
18. Římal V, Štěpánková H, Štěpánek J (2011) Analysis of NMR spectra in case of temperature-dependent chemical exchange between two unequally populated sites. Concepts Magnet Reson Part A 38A:117–127
19. Kozic J, Novák Z, Římal V, Profant V, Kuneš J, Vinšová J (2016) Conformations, equilibrium thermodynamics and rotational barriers of secondary thiobenzanilides. Tetrahedron 72:2072–2083
20. Takai A, Freas DJ, Suzuki T, Sugimoto M, Labuta J, Haruki R, Kumai R, Adachi S-I, Sakai H, Hasobe T, Matsushita Y, Takeuchi M (2017) The effect of a highly twisted CC double bond on the electronic structures of 9,9′-bifluorenylidene derivatives in the ground and excited states. Org Chem Front 4:650–657
21. Hensmann M, Booker GW, Panayotou G, Boyd J, Linacre J, Waterfield M, Campbell ID (1994) Phosphopeptide binding to the N-terminal SH2 domain of the p85α subunit of PI 3′-kinase: a heteronuclear NMR study. Protein Sci 3:1020–1030
22. Günther U, Mittag T, Schaffhausen B (2002) Probing Src homology 2 domain ligand interactions by differential line broadening. Biochemistry 41:11658–11669
23. Mittag T, Schaffhausen B, Günther UL (2004) Tracing kinetic intermediates during ligand binding. J Am Chem Soc 126:9017–9023
24. Waudby CA, Ramos A, Cabrita LD, Christodoulou J (2016) Two-dimensional NMR lineshape analysis. Sci Rep 6:1–8

25. Shinya S, Ghinet MG, Brzezinski R, Furuita K, Kojima C, Shah S, Kovrigin EL, Fukamizo T (2017) NMR line shape analysis of a multi-state ligand binding mechanism in chitosanase. J Biomol NMR 67:309–319
26. Waudby CA, Ouvry M, Davis B, Christodoulou J (2020) Two-dimensional NMR lineshape analysis of single, multiple, zero and double quantum correlation experiments. J Biomol NMR 74:95–109
27. Kovrigin EL (2020) NMR line shapes in molecular mechanisms with ligand binding and multiple conformers
28. Bain AD (2003) Chemical exchange in NMR. Prog Nucl Magn Reson Spectrosc 43:63–103
29. Kleckner IR, Foster MP (2011) An introduction to NMR-based approaches for measuring protein dynamics. Biochim Biophys Acta (BBA) - Proteins and Proteomics. Protein Dynamics: Experimental and Computational Approaches 1814:942–968
30. Furukawa A, Konuma T, Yanaka S, Sugase K (2016) Quantitative analysis of protein-ligand interactions by NMR. Prog Nucl Magn Reson Spectrosc 96:47–57
31. King EL, Altman C (1956) A schematic method of deriving the rate laws for enzyme-catalyzed reactions. J Phys Chem 60:1375–1378
32. Hill T (2012) Free energy transduction in biology: the steady-state kinetic and thermodynamic formalism. Elsevier
33. Mulquiney PJ, Kuchel PW (2003) Modelling metabolism with Mathematica: detailed examples including erythrocyte metabolism. CRC Press, Boca Raton, Fla
34. Grilli S, Lunazzi L, Mazzanti A, Casarini D, Femoni C (2001) Conformational studies by dynamic NMR. 78.1 Stereomutation of the helical enantiomers of trigonal carbon diaryl-substituted compounds: Dimesitylketone, Dimesitylthioketone, and Dimesitylethylene. J Org Chem 66:488–495
35. Grilli S, Lunazzi L, Mazzanti A, Mazzanti G (2001) Conformational studies by dynamic NMR. 79.1 Dimesityl Sulfine revisited: detection of the helical antipodes and determination of their enantiomerization pathways. J Org Chem 66:748–754
36. Biali SE, Rappoport Z (1984) Stable simple enols. 3. Static and dynamic NMR behavior of crowded triarylethenols and related compounds. Three-ring flip as the threshold mechanism for enantiomerization of crowded triarylvinyl propellers. J Am Chem Soc 106:477–496
37. Gust D, Mislow K (1973) Analysis of isomerization in compounds displaying restricted rotation of aryl groups. J Am Chem Soc 95:1535–1547
38. McConnell HM (1958) Reaction rates by nuclear magnetic resonance. J Chem Phys 28:430–431
39. Labuta J, Futera Z, Ishihara S, Kouřilová H, Tateyama Y, Ariga K, Hill JP (2014) Chiral guest binding as a probe of macrocycle dynamics and tautomerism in a conjugated tetrapyrrole. J Am Chem Soc 136:2112–2118
40. Alexander S (1962) Exchange of interacting nuclear spins in nuclear magnetic resonance. II. Chemical exchange. J Chem Phys 37:974–980
41. Kovrigin EL (2012) NMR line shapes and multi-state binding equilibria. J Biomol NMR 53:257–270
42. Feng C, Kovrigin EL, Post CB (2019) NmrLineGuru: standalone and user-friendly GUIs for Fast 1D NMR lineshape simulation and analysis of multi-state equilibrium binding models. Sci Rep 9:1–14
43. Greenwood AI, Rogals MJ, De S, Lu KP, Kovrigin EL, Nicholson LK (2011) Complete determination of the Pin1 catalytic domain thermodynamic cycle by NMR lineshape analysis. J Biomol NMR 51:21–34
44. Binsch G (1968) Topics in stereochemistry. In: Eliel EL, Allinger NL (eds) vol 3. Interscience Publishers
45. Günther UL, Schaffhausen B (2002) NMRKIN: Simulating line shapes from two-dimensional spectra of proteins upon ligand binding. J Biomol NMR 22:201–209
46. Post CB (1994) [19] Characterization of enzyme-complex formation by analysis of nuclear magnetic resonance line shapes. In: Methods in enzymology, vol 240, Part B: numerical computer methods. Academic, pp 438–446

47. Gutowsky HS, Saika A (1953) Dissociation, chemical exchange, and the Proton magnetic resonance in some aqueous electrolytes. J Chem Phys 21:1688–1694
48. Johnson CS (1965) Advances in magnetic resonance In: Waugh JS (ed) vol 1. Academic
49. Waudby CA, Alfonso I (2023) An introduction to one- and two-dimensional lineshape analysis of chemically exchanging systems. J Magn Reson Open 16–17:100102
50. Sack RA (1958) A contribution to the theory of the exchange narrowing of spectral lines. Mol Phys 1:163–167
51. Bhuvaneshwari RA, Shivamani A, Sengupta I (2024) Line shape analysis of 19F NMR-monitored chemical denaturation of a fold-switching protein RfaH reveals its slow folding dynamics. J Phys Chem B 128:465–471
52. Almeida T, Hargreaves E, Zech T, Barsukov I (2024) Dock-and-lock binding of SxIP ligands is required for stable and selective EB1 interactions, Preprint. Biochemistry
53. Connors KA (1987) Binding constants: the measurement of molecular complex stability. Wiley
54. Fišer J (1980) Ú vod do molekulov é symetrie: aplikace teorie grup v chemii. SNTL
55. Anslyn EV, Dougherty DA (2006) Modern physical organic chemistry. University Science Books
56. Parker D (1991) NMR determination of enantiomeric purity. Chem Rev 91:1441–1457
57. Rinaldi PL (1982) The determination of absolute configuration using nuclear magnetic resonance techniques. Prog Nucl Magn Reson Spectrosc 15:291–352
58. Fulwood R, Parker D (1994) 1,2-Diphenylethane-1,2-diamine: an effective NMR chiral solvating agent for chiral carboxylic acids. J Chem Soc Perkin Trans 2:57–64
59. Ebbers EJ, Ariaans GJA, Houbiers JPM, Bruggink A, Zwanenburg B (1997) Controlled racemization of optically active organic compounds: prospects for asymmetric transformation. Tetrahedron 53:9417–9476
60. Labuta J, Ishihara S, Šikorský T, Futera Z, Shundo A, Hanyková L, Burda JV, Ariga K, Hill JP (2013) NMR spectroscopic detection of chirality and enantiopurity in referenced systems without formation of diastereomers. Nat Commun 4:2188
61. Březina V (2014) Spectroscopic and theoretical study of supramolecular complexes of symmetric porphyrins with chiral guests. MA thesis. Univerzita Karlova, Matematicko-fyzikální fakulta
62. Labuta J, Ishihara S, Shundo A, Arai S, Takeoka S, Ariga K, Hill JP (2011) Chirality sensing by nonchiral porphines. Chem Eur J 17:3558–3561
63. Labuta J, Ishihara S, Payne DT, Takimoto K, Sato H, Hanyková L, Ariga K, Hill JP (2021) Estimation of enantiomeric excess based on rapid host-guest exchange. Chemosensors 9:259
64. Labuta J, Ishihara S, Ariga K, Hill JP (2014) Dynamic processes in prochiral solvating agents (pro-CSAs) studied by NMR spectroscopy. Symmetry 6:345–367
65. Shundo A, Labuta J, Hill JP, Ishihara S, Ariga K (2009) Nuclear magnetic resonance signaling of molecular chiral information using an achiral reagent. J Am Chem Soc 131:9494–9495
66. Eliel EL (1982) Prostereoisomerism (prochirality). In: Organic chemistry, Topics in current chemistry, pp 1–76
67. Hanson KR (1966) Applications of the sequence rule. I. Naming the paired ligands g, g at a Tetrahedral Atom Xggij. II. Naming the two faces of a trigonal atom Yghi. J Am Chem Soc 88:2731–2742
68. Norvaiša K, O'Brien JE, Osadchuk I, Twamley B, Borovkov V, Senge MO (2022) Importance of molecular symmetry for enantiomeric excess recognition by NMR. Chem Commun 58:5423–5426
69. Takimoto K, Ishihara S, Labuta J, Březina V, Payne DT, Hill JP, Ariga K, Sumita M, Mori S, Sato H (2020) Enantiomeric excess dependent splitting of NMR signal through dynamic chiral inversion and coligand exchange in a coordination complex. J Phys Chem Lett 8164–8169
70. Lesot P, Aroulanda C, Zimmermann H, Luz Z (2015) Enantiotopic discrimination in the NMR spectrum of prochiral solutes in chiral liquid crystals. Chem Soc Rev 44:2330–2375
71. Merlet D, Emsley JW, Lesot P, Courtieu J (1999) The relationship between molecular symmetry and second-rank orientational order parameters for molecules in chiral liquid crystalline solvents. J Chem Phys 111:6890–6896

Chapter 4
UV/vis Spectroscopy

Ultraviolet/visible (UV/vis) spectroscopy measures interaction of light with electronic structure of analytes. This structure is affected by protonation and supramolecular host-guest binding, thus, UV/vis is a good instrument to investigate these processes typically in titration experiments. Analysis of UV/vis titrations is conveniently performed using singular value decomposition as described in the next chapter.

4.1 Absorbance Measurement

The experimental setup inside a UV/vis spectrometer is shown in Fig. 4.1 (see Sect. 13 in [1] for details). The source (incandescent lamp in combination with diffraction grating) produces essentially monochromatic light beam passing through the sample in a cuvette and further to the photodetector. Light interacts with the analyte in three ways: (**i**) transmission, when the light passes though the sample without changes, (**ii**) scattering, when the photons change direction and (**iii**) absorption, when the light induces dissipative processes in the analyte and its energy is transformed to heat. The measurement is performed for all wavelengths in the accessible range, typically 300–900 nm. Incident intensity $I_0(\lambda)$ and transmitted intensity $I(\lambda)$ (Fig. 4.1) are measured by the photodetector (the former is measured without the presence of analyte) for all wavelengths in the accessible range, typically 200–900 nm. Then, the UV/vis spectrum is constructed as transmittance [2]

$$T(\lambda) = \frac{I(\lambda)}{I_0(\lambda)} \tag{4.1}$$

or absorbance

$$A(\lambda) = \log \frac{I_0(\lambda)}{I(\lambda)} = -\log T\,, \tag{4.2}$$

V. Březina, *Supramolecular Complexes of Oxoporphyrinogens with Organic Molecules*, Springer Theses, https://doi.org/10.1007/978-3-031-68646-7_4

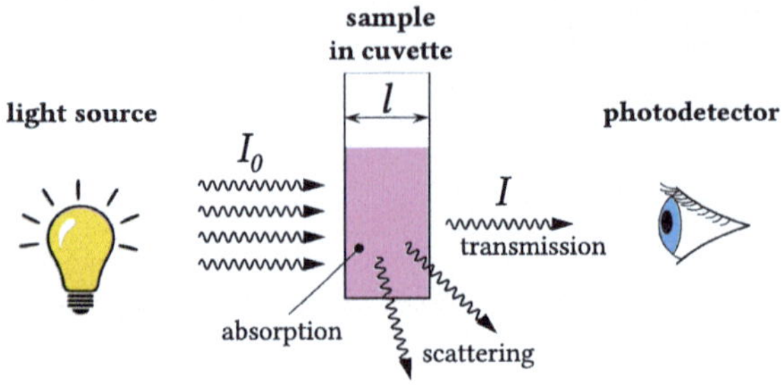

Fig. 4.1 Experimental setup of UV/vis measurement

where 'log' denotes decadic logarithm. In this work, we use absorbance for visualization of measured UV/vis spectra. Note that according to Fig. 4.1, the measured absorbance accounts not only for absorption but also for scattering of light in the sample.

Absorption of light is directly related to transitions between electronic energy levels of the analyte molecules. In this regard, HOMO-LUMO[1] transitions play the most significant role [3]. UV/vis spectra in gas state reflect splitting of electronic states due to vibrational and rotational molecular motions [4]. However, in liquid samples due to collisions with solvent, rotation of the analyte is restricted and energies of vibrations are modified in a nonuniform way, thus, rotational and vibrational lines disappear (see Chap. 14b in [1] and Chap. 7 in [5]). Typical UV/vis absorption bands of liquid samples have approximately log-normal shape [6, 7]. Scattering is caused by the presence of large molecules or particles ($\approx$ 100 nm diameter) [8]. Detailed description of UV/vis spectra of macrocyclic tetrapyrroles can be found in [9, 10].

4.2 Beer-Lambert Law of Absorption

Intensity of light transmitted through the sample decreases exponentially as formulated by the Beer-Lambert law of absorption

$$I = I_0 10^{-\epsilon l [\mathbf{X}]}\,, \tag{4.3}$$

where l is the cuvette length, $[\mathbf{X}]$ is concentration of the absorbing molecules and ϵ is the molar decadic absorption coefficient (also called extinction coefficient [2]).[2] The law yields a simple expression for absorbance,

$$A = \epsilon l [\mathbf{X}]\,. \tag{4.4}$$

[1] Highest occupied molecular orbitals and lowest unoccupied molecular orbitals, respectively.

[2] Instead of A or ϵ, absorption cross section $\sigma = \frac{1}{l N_\mathrm{A} [\mathbf{X}]} \ln \frac{I_0}{I} = \frac{1}{l N_\mathrm{A} [\mathbf{X}]} \frac{A}{\log \mathrm{e}}$ (in m^2) is used, where e is the Euler number and N_A is the Avogadro number. It can be expressed using the molar decadic absorption coefficient $\sigma = \frac{\epsilon}{N_\mathrm{A} \log \mathrm{e}} \approx 3.82 \times 10^{-24}\, \epsilon\ \mathrm{m}^2$ [2].

This expression is readily generalized for a mixture of N absorbing components $\mathbf{X}_i$,

$$\frac{A(\lambda)}{l} = \epsilon_1(\lambda)[\mathbf{X}_1] + \epsilon_2(\lambda)[\mathbf{X}_2] + \cdots + \epsilon_N(\lambda)[\mathbf{X}_N]\,. \tag{4.5}$$

Deviations from linearity in Eq. (4.4) might occur at high analyte concentrations due to several reasons (see pp. 306–315 in [1]). For instance, analyte molecules at high concentrations can interact with each other, and thus change their optical properties compared to diluted analyte. Also, absorptivity depends on the refractive index, which might change at high analyte concentration. Another factors, such as the presence of scattering introduce nonlinearity [11]. Furthermore, instrumental deviations such as stray radiation or too wide monochromator window can corrupt the linearity in Eq. (4.4). Absorbance values of $A > 2.5$ or $A < 0.05$ possess high uncertainty due to limited accuracy of the photodetector (Table 13–4 in [1]), hence the typical analyte concentrations are 10^{-5} M and typical cuvette width is 1 cm or less.

4.3 UV/vis Titration Experiment

Titration experiments (i.e. measurement of spectra at different guest concentrations when the guest is gradually added into the host solution, discussed for NMR in Sect. 3.4.2) in UV/vis spectroscopy are a common method to investigate molecular interactions. A model example is titration of **Bz$_2$OxP** with (*R*)-CSA shown in Fig. 4.2a. Here, host-guest complex with 1:1 stoichiometry is formed, thus, increase of the absorption band of **HG** at 745 nm as well as decrease of the absorption band of **H** at 500 nm are observed. In order to quantify the host-guest binding, binding isotherm is constructed from the absorbance values, which change the most during the titration. In this case, it is the absorption maximum at 745 nm. The Beer-Lambert law for mixture of absorbing species in Eq. (4.5) implies the analytical expression for fitting of the binding isotherm in the form (page 148 in [12])

$$A_{\lambda_{\mathrm{max}}} = A_{\mathbf{H}}\frac{[\mathbf{H}]}{[\mathbf{H}]_{\mathrm{t}}} + A_{\mathbf{HG}}\frac{[\mathbf{HG}]}{[\mathbf{H}]_{\mathrm{t}}}\,, \tag{4.6}$$

where $A_j = l\epsilon_j[\mathbf{H}]_{\mathrm{t}}$. Fitting to the experimental data with fitting parameters $A_{\mathbf{H}}$, $A_{\mathbf{HG}}$ and K_1 provides excellent match as shown in Fig. 4.2b. Such analysis exploits only very small part of the total measured data. In the next chapter, we show another method based on singular value decomposition in which the whole measured spectra are analyzed at once.

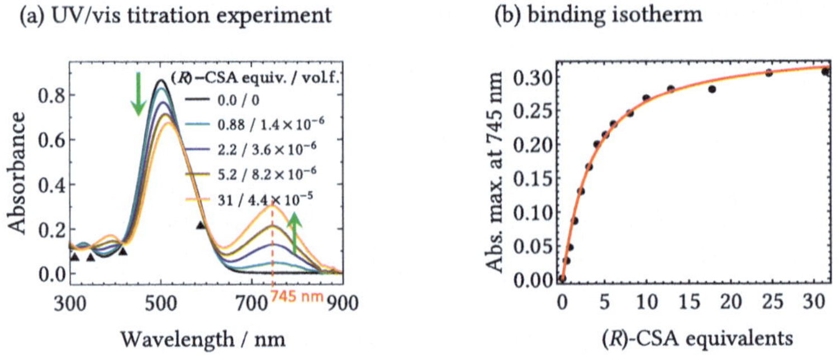

Fig. 4.2 Illustration of a titration experiment. **a** Titration of **Bz$_2$OxP** with (*R*)-CSA. **b** Binding isotherm constructed from absorbance at 745 nm (red dashed line in (a)), fitted with Eq. (4.6)

4.4 Isosbestic Points

When the spectral change involves only two absorbing species, i.e. one converts into the other, *isosbestic points* (points of same extinction) can arise [2]. These are points $\{\lambda_{\text{i.p.}}, A(\lambda_{\text{i.p.}})\}$, where the absorbance does not change, for example, during a titration experiment, see black triangles in Fig. 4.2a. The necessary condition is that the total concentration of all absorbing species $[\mathbf{X}]_t$ remains constant (or the spectra are rescaled to constant concentration, see footnote 7 on Sect. 5.2.2). Let us denote all N_{spc} absorbing species present in the sample as $\mathbf{X}_i$ with total concentration $[\mathbf{X}]_t = [\mathbf{X}_1] + [\mathbf{X}_2] + \cdots + [\mathbf{X}_N]$ and suppose interconversion only between $\mathbf{X}_1$ and $\mathbf{X}_2$. Then

$$
\begin{aligned}
\frac{A(\lambda_{\text{i.p.}})}{l} &= \epsilon_1(\lambda_{\text{i.p.}})[\mathbf{X}_1] + \epsilon_2(\lambda_{\text{i.p.}})[\mathbf{X}_2] + \cdots + \epsilon_{N_{\text{spc}}}(\lambda_{\text{i.p.}})[\mathbf{X}_{N_{\text{spc}}}] \\
&= \epsilon_1(\lambda_{\text{i.p.}})\left\{[\mathbf{X}]_t - [\mathbf{X}_2] - \cdots - [\mathbf{X}_{N_{\text{spc}}}]\right\} + \epsilon_2(\lambda_{\text{i.p.}})[\mathbf{X}_2] + \cdots + \epsilon_{N_{\text{spc}}}(\lambda_{\text{i.p.}})[\mathbf{X}_{N_{\text{spc}}}] \\
&= \underbrace{\epsilon_1(\lambda_{\text{i.p.}})[\mathbf{X}]_t}_{\text{const.}} + \underbrace{\left\{\epsilon_2(\lambda_{\text{i.p.}}) - \epsilon_1(\lambda_{\text{i.p.}})\right\}[\mathbf{X}_2]}_{0} + \underbrace{\cdots + \left\{\epsilon_{N_{\text{spc}}}(\lambda_{\text{i.p.}}) - \epsilon_1(\lambda_{\text{i.p.}})\right\}[\mathbf{X}_{N_{\text{spc}}}]}_{\text{const.}},
\end{aligned}
\tag{4.7}
$$

and all the right-hand side terms are constant or equal to zero. Therefore, isosbestic points can be formed where the individual spectra of the interconverting absorbing species cross (marked by black triangles in Fig. D.1g in Appendix D). Identification of isosbestic points in the data immediately implies spectral change involving two species.

4.5 Halochromism and Solvatochromism

Chromic phenomena refer to color change of a dye due to external stimuli. There are several types of chromism including response to light (photochromism), temperature (thermochromism), electricity (electrochromism) and many others [13]. Two chromic phenomena are relevant for supramolecular binding of oxoporphyrinogens. The first is *halochromism*, chromic response due to protonation (interaction with acid), deprotonation (base), or interaction with ions (salt). The effect of acids is sometimes called acidochromism. Halochromic substances have an immediate application as pH indicators, for example, the Universal Indicator is a mixture of several dyes including phenolphtalein, methyl red and others [14]. Halochromic spectral changes stem from the formation of new absorption bands upon binding (i.e. in the host-guest complex). Both free and bound species are present in the sample and their relative populations determine the resulting color.

The second chromic event related to supramolecular binding, *solvatochromism*, reflects changes in solvent polarity or weak interactions with solvent molecules (e.g. hydrogen bonding) resulting in shift of the absorption bands, see Sect. 6.2 in [15]. Unlike halochromism, solvatochromism does not produce new absorption bands but produces shift of the absorption maxima, see green and blue arrows in Fig. 4.3, respectively.

Mechanism of solvatochromism consists in different stabilization of ground (E_g) and excited (E_e) energy levels of molecular orbitals [15]. The energy levels of solute are lowered in polar solvent (compared to nonpolar solvent) due to solute-solvent interaction based on permanent and induced electrical dipoles. Polar solvent decreases states with high dipole moment to higher extent than states with low dipole moment. Figure 4.4 illustrates the two types of solvatochromism: (**i**) positive solvatochromism, where the excited state is stabilized more than the ground state (i.e. $\mu_e > \mu_g$) inducing red (bathochromic) shift with increasing polarity and (**ii**) negative solvatochromism, where the excited state is stabilized less than the ground state (i.e. $\mu_e < \mu_g$) inducing blue (hypsochromic) shift with increasing polarity. Note that

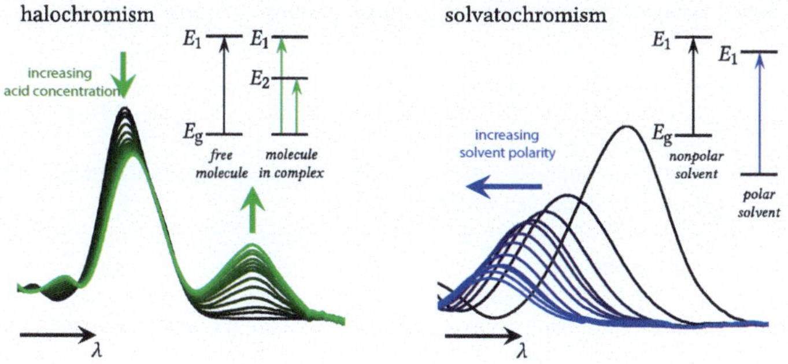

Fig. 4.3 Difference between halochromic and solvatochromic spectral changes illustrated with actual UV/vis data. These particular experiments are described and analyzed in Sect. 5.2

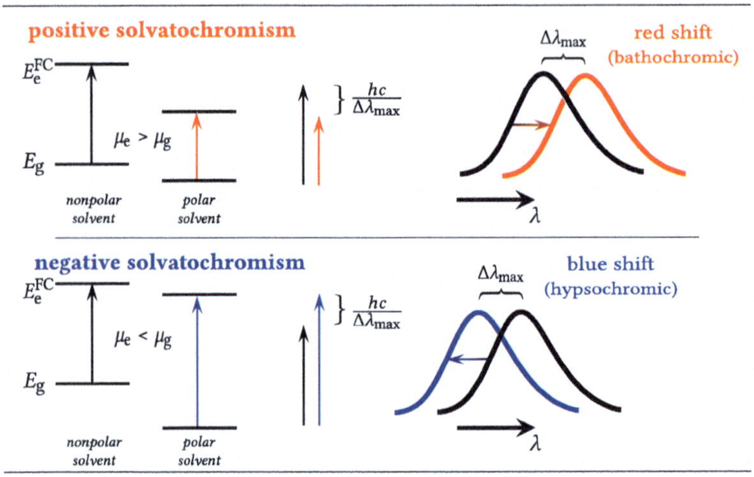

Fig. 4.4 Illustration of solvatochromic effects as solvent polarity influence on energy difference between ground state E_g and the Franck-Condon excited state E_e^{FC}

the excited state is in fact Franck-Condon state (E_e^{FC}), i.e. excited electronic states with the position of nuclei and its solvent cage corresponding to the ground state since the optical transitions are much shorter (ca. 10^{-15} s) than vibrational movemet of nuclei (ca. 10^{-12} s). A prominent example of a solvatochromic dye is Reichardt's dye [16, 17], used for construction of empirical $E_T(30)$ polarity scale for solvents.

References

1. Skoog DA, Holler FJ, Crouch SR (2017) Principles of instrumental analysis. Cengage Learning, Jan.
2. Braslavsky SE (2007) Glossary of terms used in photochemistry, 3rd edition (IUPAC Recommendations 2006). Pure Appl Chem 79:293–465
3. Soderberg T (2019) Organic chemistry with a biological emphasis volume I. Chemistry Publications
4. Voigt S, Orphal J, Burrows JP (2002) The temperature and pressure dependence of the absorption cross-sections of NO_2 in the 250–800 nm region measured by Fourier-transform spectroscopy. J Photochem Photobiol Chem 149:1–7
5. Demtröder W (2006) Atoms, molecules, and photons: an introduction to atomic-, molecular-, and quantum-physics. Springer, Berlin, New York
6. Siano DB, Metzler DE (1969) Band shapes of the electronic spectra of complex molecules. J Chem Phys 51:1856–1861
7. Koeppe B, Tolstoy PM, Limbach H-H (2011) Reaction pathways of proton transfer in hydrogen-bonded phenol-carboxylate complexes explored by combined UV-Vis and NMR spectroscopy. J Am Chem Soc 133:7897–7908
8. Dawson PL, Acton JC (2018) 22—Impact of proteins on food color. In: Yada RY (ed) Proteins in food processing (second edition). Woodhead publishing series in food science, technology and nutrition. Woodhead Publishing, pp 599–638

9. Dolphin D (1978) The porphyrins V3: physical chemistry, part A. Academic Press
10. Wamser CC, Ghosh A (2022) The hyperporphyrin concept: a contemporary perspective. JACS Au 2:1543–1560
11. Sassaroli A, Fantini S (2004) Comment on the modified Beer-Lambert law for scattering media. Phys Med Biol 49:N255–N257
12. Connors KA (1987) Binding constants: the measurement of molecular complex stability. Wiley
13. Bamfield P, Hutchings M (2018) Chromic phenomena
14. Foster LS, Gruntfest IJ (1937) Demonstration experiments using universal indicators. J Chem Educ 14:274
15. Reichardt C, Welton T (2011) Solvents and solvent effects in organic chemistry. John Wiley & Sons
16. Reichardt C (1992) Solvatochromism, thermochromism, piezochromism, halochromism, and chiro-solvatochromism of pyridinium N-phenoxide betaine dyes. Chem Soc Rev 21:147–153
17. Machado VG, Stock RI, Reichardt C (2014) Pyridinium N-phenolate betaine dyes. Chem Rev 114:10429–10475

Chapter 5
Singular Value Decomposition

Singular value decomposition (SVD) is a powerful tool for data analysis. In terms of mathematics, it is a matrix decomposition with good numerical stability and guaranteed existence. It provides a tool for dimensionality reduction, i.e. identification of patterns in high-dimensional data stored in a single matrix and filtering out noise. SVD is closely related to a statistical method called *principal component analysis* (PCA) [1], the terms SVD and PCA are sometimes used interchangeably. There are plenty of SVD- and PCA-based applications, which exploit their power to find patterns in data, for example, in imaging of gene expression in brain [2], face recognition algorithm [3], modal analysis of fluid flows [4, 5], vibration modes of controlled active microparticles [6] and others [7, 8]. Another closely related term is *factor analysis*, which is focused on finding the real factors, which can explain all the variability in the data [9]. Factor analysis stems originally from psychology where it was introduced by Spearman [10] to quantify types of intelligence.

In this chapter, we explain the use of SVD-based methods for analysis of UV/vis spectroscopic data. They serve for determination of the exact number of absorbing species, which can be done without the use of any binding model. The other purpose is determination of the absorption spectra of the individual species, which often cannot be measured separately. This task can be performed either using a particular model of chemical kinetics/binding or even without a model in some cases [11]. Applicability of the SVD-based analysis is limited to a spectroscopic measurement, where the resulting spectrum is linear combination of the individual absorbing components. This includes measurements of UV/vis or IR absorption, Raman scattering [12, 13], circular dichroism and others, but excludes e.g. UV/vis data with solvatochromic shift caused solely by solvent polarity change (see Sect. 5.2.1) and NMR spectra with presence of chemical exchange.

V. Březina, *Supramolecular Complexes of Oxoporphyrinogens with Organic Molecules*, Springer Theses, https://doi.org/10.1007/978-3-031-68646-7_5

Excellent studying material on both SVD and PCA is provided in book of Brunton and Kutz [14] including series of videolectures.[1] Various applications in chemistry from the perspective of factor analysis are discussed in book of Malinowski [9], comprehensive introduction of SVD in spectroscopy is provided in publications of Henry and Hofrichter [15, 16]. Valuable insights to these methods can be found in publications of Zimányi [17–20].

5.1 Mathematical Properties

5.1.1 Basics of SVD

Let us consider a real $m \times n$ matrix $\mathbf{A}$ consisting of experimental data, which has full rank due to the noise. SVD of $\mathbf{A}$ is defined as

$$\mathbf{A} = \mathbf{U}\mathbf{S}\mathbf{V}^{\mathsf{T}}, \tag{5.1}$$

where $\mathbf{U}$ is a square unitary[2] $m \times m$ matrix with columns called *left singular vectors*, $\mathbf{V}$ is a square unitary $n \times n$ matrix with columns called *right singular vectors*. The rectangular $m \times n$ matrix $\mathbf{S}$ is diagonal with the non-zero elements termed as *singular values*, the j-th singular value is denoted as s_j. The number of the singular values is $\min(m, n)$ depending on the shape of $\mathbf{A}$. All singular values are non-negative and follow descending order, i.e. $s_j > s_{j+1}$. For convenience, we denote[3] j-th column of $\mathbf{A}$ as $\mathbf{a}_j$, j-th column of $\mathbf{U}$ as $\mathbf{u}_j$ and j-th column of $\mathbf{V}$ as $\mathbf{v}_j$. Vectors $\mathbf{a}_j$ represent single measurements, single images, or single acquired spectra, depending on the type of data. To summarize, left singular vectors $\mathbf{u}_j$ form an orthonormal basis for the single measurements $\mathbf{a}_j$, right singular vectors multiplied by the corresponding singular value $s_j\mathbf{v}_j$ are coordinates of the basis vectors and s_j themselves indicate, how significant the j-th basis vector is for reconstruction of $\mathbf{a}_j$.

The definition in Eq. (5.1), called *full SVD* is redundant since for a "tall-skinny" matrix ($m > n$), last $m - n$ left singular vectors are multiplied by zeros of $\mathbf{S}$ as illustrated in Fig. 5.1. Let us define truncated matrices $\mathbf{U}_r = (\mathbf{u}_1, \mathbf{u}_2, \ldots, \mathbf{u}_r)$, $\mathbf{V}_r = (\mathbf{v}_1, \mathbf{v}_2, \ldots, \mathbf{v}_r)$ and $\mathbf{S}_r = \mathrm{diag}(s_1, s_2, \ldots, s_r)$ ($\mathbf{S}_r$ is a square matrix). Then, the *economy SVD* (equivalent to full SVD) is

$$\mathbf{A} = \mathbf{U}_n\mathbf{S}_n\mathbf{V}^{\mathsf{T}}. \tag{5.2}$$

[1] Available at https://www.youtube.com/playlist?list=PLMrJAkhIeNNSVjnsviglFoY2nXildDCcv.

[2] A unitary matrix $\mathbf{U}$ satistfies $\mathbf{U}\mathbf{U}^{\mathsf{T}} = \mathbf{U}^{\mathsf{T}}\mathbf{U} = 1$. Its columns form therefore an orthonormal basis as well as its rows.

[3] In this chapter uppercase bold symbols stand for matrices, lowercase bold symbols stand for column vectors.

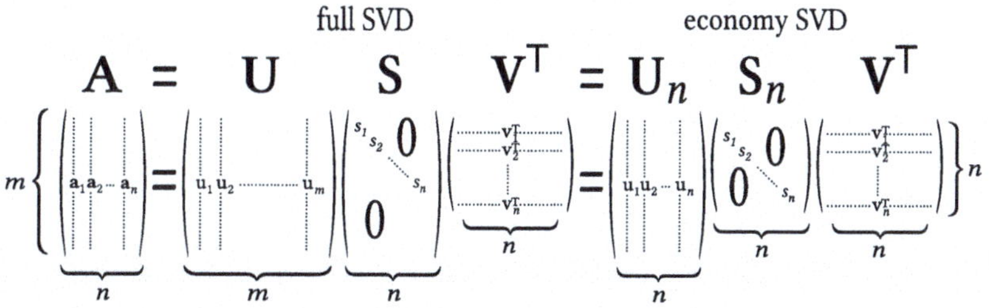

Fig. 5.1 Full and economy SVD of a "tall skinny" ($m > n$) matrix

For a "short-fat" matrix ($m < n$), the economy SVD is analogously

$$\mathbf{A} = \mathbf{U}\mathbf{S}_m\mathbf{V}_m^{\mathsf{T}} . \tag{5.3}$$

Another widely used notation for the SVD decomposition for $p = \min(m, n)$ is

$$\mathbf{A} = \mathbf{U}\mathbf{S}\mathbf{V}^{\mathsf{T}} = s_1\mathbf{u}_1\mathbf{v}_1^{\mathsf{T}} + s_2\mathbf{u}_2\mathbf{v}_2^{\mathsf{T}} + \cdots + s_p\mathbf{u}_p\mathbf{v}_p^{\mathsf{T}} , \tag{5.4}$$

where dyadic (tensor) product is used.[4] This dyadic decomposition is just another notation for the economy SVD. Note that SVD is unique up to the sign in the product $s_j\mathbf{u}_j\mathbf{v}_j^{\mathsf{T}} = s_j(-\mathbf{u}_j)(-\mathbf{v}_j^{\mathsf{T}})$, however, there exists a method to get rid of this ambiguity [21].

The most useful property of SVD is that the first r terms in Eq. (5.4) are the best possible approximation of the matrix $\mathbf{A}$ than any other matrix of rank r. Using the notation with truncated matrices,

$$\mathbf{A} \approx \mathbf{U}_r\mathbf{S}_r\mathbf{V}_r^{\mathsf{T}} , \tag{5.5}$$

with $r < \min(m, n)$. "Goodness" of SVD approximation is measured with Frobenius norm[5] $\| \bullet \|$ and it is formulated as the Eckhart-Young theorem [14, 22],

$$\min_{\mathbf{X},\, \mathrm{rank}(\mathbf{X})=r} \|\mathbf{A} - \mathbf{X}\| = \|\mathbf{A} - \mathbf{U}_r\mathbf{S}_r\mathbf{V}_r^{\mathsf{T}}\| = s_{r+1} . \tag{5.6}$$

Note the relationship with the singular value s_{j+1}.

SVD can be considered as a method for calculation of the "real" rank of a noise-containing matrix and consequently for separation of real signal from noise. If the rank of a noiseless matrix $\mathbf{A}$ is r, only the first r singular values will be significant after addition of noise, assuming the noise is not too big.

[4] Dyadic product of two column vectors $\mathbf{a}$, $\mathbf{b}$ is defined as $(\mathbf{a}\mathbf{b}^{\mathsf{T}})_{jk} = a_j b_k$.

[5] Frobenius norm of a real matrix $\mathbf{A}$ is defined as $\|\mathbf{A}\| = \sqrt{\sum_{j=1}^{m}\sum_{k=1}^{n} a_{jk}^2}$.

Eigenvalues and eigenvectors of matrices $\mathbf{AA}^\mathsf{T}$ and $\mathbf{A}^\mathsf{T}\mathbf{A}$ can be used to calculate SVD. Since the matrices $\mathbf{U}$ and $\mathbf{V}$ are unitary,

$$\mathbf{AA}^\mathsf{T} = \mathbf{USV}^\mathsf{T}\mathbf{VS}^\mathsf{T}\mathbf{U}^\mathsf{T} = \mathbf{USS}^\mathsf{T}\mathbf{U}^\mathsf{T} . \tag{5.7}$$

Therefore using $\mathbf{U}^\mathsf{T} = \mathbf{U}^{-1}$

$$\mathbf{AA}^\mathsf{T}\mathbf{u}_j = s_j^2\mathbf{u}_j , \tag{5.8}$$

which implies that the left singular vectors $\mathbf{u}_j$ are eigenvectors of $\mathbf{AA}^\mathsf{T}$ with eigenvalues s_j^2. Similarly it holds

$$\mathbf{A}^\mathsf{T}\mathbf{Av}_j = s_j^2\mathbf{v}_j , \tag{5.9}$$

so that the right singular vectors $\mathbf{v}_j$ are eigenvectors of $\mathbf{A}^\mathsf{T}\mathbf{A}$ with eigenvalues s_j^2. For numerical implementation of SVD, see book of Press et al. [23]. In *Mathematica* software, SVD is calculated as `{U,S,V}=SingularValueDecomposition[A]`, in *MATLAB*, `[U,S,V]=svd(A)` or `[U,S,V]=svd(A,'econ')` for the economy version.

5.1.2 Pseudoinverse and Linear Regression

Pseudoinverse is a generalization of matrix inverse for singular or rectangular matrices also called Moore-Penrose inverse. Unlike the inverse where $\mathbf{A}^{-1}\mathbf{A} = \mathbf{AA}^{-1} = 1$, the pseudoinverse $\mathbf{A}^\dagger$ of a real matrix must satisfy these four criteria [22, 24]

$$\mathbf{AA}^\dagger\mathbf{A} = \mathbf{A} , \tag{5.10a}$$

$$\mathbf{A}^\dagger\mathbf{AA}^\dagger = \mathbf{A}^\dagger , \tag{5.10b}$$

$$(\mathbf{AA}^\dagger)^\mathsf{T} = \mathbf{AA}^\dagger , \tag{5.10c}$$

$$(\mathbf{A}^\dagger\mathbf{A})^\mathsf{T} = \mathbf{A}^\dagger\mathbf{A} . \tag{5.10d}$$

Pseudoinverse exists for every matrix and can be expressed by an explicit formula for a full rank matrix (i.e. rank($\mathbf{A}$)$=\min(m, n)$). In the case of tall-skinny ($m > n$) matrix, the full rank ensures the existence of the inverse $(\mathbf{A}^\mathsf{T}\mathbf{A})^{-1}$, and the conditions in Eq. (5.10a, c) imply

$$\mathbf{A}^\dagger = (\mathbf{A}^\mathsf{T}\mathbf{A})^{-1}\mathbf{A}^\mathsf{T} . \tag{5.11}$$

Similarly for a short-fat ($m < n$) matrix, the conditions in Eq. (5.10b, d) yield

$$\mathbf{A}^\dagger = \mathbf{A}^\mathsf{T}(\mathbf{AA}^\mathsf{T})^{-1} . \tag{5.12}$$

Economy SVD can be used to calculate pseudoinverse when Eqs. (5.2) or (5.3) is substituted into the first pseudoinverse condition in Eq. (5.10a) yielding

$$\mathbf{A}^{\dagger} = \mathbf{V}\mathbf{S}_n^{-1}\mathbf{U}_n^{\mathsf{T}} \text{ for } m > n\,, \tag{5.13a}$$

$$\mathbf{A}^{\dagger} = \mathbf{V}_m\mathbf{S}_m^{-1}\mathbf{U}^{\mathsf{T}} \text{ for } m < n\,. \tag{5.13b}$$

Let us consider a system of linear equations defined by a full-rank $m \times n$ matrix **A**,

$$\mathbf{A}\mathbf{x} = \mathbf{b}\,, \tag{5.14}$$

which consists of m equations and a vector of n unknowns **x**. For a square matrix **A**, the solution is obtained simply as $\mathbf{x} = \mathbf{A}^{-1}\mathbf{b}$. If **A** is a tall-skinny ($m > n$) matrix, Eq. (5.14) provides an overdetermined system of equations, which cannot be solved exactly. However, an approximate solution with minimal least squares error $\|\mathbf{A}\mathbf{x} - \mathbf{b}\|^2$ can be calculated using the pseudoinverse. Substitution of Eqs. (5.2) and (5.13) into Eq. (5.14) reveals

$$\mathbf{x} = \mathbf{A}^{\dagger}\mathbf{b} = \mathbf{V}\mathbf{S}_n^{-1}\mathbf{U}_n^{\mathsf{T}}\,. \tag{5.15}$$

This solution of an overdetermined system is also called *linear regression* and it is widely used in statistics [25]. On the other hand, if **A** is a short-fat ($m < n$) matrix, Eq. (5.14) is an underdetermined system with infinitely many solutions. Among those, pseudoinverse in $\mathbf{x} = \mathbf{A}^{\dagger}\mathbf{b}$ yields a solution with minimal norm $\|x\|$.

5.2 Analysis of UV/vis Spectral Series via SVD

Series of spectral measurements, where one parameter is changed, are an important tool of chemical physics. The parameter of change is usually concentration (titration experiments), time or temperature [26]. The methods presented here are applicable if the spectral changes depend linearly on concentration of absorbing components, in UV/vis spectroscopy it is limited by validity of the Beer-Lambert law (see Sect. 4.2). UV/vis titrations are usually analyzed at a particular wavelength with the biggest spectral change, often at maximum of an absorbing band $\lambda_{\max}$ (see Sect. 4.3). However, if some of the absorbing components have small absorption coefficient $\epsilon_j(\lambda_{\max})$, their concentration is obtained with large uncertainty. A quick remedy in this case is simultaneous fitting of data at more than one wavelength. However, the more wavelengths are analyzed at once, the more parameters (absorption coefficients $\epsilon(\lambda_j)$) should be fitted. Here comes the SVD based analysis, which enables fitting data at all measured wavelengths at once while reducing the number of fitting parameters. Moreover, it allows elimination of the experimental noise and reconstruction of the individual absorbing species, which is not possible with the simple one-wavelength approach. SVD terminology is adjusted for application in UV/vis spectra analysis. The used terms are listed in the following Table 5.1 including not yet defined quantities.

Table 5.1 Terminology for SVD-based analysis of UV/vis spectral series

	Mathematical description	Application on spectral series
$\mathbf{u}_j$	Left singular vector	Basis spectrum (factor (for $j \leq N_{spc}$))
$\mathbf{v}_j$	Right singular vector	Amplitude vector
s_j	Singular value	Singular value
N_{exp}	n	No. of measured experiments
N_{pts}	m	No. of points in one experiment
N_{spc}		No. of individual absorption species (Factor dimension)
$\mathbf{Z}$		Matrix of absorbing components spectra
$\mathbf{z}_j$		Spectrum of j-th absorbing component (Columns of $\mathbf{Z}$)
$\mathbf{P}_{exp}$		Population matrix constructed from exp. data
$\mathbf{P}_{mod}$		Population matrix calculated from a model
$\mathbf{p}_j$		Populations of j-th component (Columns of $\mathbf{P}$)
$s_j\mathbf{v}_j$		Combination coefficients
$\mathbf{T}$		Transformation matrix (Often denoted as rotation matrix $\mathbf{R}$)
$\mathbf{t}_j$		Component coordinates in comb. coeff. space (Columns of $\mathbf{T}$)

The SVD-based analysis is illustrated with two experiments in this chapter: (**i**) titration of **Bz$_2$OxP** with (*R*)-CSA in Fig. 5.2a. Halochromic spectral change in these data comprises of growing a new absorption band at 750 nm and decrease of a 500 nm band (green arrows). Detailed analysis of this measurement is given in Appendix D.1. (**ii**) Reichardt's dye in solvent mixture of varying polarity in Fig. 5.2b shows solvatochromic shift (blue arrow) that cannot be modeled as linear combination of individual absorbing species (Sect. 4.5).

For the purpose of SVD analysis, the measured spectra $\mathbf{a}_j$ are arranged into columns of the absorbance matrix $\mathbf{A}$, see Fig. 5.3. All spectra must be in the same wavelength range. The abstract basis spectra $\mathbf{u}_j$ will be transformed into the real spectra of absorbing components $\mathbf{z}_j$, the amplitude vectors $\mathbf{v}_j$ will be transformed into vectors of populations $\mathbf{p}_j$. Products $s_j\mathbf{v}_j$ are denoted as *combination coefficients* [17] as they combine the basis spectra $\mathbf{u}_j$ to reproduce the matrix $\mathbf{A}$ (Eq. (5.4)). Note that the basis spectrum $\mathbf{u}_1$ is equal to mean of all measured spectra (possibly multiplied by -1 due to the aforementioned ambiguity in SVD), therefore, it never crosses the x-axis. On the other hand, all other $\mathbf{u}_j$ must always cross the axis in order to preserve orthogonality to $\mathbf{u}_1$.

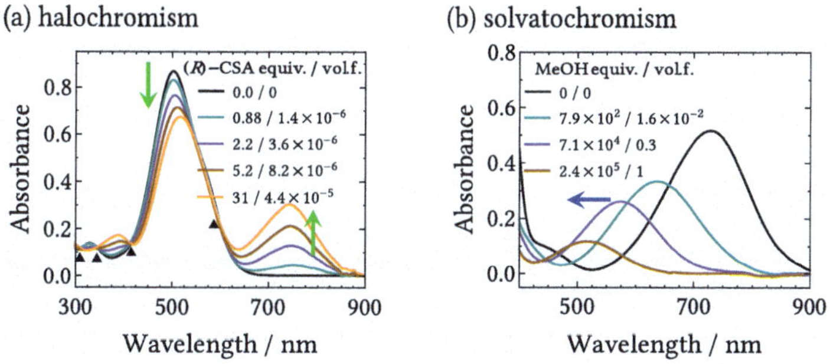

Fig. 5.2 UV/vis spectra with two different types of spectral changes, (**a**) halochromic spectral change, titration of **Bz$_2$OxP** (7μM (initial), $CDCl_3$) with (*R*)-CSA, dilution during experiment compensated by rescaling to the initial concentration, black triangles denote the isosbestic points (**b**) solvatochromic spectral change, Reichardt's dye (55μM, $CDCl_3$) in MeOH/chloroform mixture (MeOH increases solvent polarity)

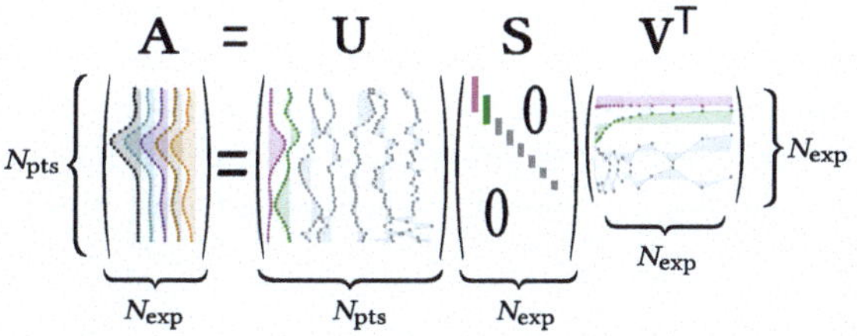

Fig. 5.3 Illustration of the SVD of UV/vis spectral series. Constructed from the data in Fig. 5.2a, spectra in columns $\mathbf{a}_j$ are measured at different acid concentration. Gray values correspond to noise

5.2.1 *Linear Combination of Absorbing Components*

The Beer-Lambert law for mixture of absorbing components in Eq. (4.5) can be expressed in matrix form as $\mathbf{a}_i = \sum_j l\boldsymbol{\epsilon}_j(\mathbf{c}_j)_i$, where $\mathbf{a}_i$ is i-th column of the data matrix $\mathbf{A}$, $\boldsymbol{\epsilon}_j$ is the absorption coefficient of j-th absorbing component and $(\mathbf{c}_j)_i$ is the concentration of j-th absorbing component in i-th measured spectrum (i.e. $\mathbf{A} = l\mathcal{E}\mathbf{C}^\mathsf{T}$). If the total concentration of absorbing components is $[\mathbf{X}]_\mathrm{t} = [\mathbf{X}_1] + [\mathbf{X}_2] + \cdots + [\mathbf{X}_{N_\mathrm{spc}}]$, the concentration $\mathbf{c}_j$ can be rescaled to relative population $\mathbf{p}_j = \mathbf{c}_j/[\mathbf{X}]_\mathrm{t}$, and $\boldsymbol{\epsilon}_j$ can be rescaled to $\mathbf{z}_j = l\boldsymbol{\epsilon}_j[\mathbf{X}]_\mathrm{t}$ (i.e. absorbance at the total concentration).[6] Thus, the total absorbance $\mathbf{A}$ can be expressed in terms of absorbance of the absorbing components $\mathbf{Z}$ and their populations $\mathbf{P}$,

[6] For illustration, spectrum formed by two absorbing components at concentrations $[\mathbf{X}_1] + [\mathbf{X}_2] = [\mathbf{X}]_\mathrm{t}$ and the first measured spectrum $\mathbf{a}_1$ can be expressed as

$$\mathbf{a}_1 = l\epsilon_1[\mathbf{X}_1] + l\epsilon_2[\mathbf{X}_2] = \underbrace{l\epsilon_1[\mathbf{X}]_\mathrm{t}}_{\mathbf{z}_1}\underbrace{\frac{[\mathbf{X}_1]}{[\mathbf{X}]_\mathrm{t}}}_{(\mathbf{p}_1)_1} + \underbrace{l\epsilon_2[\mathbf{X}]_\mathrm{t}}_{\mathbf{z}_2}\underbrace{\frac{[\mathbf{X}_2]}{[\mathbf{X}]_\mathrm{t}}}_{(\mathbf{p}_2)_1}$$

.

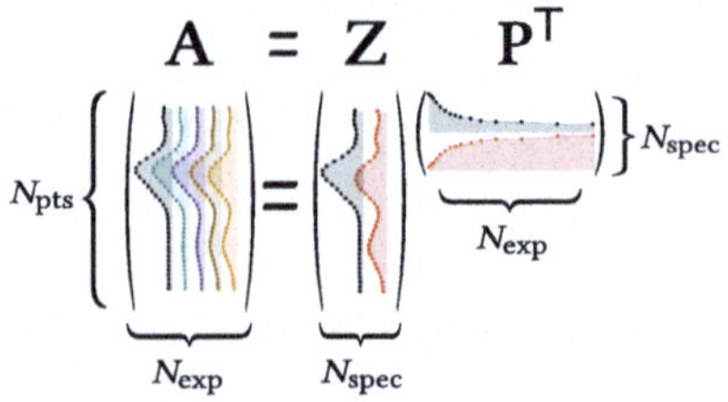

Fig. 5.4 Decomposition of halochromic spectra into two absorbing components. Constructed from the data in Fig. 5.2a, spectra in columns $\mathbf{a}_j$ are measured at different acid concentration

$$\mathbf{A} = \mathbf{Z}\mathbf{P}^{\mathsf{T}}\,, \tag{5.16}$$

see Fig. 5.4. Transformation from $\mathbf{A} = l\mathcal{E}\mathbf{C}^{\mathsf{T}}$ to Eq. (5.16) is convenient for UV/vis data processing.

In order to ensure that the populations sum to unity in every column of $\mathbf{P}$, $[\mathbf{X}]_{\mathrm{t}}$ should be constant throughout the experiment. Otherwise the spectra should be rescaled to the initial concentration.[7] In our example of halochromic spectra in Fig. 5.2a, the total concentration decreased since the sample was diluted by adding stock solution during the experiment. The spectra were indeed rescaled to the initial concentration since the isosbestic points appeared. Rescaling has little impact on the SVD except for the amplitude vector $\mathbf{v}_1$, which is sensitive to total concentration, see Appendix D.1.

5.2.2 *Determining the Number of Absorbing Species*

One of the key moments in SVD-based analysis is determination of N_{spc}, the number of individual absorbing species in the sample. This task is analogous to finding rank of the signal in a matrix containing signal and noise. The amplitude of s_j is proportional to the significance of $\mathbf{u}_j$ and $\mathbf{v}_j$ in the reconstruction of $\mathbf{A}$ according to Eq. (5.5). Thus, inspection of logarithmic plot of the singular values is a great indicator for obtaining N_{spc}. Figure 5.5a, obtained from our halochromic example (Fig. 5.2a), shows an obvious gap after the second singular value, suggesting two-component spectral change. On the contrary, our solvatochromic example (Fig. 5.2b) does not show any gap between singular values, see Fig. 5.5b. This nonlinear spectral change with bands shifts requires many-dimensional SVD approximation to account

---

[7] Assume that $(\mathbf{c}_{\mathrm{t}})_j$ is defined as the total concentration at the j-th measurement, and $(\mathbf{c}_{\mathrm{t}})_1 = (\mathbf{c}_1)_1 + (\mathbf{c}_2)_1$. Then, the measured spectra are rescaled to the initial concentration using

$$\mathbf{a}_1^{\mathrm{resc}} = \mathbf{a}_1^{\mathrm{orig}}\,,$$

$$\mathbf{a}_j^{\mathrm{resc}} = \mathbf{a}_j^{\mathrm{orig}}\frac{(\mathbf{c}_{\mathrm{t}})_1}{(\mathbf{c}_{\mathrm{t}})_j} = \underbrace{l\epsilon_1(\mathbf{c}_{\mathrm{t}})_1}_{\mathbf{z}_1}\underbrace{\frac{(\mathbf{c}_1)_j}{(\mathbf{c}_{\mathrm{t}})_j}}_{(\mathbf{p}_1)_j} + \underbrace{l\epsilon_2(\mathbf{c}_{\mathrm{t}})_2}_{\mathbf{z}_2}\underbrace{\frac{(\mathbf{c}_2)_j}{(\mathbf{c}_{\mathrm{t}})_j}}_{(\mathbf{p}_2)_j} \quad \text{for } j = 2, \ldots N_{\mathrm{exp}}\,.$$

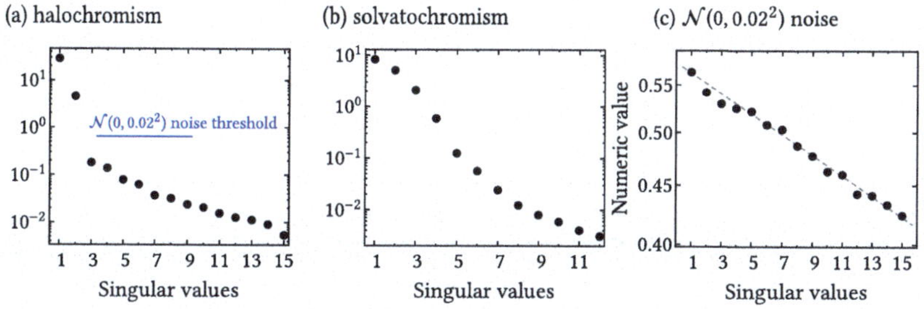

Fig. 5.5 Log-plot of singular values with **a** halochromic and **b** solvatochromic spectral changes, **c** uncorrelated noise, approximately following (gray dashed) straight line. Spectral data for (**a**), (**b**) are given in Fig. 5.2. Noise data for (**c**) have normal distribution $\mathcal{N}(0, 0.02^2)$ (mean = 0, std. = 0.02) and the same dimensions as (**a**), the noise amplitude was estimated from Fig. 5.5a. The noise threshold, blue line in (**a**), is given by the highest singular value in (**c**). Values in (**a**) and (**b**) were obtained from the spectra shown in Fig. 5.2a, b, respectively

for SVD of uncorrelated Gaussian noise reveals typical exponential decay (straight line in logarithmic plot), see Fig. 5.5c. Similar decay for the noise component is exhibited in Fig. 5.5a (s_3–s_{15}) and Fig. 5.5b (ca. s_5–s_{12}). The highest singular value in (c) provides a crude estimate of noise threshold in Fig. 5.5a (blue line, noise level estimate based on the Fig. 5.2a).

The cumulative sum $s_1^2 + s_2^2 + \cdots + s_j^2$ is proportional to the amount of data from **A** reconstructed by j-dimensional SVD approximation according to Eq. (5.5) (or the amount of reconstructed variance in PCA, cf. Sect. 5.3.2). Thus, the proportion of reconstructed data, also called the *cumulative energy* [14], is defined as

$$\text{cum.en.}(j) = \frac{\sum_{k=1}^{j} s_k^2}{\sum_{l=1}^{N_{\text{exp}}} s_l^2}. \tag{5.17}$$

In our halochromic example, the cumulative energy for $j = 2$ already accounts for almost 100% of data, see Fig. 5.6a. Another quantity, *residual standard deviation* (RSD) determines the error of j-dimensional SVD approximation as (Eq. (4.44) in Malinowski [9])

$$\text{RSD}(j) = \sqrt{\frac{\sum_{k=j+1}^{N_{\text{exp}}} s_k^2}{N_{\text{pts}}(N_{\text{exp}} - j)}}. \tag{5.18}$$

If the experimental error is known, it can be directly compared to the RSD value and serve as a criterion for N_{spc} determination. RSD of our halochromic example (Fig. 5.2a) is given in Fig. 5.6b. There is significant drop of RSD for $j = 2$ under the estimated experimental error threshold (std = 0.02), which is another indicator of the presence of two absorbing species.

If the signal-to-noise ratio is too low, it is hard to determine N_{spc} just from the visual inspection of singular values, cumulative energy or RSD. If the noise level is known,

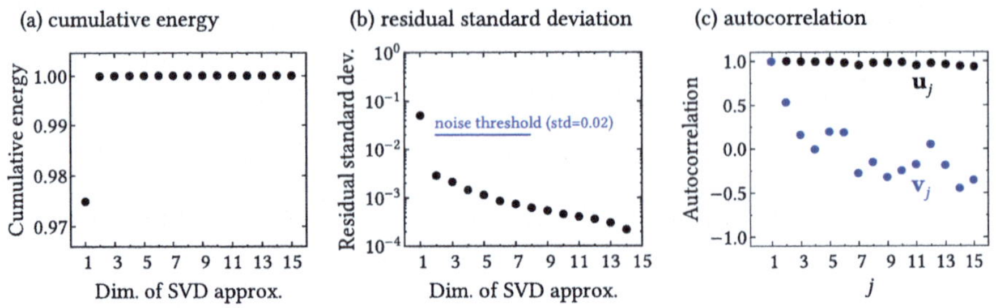

Fig. 5.6 Various indicators for N_{spc} determination for our halochromic data (Fig. 5.2a), **a** cumulative energy and **b** residual standard deviation based on the singular values, **c** autocorrelation coefficients calculated for neighbouring points

the noise threshold can be crudely estimated (Fig. 5.5a, c). Estimation of the noise threshold for white noise is given by Gavish and Donoho [27] even for unknown noise level. Determination of N_{spc} in UV/vis spectra is limited by the noise from the detector and illumination lamp (dependent on λ [18]) or by imperfect baseline subtraction. Shifts of the absorption bands due to solvatochromic or temperature effects or deviations from the Beer-Lambert law (Sect. 4.2) pose other challenges to SVD analysis. Complications also arise if vectors $\mathbf{z}_j$ or vectors $\mathbf{p}_j$ are approximately linearly dependent [18].

In the presence of uncorrelated noise, the noise-containing vectors $\mathbf{u}_j$ and $\mathbf{v}_j$ for $j > N_{\text{spc}}$ are often less smooth than the signal-containing vectors. The "smoothness" can be quantified as autocorrelation [15], i.e. correlation of a vector $\mathbf{u}_j$ (or $\mathbf{v}_j$) with the same vector shifted by one data point (or more data points, if the noise is correlated at short distances [18]), in particular,

$$\text{autocorr.}(\mathbf{u}_j) = \sum_{k=1}^{N_{\text{pts}}-1} (\mathbf{u}_j)_k (\mathbf{u}_j)_{k+1} , \tag{5.19}$$

$$\text{autocorr.}(\mathbf{v}_j) = \sum_{k=1}^{N_{\text{exp}}-1} (\mathbf{v}_j)_k (\mathbf{v}_j)_{k+1} . \tag{5.20}$$

In our halochromic example (Fig. 5.2a), the basis spectra are smooth as the autocorrelation is close to unity, Fig. 5.6c. This is probably a consequence of data smoothing during the data processing. On the other hand, only the first two amplitude vectors have a big positive autocorrelation, which can be an indicator for $N_{\text{spc}} = 2$. Henry and Hofrichter [15] introduced a 'rotation procedure', which can transform set of p amplitude vectors (although not the first p vectors) to another set of p orthonormal vectors with extremal autocorrelation. Thus, some of the new vectors have bigger autocorrelation on the expense of others, which helps to "push" the signal forward. Nevertheless, we are not going to use this procedure in this work.

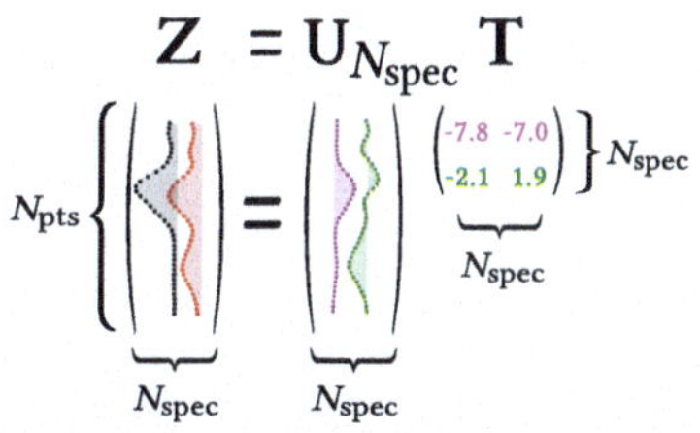

Fig. 5.7 Reconstruction of real spectra from abstract basis spectra $\mathbf{u}_j$. Constructed from the data in Fig. 5.2a

5.2.3 *Transformation from Abstract to Real Spectra*

If all spectra of individual absorbing species can be measured in isolation, their populations can be obtained without SVD as described in Sect. 5.2.5. In the other case, SVD has to be applied together with a model for concentrations of the chemical species (e.g. model based on host-guest binding or chemical kinetics).

First $N_{\rm spc}$ basis spectra can be transformed to the spectra of absorbing components using the transformation matrix $\mathbf{T}$,[8]

$$\mathbf{Z} = \mathbf{U}_{N_{\rm spc}}\mathbf{T}\,, \tag{5.21}$$

see Fig. 5.7.

Analogously, populations are reconstructed using the inverse matrix $\mathbf{T}^{-1}$,

$$\mathbf{P}^{\sf T} = \mathbf{T}^{-1}\mathbf{S}_{N_{\rm spc}}\mathbf{V}^{\sf T}_{N_{\rm spc}} \tag{5.22}$$

since $\mathbf{A} = \mathbf{U}_{N_{\rm spc}}\mathbf{T}\mathbf{T}^{-1}\mathbf{S}_{N_{\rm spc}}\mathbf{V}^{\sf T}_{N_{\rm spc}} = \mathbf{Z}\mathbf{P}^{\sf T}$, see Fig. 5.8. At this point, we have two ways of expressing the matrix $\mathbf{A}$,

$$\mathbf{A} = \mathbf{U}_{N_{\rm spc}}\mathbf{S}_{N_{\rm spc}}\mathbf{V}^{\sf T}_{N_{\rm spc}}\,, \tag{5.23}$$

and

$$\mathbf{A} = \mathbf{Z}\mathbf{P}^{\sf T} = \mathbf{U}_{N_{\rm spc}}\mathbf{T}\mathbf{P}^{\sf T}\,. \tag{5.24}$$

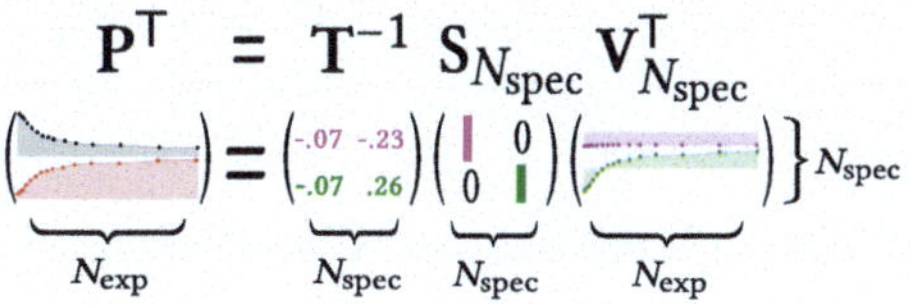

Fig. 5.8 Reconstruction of real populations from abstract amplitude vectors $\mathbf{v}_j$. Constructed from the data in Fig. 5.2a

[8] It is sometimes called 'rotation procedure' [9] (denoting the transformation matrix as $\mathbf{R}$), however, the transformation is not unitary.

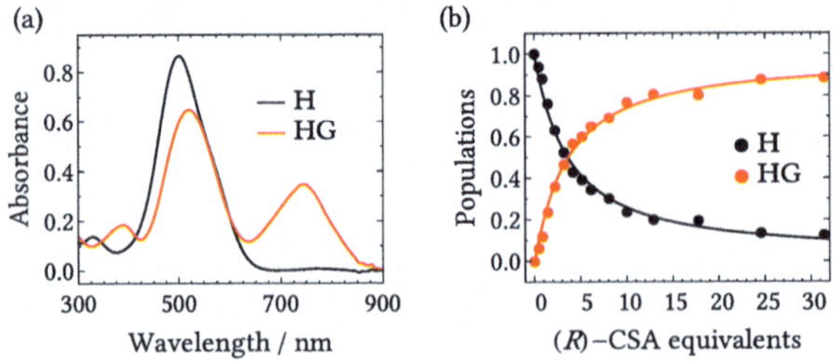

Fig. 5.9 Reconstructed (**a**) spectra of absorbing species **Z** and (**b**) their populations **P** throughout the halochromic spectral change. Points stand for $\mathbf{P}_{\text{exp}}$ calculated from Eq. (5.22), lines for $\mathbf{P}_{\text{mod}}$ calculated from a 1:1 host-guest model (Eq. (2.26))

Comparing these two expressions, we get

$$\mathbf{S}_{N_{\text{spc}}}\mathbf{V}_{N_{\text{spc}}}^{\mathsf{T}} = \mathbf{T}\mathbf{P}^{\mathsf{T}} . \tag{5.25}$$

The left-hand side of this equation is fixed and based solely on SVD. The populations **P** follow a system-dependent model derived from thermodynamics or chemical kinetics, see Chap. 2. Hence, the right-hand side of Eq. (5.25) should be adjusted by least squares fitting of elements of **T** and model parameters for **P** (binding constants or reaction rate coefficients). Thus, we distinguish populations $\mathbf{P}_{\text{exp}}$ obtained from the experiment after calculating **T** (Eq. (5.22), typically noisy) from populations $\mathbf{P}_{\text{mod}}$ obtained from a model (smooth, could be interpolated between experimental points), see Fig. 5.9b. In our halochromic example (Fig. 5.2a), the host-guest binding model with 1:1 stoichiometry (Eq. (2.26)) was used, and the function

$$\chi^2 = \|\mathbf{S}_{N_{\text{spc}}}\mathbf{V}_{N_{\text{spc}}}^{\mathsf{T}} - \mathbf{T}\mathbf{P}_{\text{mod}}^{\mathsf{T}}(K_{\mathbf{HG}})\|^2 \tag{5.26}$$

was minimized.

The reconstructed spectra are shown in Fig. 5.9a. The first measured spectrum $\mathbf{a}_1$ with no acid added was identified as the first absorbing component $\mathbf{z}_1$, therefore, the first column of **T** was calculated from Eqs. (5.21) and (5.15) as $\mathbf{t}_1 = \mathbf{U}_{N_{\text{spc}}}^{\dagger}\mathbf{a}_1^{\mathsf{T}}$ or simply read out from the first column of $\mathbf{S}_{N_{\text{spc}}}\mathbf{V}_{N_{\text{spc}}}^{\mathsf{T}}$ (combination coefficients of the first measured spectrum, see next section). Quality of the fit can be checked by comparison of populations $\mathbf{P}_{\text{mod}}$ and $\mathbf{P}_{\text{exp}}$, see Fig. 5.9b.

5.2.4 Combination Coefficients

Combination coefficients $s_j\mathbf{v}_j$ provide an illustrative "map" of the course of the experiment. Each point in plot of combination coefficients corresponds to a particular measurement. This plot is linear for our halochromic example (Fig. 5.2a) as shown in Fig. 5.10a (full circles). This is a rule for every spectral change involv-

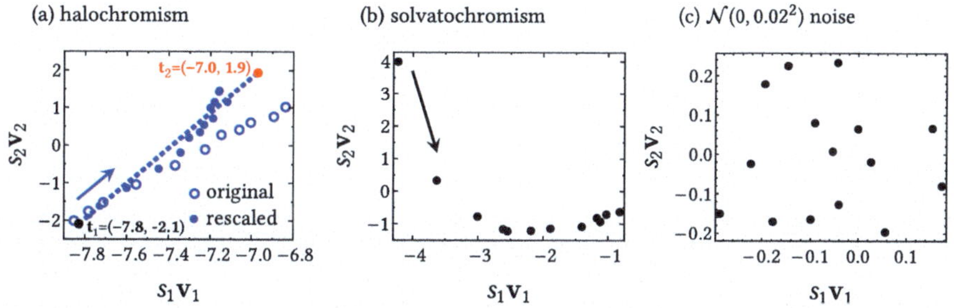

Fig. 5.10 Plot of first two combination coefficients for **a** halochromic (data in Fig. 5.2a), **b** solvatochromic (data in Fig. 5.2b) spectral change and **c** spectra generated as $\mathcal{N}(0, 0.02^2)$ noise with dimensions of (a). Course of the experiment is indicated by arrows. In (a) black (red) circle corresponds to combination coefficients of the first (second) real component **H** (**HG**), dashed line denotes the linear path between the two components in two-component spectral change

ing exactly two absorbing species with constant dye concentration (or rescaled to constant dye concentration, see footnote 7 on Sect. 5.2.2). In our case, also the original spectra measured with gradual sample dilution show linear dependence (empty circles) because the dilution was uniform. The change of total dye concentration is mostly captured by the first combination coefficient $s_1\mathbf{v}_1$, see Fig. D.1f in Appendix D.1. Solvatochromic spectral change (Fig. 5.2b) shows systematic shift without the linear dependence, see Fig. 5.10b. Spectra consisting only of noise do not show any systematic shift, see Fig. 5.10c.

If a measured spectrum $\mathbf{a}_j$ is the spectrum of an isolated individual absorbing species in two-component system, e.g. $\mathbf{a}_j = \mathbf{z}_1$, then Eq. (5.4) implies $\mathbf{z}_1 = \mathbf{u}_1 s_1 (\mathbf{v}_1^\mathsf{T})_j + \mathbf{u}_2 s_2 (\mathbf{v}_2^\mathsf{T})_j$. Hence, the combination coefficients are identical to T_{11} and T_{21} (cf. Eq. (5.21)) in this case. In other words, j-th column of $\mathbf{T}$ (i.e. $\mathbf{t}_j$) constitutes coordinates of the j-th absorbing component in the combination coefficient plot. Coordinates of the first and second absorbing species are shown in Fig. 5.10a (compare the coordinates with Fig. 5.7). Since the measured spectra are linear combination of $\mathbf{z}_1$ and $\mathbf{z}_2$, all experimental points lie on a curve $(T_{11}, T_{21})^\mathsf{T} p_1 + (T_{12}, T_{22})^\mathsf{T}(1 - p_1) = \mathbf{t}_1 p_1 + \mathbf{t}_2(1 - p_1)$, where $p_1 \in [0, 1]$ is the population of the first component (dashed line in Fig. 5.10a). This is parametric equation for a line segment implying linear dependence in the two-component spectral change.

This linear dependence can be generalized to higher number of absorbing components. For example, spectral data set with $N_{\mathrm{spc}} = 3$ implies that the combination coefficients lie on a 2D plane inside a triangle given by $\mathbf{t}_1$, $\mathbf{t}_2$ and $\mathbf{t}_3$ [18]. Also in such system, a spectral change two absorbing components (population of the third one is constant) implies that the corresponding combination coefficients lie on a straight line. Generally, N_{spc} components form an $(N_{\mathrm{spc}} - 1)$-dimensional object (a convex hull) in N_{spc}-dimensional space of combination coefficients, although small overlap with other dimensions occurs due to the noise. Combination coefficients corresponding to a spectral change involving r components form an $(r - 1)$-dimensional object in N_{spc}-dimensional space of combination coefficients.

5.2.5 Decomposition to the Absorbing Components Without SVD

If all spectra of individual components $\mathbf{Z}$ can be measured separately, transformation from abstract to real vectors is not necessary and their populations can be calculated without SVD as

$$\mathbf{P}_{\text{exp}}^{\text{T}} = \mathbf{Z}^{\dagger}\mathbf{A}\,, \tag{5.27}$$

cf. Eq. (5.15). These populations $\mathbf{P}_{\text{exp}}$ obtained directly from the experiment serve as binding isotherms and can be fitted with $\mathbf{P}_{\text{mod}}$ using a binding or kinetic model to obtain binding constants. In our publication [28], we used this type of analysis and called it 'decomposition to spectra of individual species' (DSIS). Instead of Eq. (5.27) the equivalent approach of least squares fitting was used and we minimized

$$\chi^2 = \|\mathbf{A} - \mathbf{Z}\mathbf{P}_{\text{exp}}^{\text{T}}\|^2 \tag{5.28}$$

with all elements of $\mathbf{P}_{\text{exp}}^{\text{T}}$ being fitting parameters. In contrary to pseudoinverse, least squares fitting allows for constrains such as $\sum \mathbf{p}_j = 1$ or $0 \le \mathbf{p}_j \le 1$.

5.2.6 SVD Analysis–Summary

The SVD-based analysis comprises of four basic steps described in Scheme 5.1.

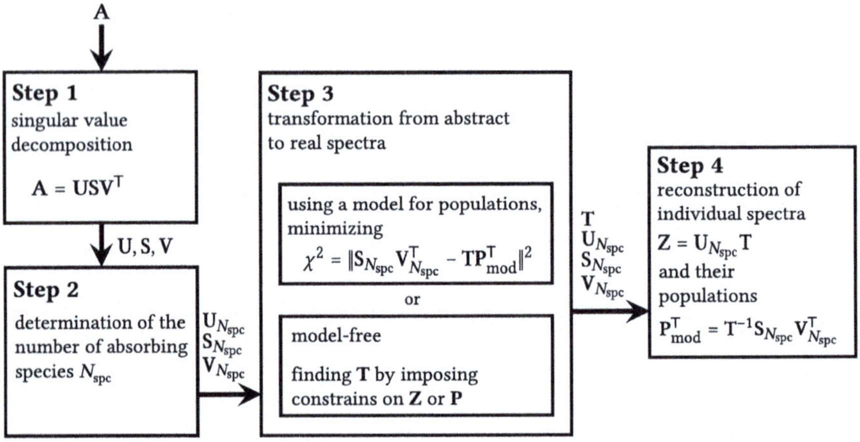

Scheme 5.1 Steps in SVD-based analysis of UV/vis spectral series. Model-free determination of T is explained in [11]

5.3 Analogy with Principal Component Analysis

5.3.1 Calculation of Principal Components

Principal component analysis (PCA) is a statistical tool identifying directions of maximal variance in multidimensional data sets and consequently finding patterns (clustering of datapoints) in high-dimensional data. We show in Sect. 5.3.2 that PCA is mathematically identical with SVD and can be regarded as statistical interpretation of SVD. In the PCA literature it is common to construct the analyzed matrix as rows of measurements in contrast to our SVD column-wise convention. We follow the usual PCA convention since the common programs as *Mathematica* or *MATLAB* use it too, and we denote the row-wise data matrix $\mathbf{B} \equiv \mathbf{A}^\mathsf{T}$ (in this section $\mathbf{B}$ is a $m \times n$ matrix). As shown in Eq. (5.29), measurements $\mathbf{b}_j^\mathsf{T}$ (e.g. measured absorption spectra) are arranged in rows. Columns $\mathbf{x}_j$ denote one particular feature (e.g. measured absorbance at one particular wavelength). In statistical application of PCA, the columns $\mathbf{x}_j$ stand for particular analyzed features such as age, height and weight, whereas rows $\mathbf{b}_j^\mathsf{T}$ denote entries for a particular person.

$$\mathbf{B} = \begin{pmatrix} \dots \; \mathbf{b}_1^\mathsf{T} \; \dots \\ \dots \; \mathbf{b}_2^\mathsf{T} \; \dots \\ \vdots \\ \dots \; \mathbf{b}_m^\mathsf{T} \; \dots \end{pmatrix} = \begin{pmatrix} \vdots & \vdots & & \vdots \\ \mathbf{x}_1 & \mathbf{x}_2 & \dots & \mathbf{x}_n \\ \vdots & \vdots & & \vdots \end{pmatrix} . \tag{5.29}$$

Unlike SVD, PCA requires mean-subtracted data $\overline{\mathbf{B}}$ (denoted with bar) defined as

$$\overline{\mathbf{B}} \equiv \mathbf{B} - \begin{pmatrix} 1 \\ 1 \\ \vdots \\ 1 \end{pmatrix} (\overline{x}_1, \overline{x}_2, \dots, \overline{x}_n) , \tag{5.30}$$

where $\overline{x}_j = 1/n \sum_{k=1}^{m} (\mathbf{x}_j)_k$ is arithmetic mean of j-th feature.

PCA is searching for directions $\boldsymbol{\alpha}_j$ in which a linear transformation of mean-centered feature vectors $\overline{\mathbf{x}}_j = \mathbf{x}_j - \overline{x}_j$ captures maximum of measured variance. These direction vectors have unit norm and zero mutual covariance,[9] i.e. they are uncorrelated. The transformed vectors are denoted as *principal components* (PCs), and the transformation from mean-centered feature vectors reads

[9] Covariance is defined as $\mathrm{Cov}(\mathbf{x}, \mathbf{y}) = 1/(n-1) \sum_{j=1}^{n} (x_j - \overline{x})(y_j - \overline{y})$ and variance as $\mathrm{Var}(\mathbf{x}) = \mathrm{Cov}(\mathbf{x}, \mathbf{x})$. Covariance calculated from measured data is also called 'sample covariance' because the true value is not known. To obtain unbiased value, $n-1$ should be in the denominator rather than n.

$$\underbrace{\begin{pmatrix} \vdots & \vdots & & \vdots \\ \mathrm{PC}_1 & \mathrm{PC}_2 & \ldots & \mathrm{PC}_n \\ \vdots & \vdots & & \vdots \end{pmatrix}}_{m\times n} = \underbrace{\begin{pmatrix} \vdots & \vdots & & \vdots \\ \overline{\mathbf{x}}_1 & \overline{\mathbf{x}}_2 & \ldots & \overline{\mathbf{x}}_n \\ \vdots & \vdots & & \vdots \end{pmatrix}}_{m\times n} \underbrace{\begin{pmatrix} \vdots & \vdots & & \vdots \\ \boldsymbol{\alpha}_1 & \boldsymbol{\alpha}_2 & \ldots & \boldsymbol{\alpha}_n \\ \vdots & \vdots & & \vdots \\ \vdots & \vdots & & \vdots \end{pmatrix}}_{n\times n} . \tag{5.31}$$

PCs are automatically mean-centered due to the centering of feature vectors. PCs are sometimes denoted as *scores* ($m \times n$ matrix) and the directions $\boldsymbol{\alpha}_j$ as *loadings* ($n \times n$ matrix).

Search for PCs is performed [1] by maximization of $\mathrm{Var}(\mathrm{PC}_j)$ using Lagrange multipliers with normalization conditions $\boldsymbol{\alpha}_j^{\mathsf{T}}.\boldsymbol{\alpha}_j = 1$ and zero covariance between different PCs, i.e. $\mathrm{Cov}(\mathrm{PC}_j, \mathrm{PC}_k) = 0$ (equivalent to $\boldsymbol{\alpha}_j^{\mathsf{T}}.\boldsymbol{\alpha}_k = 0$). It turns out that the solution is equivalent to finding eigenvectors of the covariance matrix $1/(n-1)\overline{\mathbf{B}}^{\mathsf{T}}\overline{\mathbf{B}}$,

$$\frac{1}{n-1}\overline{\mathbf{B}}^{\mathsf{T}}\overline{\mathbf{B}}\boldsymbol{\alpha}_j = \lambda_j \boldsymbol{\alpha}_j . \tag{5.32}$$

Since the variance captured by j-th PC is equal to its corresponding eigenvalue λ_j (as shown below), the PCs and corresponding loadings $\boldsymbol{\alpha}_j$ are sorted in descending order according to their eigenvalues,[10] $\lambda_1 > \lambda_2 > \cdots > \lambda_n$. The covariance matrix $1/(n-1)\overline{\mathbf{B}}^{\mathsf{T}}\overline{\mathbf{B}}$ is symmetric (Hermitian), therefore, its eigenvalues are real and its eigenvectors are orthogonal.

5.3.2 Comparison to SVD

SVD of the mean-subtracted data matrix is $\overline{\mathbf{B}} = \mathbf{USV}^{\mathsf{T}}$. Recalling the calculation of SVD vectors as eigenvalues (Eq. (5.9)), we already know that

$$\overline{\mathbf{B}}^{\mathsf{T}}\overline{\mathbf{B}}\mathbf{v}_j = s_j^2 \mathbf{v}_j . \tag{5.33}$$

Comparison to Eq. (5.32) implies that $\lambda_j = s_j^2/(n-1)$ and $\boldsymbol{\alpha}_j = \mathbf{v}_j$. Furthermore, comparing SVD in the form $\mathbf{US} = \overline{\mathbf{B}}\mathbf{V}$ to Eq. (5.31) yields $\mathrm{PC}_j = \mathbf{u}_j s_j$.

This analogy with SVD allows for a quick proof that the variance captured by PC_j is λ_j, in particular $\mathrm{Var}(\mathrm{PC}_j) = 1/(n-1)\mathrm{PC}_j\,\mathrm{PC}_j^{\mathsf{T}} = s_j^2/(n-1)\mathbf{u}_j\mathbf{u}_j^{\mathsf{T}} = s_j^2/(n-1) = \lambda_j$. Relative amount of captured variance by first j PCs is then calculated as $\sum_{k=1}^{j}\lambda_k / \sum_{l=1}^{n}\lambda_l$ (cf. the cumulative energy in Eq. (5.17)). Due to the analogy of PCA and SVD, PCA of real data sets can be calculated using numerical algorithms for SVD.

[10] Since the experimental data contain noise, usually no pair of eigenvalues have the same value.

5.3.3 Application

Standard application of PCA is visualization of high-dimensional data in low-dimensional plot of principal components. This allows to observe clustering of data points, revealing similarities or differences in various data sets. We illustrate PCA analysis with three artificial data sets shown in Fig. 5.11a. Age, height and weight are the observed features ($n = 3$), and each data set has 150 entries (i.e. people, $m = 150$). PCA is calculated from all three data sets combined. Mutual plot of PC_1 and PC_2 (together capturing 92.9% of variance) in Fig. 5.11b projects the 3D data to 2D plot and shows that data sets 1 and 2 mostly overlap, while data set 3 is significantly different. The corresponding loadings α_1 and α_2 represent directions in the combined data set, where the data are maximally "stretched out". Mutual plot of PC_2 and PC_3 in Fig. 5.11c confirms that PC_3 captures the smallest variance. Note that k-th data point $\overline{\mathbf{b}}_k$ can be projected to j-th PC using the loading α_j according to $(\mathrm{PC}_j)_k = \overline{\mathbf{b}}_k^{\mathsf{T}}.\boldsymbol{\alpha}_j$ as follows from Eq. (5.31). In real life applications of PCA, high-dimensional data are processed (i.e. with large number of features), for example, when thousands of genetic markers are analyzed to observe clustering based on common ancestry [29]. Recent developments of PCA are reviewed by Jolliffe and Cadima [30]. In this work, we use PCA to analyze combination coefficients obtained from SVD, the procedure is described in detail in Appendix D.5.

In *Mathematica* software, PCs are obtained by `PC=PrincipalComponents[B]`, in *MATLAB*, the command `[alpha,PC,lambda]=pca(B)` is used to obtain loadings, PC scores and the variances. Both applications use the row-wise data matrix according to Eq. (5.31) and subtract the mean automatically.

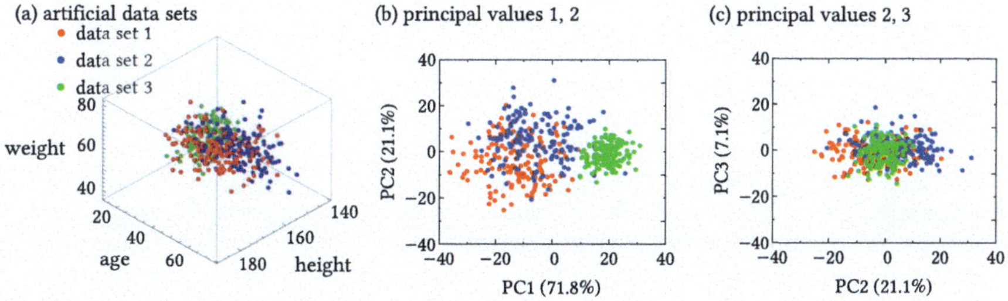

Fig. 5.11 Illustration of PCA, **a** three artificially constructed data sets with three features (normally distributed), each of 150 entries. **b** Mutual plot of first two principal components capturing 92.9% of the variance, **c** mutual plot of principal components 2 and 3

References

1. Jolliffe IT (2002) Principal component analysis, second, springer series in statistics. Springer-Verlag, New York
2. Brown VM, Ossadtchi A, Khan AH, Yee S, Lacan G, Melega WP, Cherry SR, Leahy RM, Smith DJ (2002) Multiplex three-dimensional brain gene expression mapping in a mouse model of Parkinson's disease. Genome Res 12:868–884
3. Zeng G (2007) Facial recognition with singular value decomposition. In: Elleithy K (ed) Advances and innovations in systems, computing sciences and software engineering, pp 145–148
4. Taira K, Brunton SL, Dawson STM, Rowley CW, Colonius T, McKeon BJ, Schmidt OT, Gordeyev S, Theofilis V, Ukeiley LS (2017) Modal analysis of fluid flows: an overview. AIAA J 55:4013–4041
5. Scherl I, Strom B, Shang JK, Williams O, Polagye BL, Brunton SL (2020) Robust principal component analysis for modal decomposition of corrupt fluid flows. Phys Rev Fluids 5:054401
6. Khadka U, Holubec V, Yang H, Cichos F (2018) Active particles bound by information flows. Nat Commun 9:3864
7. Holliday TW, Friedl L (2013) Hominoid humeral morphology: 3D morphometric analysis. Am J Phys Anthropol 152:506–515
8. Martinón-Torres M, d'Errico F, Santos E, Álvaro Gallo A, Amano N, Archer W, Armitage SJ, Arsuaga JL, Bermúdez de Castro JM, Blinkhorn J, Crowther A, Douka K, Dubernet S, Faulkner P, Fernández-Colón P, Kourampas N, González García J, Larreina D, Le Bourdonnec F-X, MacLeod G, Martín-Francés L, Massilani D, Mercader J, Miller JM, Ndiema E, Notario B, Pitarch Martí A, Prendergast ME, Queffelec A, Rigaud S, Roberts P, Shoaee MJ, Shipton C, Simpson I, Boivin N, Petraglia MD (2021) Earliest known human burial in Africa. Nature 593:95–100
9. Malinowski ER (2002) Factor analysis in chemistry. Wiley, Mar.
10. Spearman C (1904) "General intelligence," objectively determined and measured. Am J Psychol 15:201–292
11. Payne DT, Labuta J, Futera Z, Březina V, Hanyková L, Chahal MK, Hill JP (2022) Molecular rotor based on an oxidized resorcinarene. Org Chem Front 9:39–50
12. Hanuš J, Chmelová K, Štěpánek J, Turpin P-Y, Bok J, Rosenberg I, Točík Z (1999) Raman spectroscopic study of triplex-like complexes of polyuridylic acid with the isopolar, non-isosteric phosphonate analogues of diadenosine monophosphate. J Raman Spectrosc 30:667–676
13. Palacký J, Mojzeš P, Bok J (2011) SVD-based method for intensity normalization, background correction and solvent subtraction in Raman spectroscopy exploiting the properties of water stretching vibrations. J Raman Spectrosc 42:1528–1539
14. Brunton SL, Kutz JN (2019) Data-driven science and engineering: machine learning, dynamical systems, and control. Cambridge University Press
15. Henry ER, Hofrichter J (1992) [8] Singular value decomposition: application to analysis of experimental data. In: Methods in enzymology, vol 210. Numerical computer methods. Academic Press, pp 129–192
16. Henry ER (1997) The use of matrix methods in the modeling of spectroscopic data sets. Biophys J 72:652–673
17. Zimányi L, Kulcsár Á, Lanyi JK, Sears DF, Saltiel J (1999) Singular value decomposition with self-modeling applied to determine bacteriorhodopsin intermediate spectra: Analysis of simulated data. PNAS 96:4408–4413
18. Zimányi L (2004) Analysis of the Bacteriorhodopsin Photocycle by Singular Value Decomposition with Self-Modeling: A Critical Evaluation Using Realistic Simulated Data. J. Phys. Chem. B 108:4199–4209
19. Zimányi L, Saltiel J, Brown LS, Lanyi JK (2006) A Priori Resolution of the Intermediate Spectra in the Bacteriorhodopsin Photocycle: The Time Evolution of the L Spectrum Revealed. J. Phys. Chem. A 110:2318–2321

20. Zimányi L, Thekkan S, Eckert B, Condren AR, Dmitrenko O, Kuhn LR, Alabugin IV, Saltiel J (2020) Determination of the pKa Values of trans-Resveratrol, a Triphenolic Stilbene, by Singular Value Decomposition. Comparison with Theory. J. Phys. Chem. A 124:6294–6302
21. Bro R, Acar E, Kolda TG (2008) Resolving the sign ambiguity in the singular value decomposition. Journal of Chemometrics 22:135–140
22. G. H. Golub and C. F. V. Loan, *Matrix Computations* (JHU Press, 2013)
23. Press WH, Teukolsky SA, Vetterling WT, Flannery BP (2007) Numerical Recipes 3rd Edition: The Art of Scientific Computing. Cambridge University Press, Sept
24. Penrose R (1955) A generalized inverse for matrices. Mathematical Proceedings of the Cambridge Philosophical Society 51:406–413
25. N. R. Draper and H. Smith, *Applied Regression Analysis* (John Wiley & Sons, Aug. 2014)
26. K. A. Connors, *Binding constants: the measurement of molecular complex stability* (Wiley, May 1987)
27. Gavish M, Donoho DL (2014) The Optimal Hard Threshold for Singular Values is 4/. IEEE Transactions on Information Theory 60:5040–5053
28. Březina V, Ishihara S, Lang J, Hanyková L, Ariga K, Hill JP, Labuta J (2018) Structural Modulation of Chromic Response: Effects of Binding-Site Blocking in a Conjugated Calix[4]pyrrole Chromophore. ChemistryOpen 7:323–335
29. Ioannidis AG, Blanco-Portillo J, Sandoval K, Hagelberg E, Miquel-Poblete JF, Moreno-Mayar JV, Rodríguez-Rodríguez JE, Quinto-Cortés CD, Auckland K, Parks T, Robson K, Hill AVS, Avila-Arcos MC, Sockell A, Homburger JR, Wojcik GL, Barnes KC, Herrera L, Berríos S, Acuña M, Llop E, Eng C, Huntsman S, Burchard EG, Gignoux CR, Cifuentes L, Verdugo RA, Moraga M, Mentzer AJ, Bustamante CD, Moreno-Estrada A (2020) Native American gene flow into Polynesia predating Easter Island settlement. Nature 583:572–577
30. Jolliffe IT, Cadima J (2016) Principal component analysis: a review and recent developments. Philosophical Transactions of the Royal Society A: Mathematical, Physical and Engineering Sciences 374:20150202

Chapter 6
Chromic and Binding Properties of OxP Derivatives

As discussed in Sect. 4.5, chromic changes in dye molecules can be caused by protonation (halochromism) or by change of solvent properties (solvatochromism). In this chapter we show our results about halochromism and solvatochromism of the oxoporphyrinogen derivatives **OxP**, **Bz_2OxP** and **Bz_4OxP**. Aside from the complete description of color variation, particular protonated species are identified and the binding constants were determined, where possible. To remind the structures of oxoporphyrinogens, see Fig. 6.1. The **OxP** molecule possesses two calix[4]pyrrole-type binding sites (shortly NH/amine binding sites, black arrows) and four protonation sites at the carbonyl goups (C = O groups, green arrows). N-benzylation at the macrocyclic nitrogen atoms of **OxP** leads to stepwise elimination of the two NH binding sites (red crosses) and subsequent serial variation of the chromic properties of the N-benzylated products. The color related properties were studied by UV/vis spectroscopy, however, to verify our conclusions related to protonation, NMR titration experiments (i.e. measurement of spectra at different guest concentrations when the guest is gradually added into the host solution) were conducted. Following sections on chromism and binding are mostly based on our article [1] including parts of text, most figures were adopted from [1] with modifications. These were reproduced with permission of Wiley-VCH Verlag. Several NMR spectra analyzed here and some preliminary results were presented in the master's thesis [2]. Experimental details about the measurements are given in Appendix C.2.

6.1 Halochromism of OxP

Halochromic response of **OxP** was reported for the first time by Shundo et al. [4]. The cause of color variation upon addition of trifluoroacetic acid (TFA) was the presence of di- and tetraprotonated **OxP** species. The protonation at the hemiquinonoid carbonyl sites was proven by direct observation of the intensity of the C-O-H infrared

V. Březina, *Supramolecular Complexes of Oxoporphyrinogens with Organic Molecules*, Springer Theses, https://doi.org/10.1007/978-3-031-68646-7_6

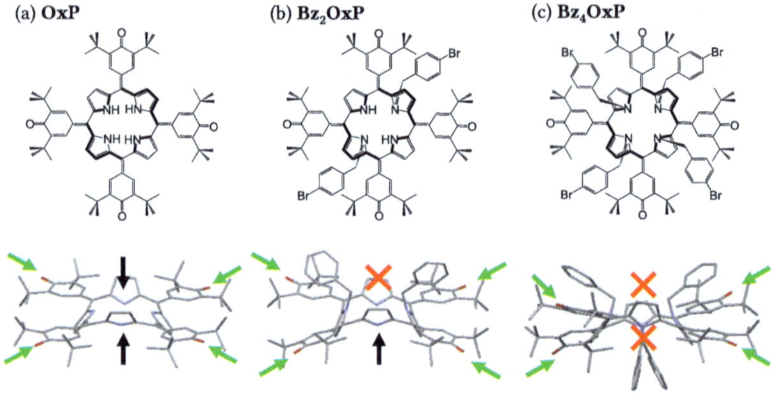

Fig. 6.1 Chemical structures (top) and X-ray crystal structures (bottom) of **a** oxoporphyrinogen (**OxP**), **b** bis(4-bromobenzyl)oxoporphyrinogen (**Bz_2OxP**), **c** tetrakis(4-bromobenzyl)oxoporphyrinogen (**Bz_4OxP**) used in this work. X-ray crystal structure in panels (**b**, **c**) are of the non-brominated derivatives [3]. Black arrows: accessible calix[4]pyrrole-type binding sites, red crosses: blocked binding sites, green arrows: protonation carbonyl sites. Adapted from [1] with permission of Wiley-VCH Verlag

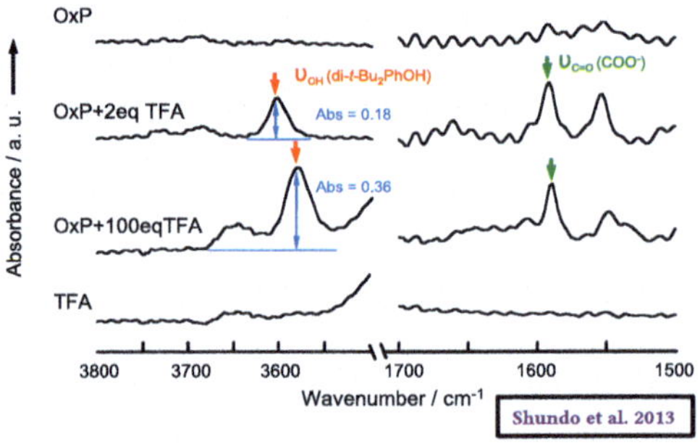

Fig. 6.2 Solution (CH_2Cl_2, 0 °C) state FTIR spectra of **OxP** with various amounts of TFA (0, 2 and 100 equiv. of TFA) and pure TFA. Absorbance of the **OxP** hydroxyl stretching peak is indicated. Appearance of the OH stretching signal is due to first diprotonation (2 equiv. of TFA) then further protonation at higher concentration of TFA. Reproduced from [4] with permission from the Royal Society of Chemistry

vibration, see Fig. 6.2. Since the binding steps **HG→HG$_2$** and **HG$_3$ →HG$_4$** are strongly cooperative (i.e. high interaction parameters α_{12} and α_{34}, see Sect. 2.2.4), mono- and tri- protonated **OxP** species are virtually absent. In our publication [1], we reanalyzed the UV/vis data from the study [4] to assess quantitatively the halochromic properties of **OxP**.

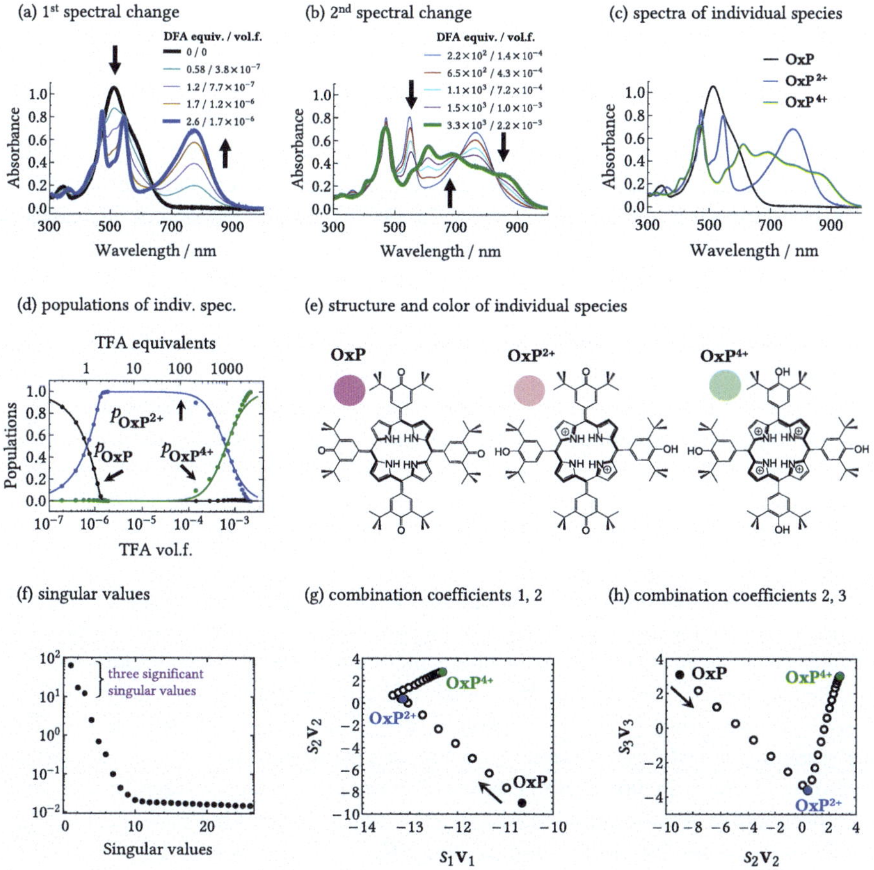

Fig. 6.3 UV/vis titration of **OxP** (10^{-5} M, CH_2Cl_2) with TFA and corresponding analyses. **a**, **b** First and second sets of spectral changes occurring during titration. These data were first published in [4]. **c** Spectra of individual absorbing species **Z** identified as the start and endpoints of each spectral change (bold lines in panels (**a**, **b**)). **d** Populations $\mathbf{P}_{\text{exp}}$ of absorbing species **Z** as obtained by DSIS (solid circles). Fitting of $\mathbf{P}_{\text{exp}}$ to 1:4 binding model provided model populations $\mathbf{P}_{\text{mod}}$ (solid lines). **e** Chemical structures and colors of absorbing species (colors calculated from the spectra shown in panel (**c**), details in Appendix C.2). SVD analysis confirms $N_{\text{spc}} = 3$ via plots of **f** singular values and (**g**, **h**) combination coefficients space. Adapted from [1] with permission of Wiley-VCH Verlag

Titration spectra[1] in Fig. 6.3a, b show two consecutive spectral changes. The presence of isosbestic points (see Sect. 4.4) indicates consecutive transitions (i.e. only two species are present at a time) **OxP**→$\mathbf{OxP^{2+}}$ and $\mathbf{OxP^{2+}}$→$\mathbf{OxP^{4+}}$ corresponding to the first and second spectral change, respectively. Singular value decomposition

---

[1] The amount of added guest is described by guest equivalents defined as equiv. $= [\mathbf{G}]_t/[\mathbf{H}]_t$. In addition, volume fraction (vol.f./ φ) is used, especially for later description of solvatochromism (polarity of the sample depends on vol.f. of its components).

(SVD) analysis of the spectra confirms the presence of three absorbing species based on three significant singular values as shown in Fig. 6.3f (see Sect. 5.2.2). Mutual plots of combination coefficients in Fig. 6.3g, h show two linear segments connecting all three component coordinates. Since arrangement of combination coefficients on a straight line implies that exactly two components are involved in the particular spectral change (see Sect. 5.2.4), this plot further confirms the aforementioned consecutive protonation scheme. In Fig. 6.3g, the **OxP^{2+}** coordinates do not lie exactly at the intersection due to small change in total dye concentration since the combinaton coefficient $s_1\mathbf{v}_1$ is more sensitive to this change than $s_2\mathbf{v}_2$ and $s_3\mathbf{v}_3$, see details in Appendix D.2.

As the two spectral changes are separated, the spectra of individual absorbing species, Fig. 6.3c, were identified at the end of each spectral change, see thick lines in Fig. 6.3a, b. Since the observed spectral changes are purely of halochromic nature (see Sect. 5.2.1), the measured UV/vis spectra are linear combination of the absorbing components spectra, $\mathbf{A} = \mathbf{Z}\mathbf{P}^{\mathrm{T}}_{\mathrm{exp}}$. Analysis from Sect. 5.2.5 referred to as 'decomposition into the spectra of individual species' (DSIS) can be applied. DSIS of $\mathbf{a}_{\varphi_j}$, measured spectrum at volume fraction φ_j, is formulated as

$$\mathbf{a}_{\varphi_j} = p_{\mathbf{OxP}}(\varphi_j)\,\mathbf{z}_{\mathbf{OxP}} + p_{\mathbf{OxP^{2+}}}(\varphi_j)\,\mathbf{z}_{\mathbf{OxP^{2+}}} + p_{\mathbf{OxP^{4+}}}(\varphi_j)\,\mathbf{z}_{\mathbf{OxP^{4+}}}\,, \tag{6.1}$$

where $\mathbf{z}_i$ is the spectrum of i-th absorbing component and $\mathbf{p}_i(\varphi_j)$ its population at TFA volume fraction φ_j. Populations $\mathbf{P}_{\mathrm{exp}}$ obtained by DSIS (with constrains $\mathbf{p}_j > 0$ and $\sum \mathbf{p}_j = 1$) are shown in Fig. 6.3d (solid circles). The quality of DSIS is excellent, as the error $\|\mathbf{A} - \mathbf{Z}\mathbf{P}^{\mathrm{T}}_{\mathrm{exp}}\|$ is small, see Fig. D.3 in Appendix D.2. Decomposition according to Eq. (6.1) would not be possible in the presence of solvatochromic shift.

Subsequently, we attempted to describe the interaction of **OxP** with TFA in terms of a host- guest binding model minimizing $\|\mathbf{P}_{\mathrm{exp}} - \mathbf{P}_{\mathrm{mod}}\|$, where $\mathbf{P}_{\mathrm{mod}}$ is a matrix of concentrations reconstructed using a particular binding model. Here, the first spectral change saturates at 2 guest equiv. implying double protonation. The second spectral change is saturated at relatively low concentration of TFA ($\varphi = 0.002$ in CH_2Cl_2), thus, equations governing the chemical equilibria can be used without correction for activity (i.e. assuming $\gamma_j = 1$, see Sect. 2.1.1). A 1:4 host-guest binding model (derived in Sect. 2.2.2 with implementation in Appendix A.4) was adopted to obtain the equilibrium binding constants listed in Table 6.1, Sect. 6.9. The resulting fit is shown in Fig. 6.3d (solid lines). The diprotonated form is stabilized by binding of two counteranions (on the top and on the bottom, Fig. 6.1a) at the central NH groups. Another source of stabilization is placement of the protonation at opposite carbonyl groups [5] (protonation sites can change during a tautomerization process, see Sect. 7.1.1). The good quality of the data fit also confirms the low stability of any mono- and triprotonated **OxP** species, which are essentially absent due to the ease of further protonation (analogously to tetraphenylporphyrin monocation [6–8]). The chemical structures of protonated species including the changes in conjugation are given in Fig. 6.3e. Results of UV/vis titration are confirmed by NMR, see Sect. 6.5.

Table 6.1 Stepwise binding constants[a] (in M^{-1}) and corresponding interaction parameters[b] (measure of cooperativity) determined by fitting of the NMR and UV/vis titration experiments. Decadic logarithms ('log') of parameters are given for **OxP**. Details about the analysis and errors are given in Appendix C.5

	DFA/TFA		CSA	
	NMR	UV/vis	NMR	UV/vis
OxP				
$\log K_1$	4.5–8.7	2.3–6.0[c]	4.0–5.0	
$\log K_1 K_2$	8.0–16	13–15[c]	10–12	
$\log K_3$	2.5–4.0	< 1.0[c]	< 5.0	N/A
$\log K_3 K_4$	4.0–6.0	3.6–5.0[c]	2.3–2.5	
$\log \alpha_{12}$	–	> 3.7[c]	–	
$\log \alpha_{34}$	–	> 5[c]	–	
Bz₂OxP K_1	$(1.3 \pm 0.3) \times 10^3$	$(1.8 \pm 0.2) \times 10^3$	$(8.0 \pm 5.0) \times 10^4$	$(5.0 \pm 1.0) \times 10^4$
Bz₄OxP K_1	10 ± 9	N/A	No spectral change	No spectral change

[a]Binding constants for processes at higher acid vol.f. could not be obtained because of the solubility threshold of CSA, the unknown activity coefficients of the species in solution and peak shifts with polarity in NMR measurements. Due to the nature of the binding processes and the models used, only the range of the binding constants could be determined in the case of **OxP**. Also, due to the strong cooperativity, products of binding constants, i.e. $K_1 K_2$, $K_3 K_4$, are used to describe the binding steps **H** → **HG₂** and **HG₂** → **HG₄**, respectively (cf. Eq. (2.30)). [b]Interaction parameters were obtained as $\alpha_{12} = \frac{8}{3}\frac{K_2}{K_1}$ and $\alpha_{34} = \frac{8}{3}\frac{K_4}{K_3}$ (see Table 2.3). [c]Measured with TFA instead of DFA

6.2 Halochromism of $\mathbf{Bz_2OxP}$

$\mathbf{Bz_2OxP}$ differs from **OxP** in that only single counteranion stabilization is allowed due to its double N-alkylation and consequent blocking of the top central binding site shown in Fig. 6.1b. For this reason, a stable monoprotonated **$\mathbf{Bz_2OxP^+}$** form is readily formed. UV/vis titration of **$\mathbf{Bz_2OxP}$** with difluoroacetic acid[2] (DFA) reveals three consecutive spectral changes, Fig. 6.4a, b, c. That indicates presence of four individual absorbing species, whose spectra are identified as endpoints of the respective spectral changes, Fig. 6.4d, bold lines in Fig. 6.4a, b, c. The second spectrum was identified as **$\mathbf{Bz_2OxP^+}$**, the presence of a monoprotonated species is supported by fitting of its binding constant in both UV/vis (see below) and NMR titrations (see Sect. 6.5). Assignment of the remaining individual species as the two- and fourfold protonation products is based on their significant similarities to $\mathbf{z}_{OxP^{2+}}$ and $\mathbf{z}_{OxP^{4+}}$ (cf. Figs. 6.3c and 6.7). Similarly to **OxP**, triprotonated species is virtually absent implying strong cooperativity in the **$\mathbf{HG_3}$** →**$\mathbf{HG_4}$** binding step. SVD analysis also confirms $N_{spc} = 4$ based on four significant singular values (see Fig. D.4) and the combination coefficient space (see Fig. 6.4f), which shows consecutive spectral changes as linear segments connecting the component coordinates. The chemical structures of

[2] DFA was used because of its nonexchanging hydrogen utilized in NMR spectroscopy analysis, which is lacking in TFA.

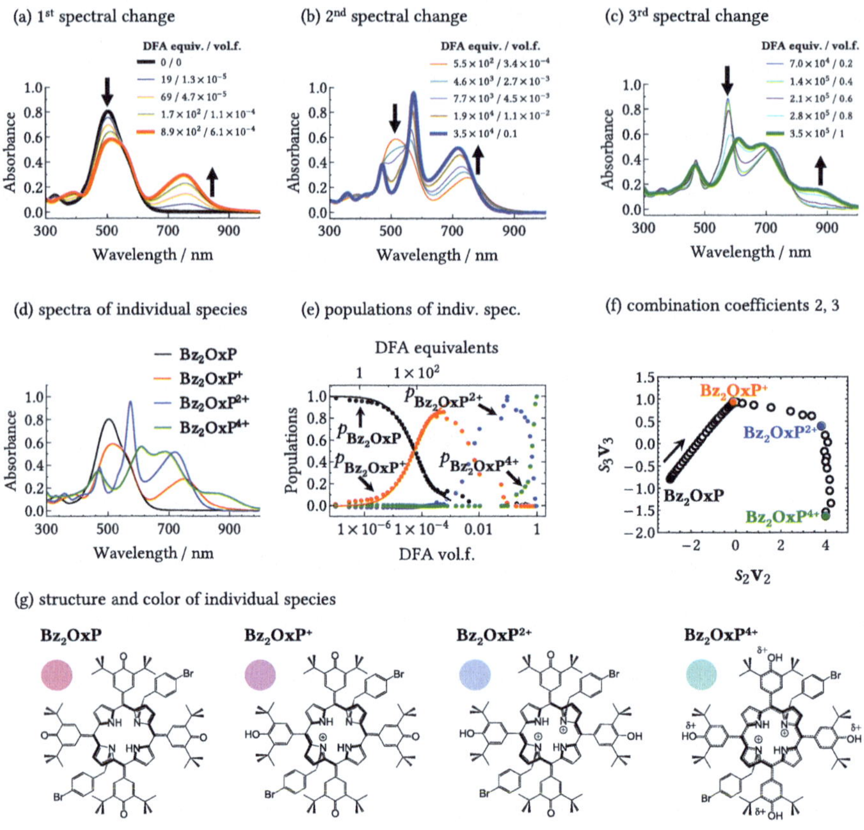

Fig. 6.4 UV/vis titration of **Bz$_2$OxP** (7×10^{-6} M, dilution during measurement compensated by rescaling to initial dye concentration, $CDCl_3$) with DFA and corresponding analyses. (**a**, **b**, **c**) Three sets of spectral changes occurring during titration. **d** Spectra of individual absorbing species **Z** identified as the start and endpoints of each spectral change (bold lines in panels (**a**, **b**, **c**)). **e** Populations $\mathbf{P}_{\mathrm{exp}}$ of absorbing species **Z** as obtained by DSIS (solid circles). Fitting of $\mathbf{P}_{\mathrm{exp}}$ to 1:1 host-guest binding model provided model populations $\mathbf{P}_{\mathrm{mod}}$ for the first protonation (solid lines). **f** SVD analysis confirms via plot of combination coefficients space, that $N_{\mathrm{spc}} = 4$ and consecutive spectral changes include only two species at once. **g** Chemical structures and colors of absorbing species (colors calculated from the spectra shown in panel (**d**), details in Appendix C.2). Adapted from [1] with permission of Wiley-VCH Verlag

the individual species and their respective colors are given in Fig. 6.4g. Analogously to the **OxP** case, linear combination of these four components accounts for all experimentally observed spectra, thus, DSIS analysis was performed using

$$\mathbf{a}_{\varphi_j} = p_{\mathbf{Bz_2OxP}}(\varphi_j)\,\mathbf{z}_{\mathbf{Bz_2OxP}} + p_{\mathbf{Bz_2OxP^+}}(\varphi_j)\,\mathbf{z}_{\mathbf{Bz_2OxP^+}} + p_{\mathbf{Bz_2OxP^{2+}}}(\varphi_j)\,\mathbf{z}_{\mathbf{Bz_2OxP^{2+}}} + p_{\mathbf{Bz_2OxP^{4+}}}(\varphi_j)\,\mathbf{z}_{\mathbf{Bz_2OxP^{4+}}}\,. \tag{6.2}$$

Populations $\mathbf{P}_{\text{exp}}$ obtained from DSIS are shown in Fig. 6.4e (solid circles). The reconstruction $\mathbf{ZP}^{\mathsf{T}}_{\text{exp}}$ excellently matches the experimental spectra **A** (see Fig. D.5) confirming that **Bz$_2$OxP** is also a halochromic dye.

In order to describe the interactions of **Bz$_2$OxP** with DFA, the first spectral change (up to $\varphi = 0.001$) was fitted using a 1:1 host-guest binding model, see Fig. 6.4e, solid lines. The quality of fit was very high, see Fig. D.5. Quantities of acid greater than $\varphi = 0.001$ change the polarity of the solvent so that activity cannot be approximated by concentration (Sect. 2.1.1). Consequently, the host-guest binding model is not applicable for higher protonated states of **Bz$_2$OxP** beyond monoprotonation.

UV/vis titration of **Bz$_2$OxP** with CSA shown in Appendix D.1 revealed only monoprotonated species, higher protonation states are inaccessible due to low solubility of CSA in chloroform, see Appendix C.1. NMR spectroscopy confirms protonation at the carbonyl oxygen atoms of the hemiquinonoid groups, see Sect. 6.5. Binding constants obtained from both UV/vis and NMR titrations are listed in Table 6.1, Sect. 6.9.

6.3 Chromism of Bz$_4$OxP

The title of this section avoids term 'halochromism' on purpose because **Bz$_4$OxP** shows both halo- and solvatochromism. Of the compounds studied, this derivative is the least susceptible to protonation due to complete N-alkylation and consequent blockage of both central binding sites, Fig. 6.1c. No stabilizing cation-counteranion interactions are allowed.

UV/vis titration with DFA shows three sets of spectral changes. The first one (up to $\varphi = 0.01$) is a red shift of absorbance maximum (500 nm), i.e. positive solvatochromism (see Sect. 4.5), see Fig. 6.5a (green arrow). This shift was not observed in the previous cases and is attributed to hydrogen bonding of acid molecules to the carbonyl oxygen atoms. It was shown in [9] that hydrogen bonding with water (in tetrahydrofuran as solvent) was accompanied by a similar spectral shift in the case of both di- and tetra-N-substituted oxoporphyrinogens. The solvatochromic effect is weak since the shift of the absorbance maximum is less than 10 nm as shown in Fig. 6.5e. The second and third spectral changes are of typical halochromic nature, see Fig. 6.5a, b (black arrows). Again, the corresponding spectra of individual species were identified as endpoints of the spectral changes, see Fig. 6.5c. Comparing to the individual species of other oxoporphyrinogens, we assigned the last two absorbing components as mono- and diprotonated forms (also protonated at the hemiquinonoid carbonyl sites) as shown in Fig. 6.7b, c. SVD analysis of the data set excluding initial solvatochromic shift (i.e. data corresponding to $\varphi > 0.01$) confirms consecutive interchange between three components by singular values shown in Fig. D.6a and plot of combination coefficient space shown in Fig. 6.5f.

The same halochromic data set (as in the case of SVD) was used for DSIS, decomposing the measured spectra into

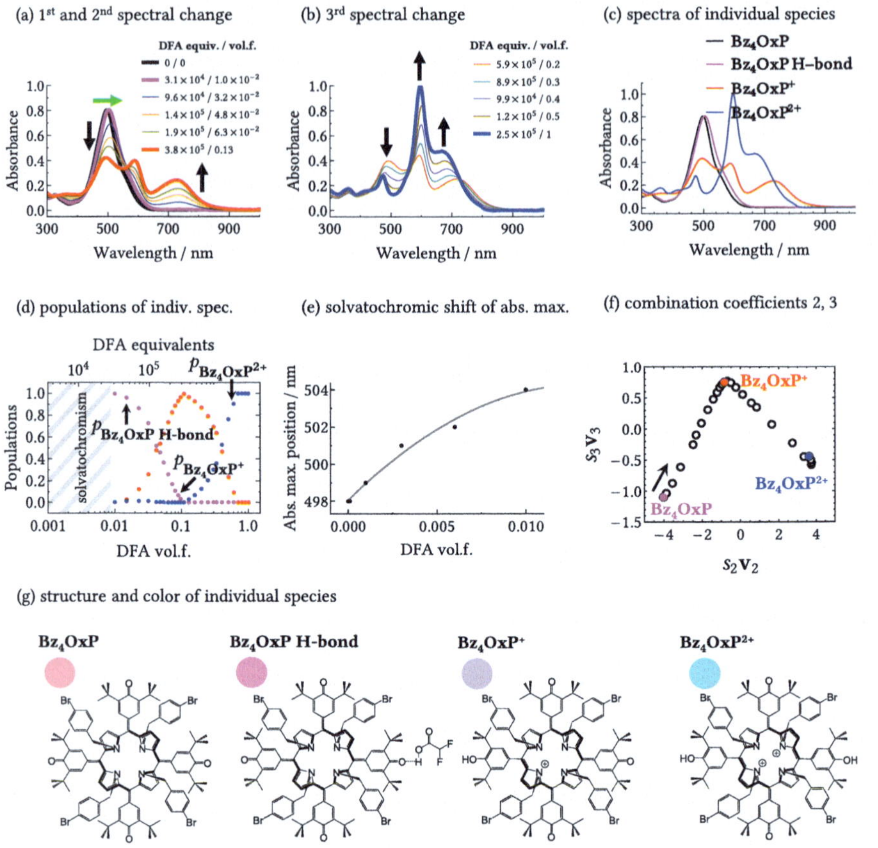

Fig. 6.5 UV/vis titration of **Bz4OxP** (5×10^{-6} M, measured spectra rescaled to initial dye concentration, $CDCl_3$) with DFA and corresponding analyses. **a** First solvatochromic (green arrow) and second (black arrow) sets of spectral changes occurring during titration, **b** third spectral change. **c** Spectra of individual absorbing species **Z** identified as the start and endpoints of each spectral change (bold lines in panels (**a**, **b**)). **d** Populations $\mathbf{P}_{\text{exp}}$ of absorbing species **Z** (free **Bz4OxP** excluded due to solvatochromic shift) as obtained by DSIS. **e** Shift of absorbance maxima, gray line is eyeguide. **f** SVD analysis confirms $N_{\text{spc}} = 3$ and consecutive spectral change via plot of combination coefficients space. **g** Chemical structures and colors of absorbing species (colors calculated from the spectra shown in panel (**c**), details in Appendix C.2). Adapted from [1] with permission of Wiley-VCH Verlag

$$\mathbf{a}_{\varphi_j} = p_{\mathbf{Bz_4OxP\ H\text{-}bond}}(\varphi_j)\,\mathbf{z}_{\mathbf{Bz_4OxP\ H\text{-}bond}} + p_{\mathbf{Bz_4OxP^+}}(\varphi_j)\,\mathbf{z}_{\mathbf{Bz_4OxP^+}} + p_{\mathbf{Bz_4OxP^{2+}}}(\varphi_j)\,\mathbf{z}_{\mathbf{Bz_4OxP^{2+}}}\,. \tag{6.3}$$

Populations of the halochromic species obtained from DSIS are shown in Fig. 6.5d. As in the previous cases, the quality of DSIS is excellent, see Fig. D.7. Populations $\mathbf{P}_{\text{exp}}$ from DSIS cannot be fitted by a simple host-guest binding model, as the high concentration of acid ($\varphi > 0.01$) causes the activity cannot be approximated by concentrations.

On the other hand, dye concentrations used in NMR titration are higher than in UV/vis and the first protonation could be achieved at lower DFA vol.f. This enables fitting of the first protonation by 1:1 host-guest binding model presented in Table 6.1, Sect. 6.9. The NMR titration also reveals a significant shift in the pyrrole β-proton resonance, Figs. 6.12c and E.7, which strongly suggests that upon protonation the positive charges are accommodated at the central nitrogen atoms of **Bz$_4$OxP** as shown in Fig. 6.5g.

6.4 Mechanism of Halochromism

Most simple dye systems operate solely on the basis of protonation, or solvent polarity. Molecular design allows for introduction of some complexing moiety to introduce specificity of the analyte-dye interaction [10, 11]. In this context, halochromism of the oxoporphyrinogen derivatives investigated here is strongly affected by the presence of central calix[4]pyrrole-type binding site.

The conjugation pathway of **OxP** can be varied upon protonation so the charge is shared over electronegative heteroatoms (i.e. N and O). After double protonation, even reorganization of the chromophore structure through tautomerization is possible. In fact, delocalization of charge may be one of the factors allowing the existence of the highly charged tetracation. The effect of N-alkylation of **OxP** is twofold: first it masks the central binding site(s), which prevents counteranion binding; second, it sterically restricts redistribution of the conjugated structure of the chromophore, which might otherwise further stabilize protonated species.

As shown in Fig. 6.6, the binding site blocking shifts approximate lower threshold of colorimetric response from $\varphi = 10^{-6}$ for **OxP** to $\varphi = 10^{-3}$ for **Bz$_2$OxP** and further to $\varphi = 0.01$ for **Bz$_4$OxP**. Such big change in sensitivity is caused by removing not only one, but two sources of stabilization, as mentioned above. Also the acid vol.f., at which all spectral changes saturate, depends strongly on the N-alkylation. The population plots in previous sections show, that **OxP** protonation saturates approximately at $\varphi = 0.002$, for **Bz$_2$OxP** at $\varphi = 1$ and for **Bz$_4$OxP** at $\varphi = 0.7$ (only double protonation).

Spectra of the corresponding mono-, di- and tetraprotonated species of each **OxP** derivative are similar in form as shown in Fig. 6.7. This assumption was exploited to assign the degrees of protonation of each absorbing species. Diprotonated species, present in each derivative, show gradual shifts of absorption bands correlated with the degree of N-alkylation (red arrows in Fig. 6.7c).

There exists strong positive cooperativity between the first and second protonation steps in **OxP** because of the stabilization by two counterions (Table 6.1, similar to the case for double protonation of free-base porphyrins [7]). Therefore, monoprotonated spectrum of **OxP** is not observable. Since **Bz$_2$OxP** is stabilized by only one counteranion, its monoprotonated form is stable. Surprisingly, another strong cooperativity arises between the third and fourth protonation step in both **OxP** and **Bz$_2$OxP**. As a result, their triprotonated form cannot be observed either.

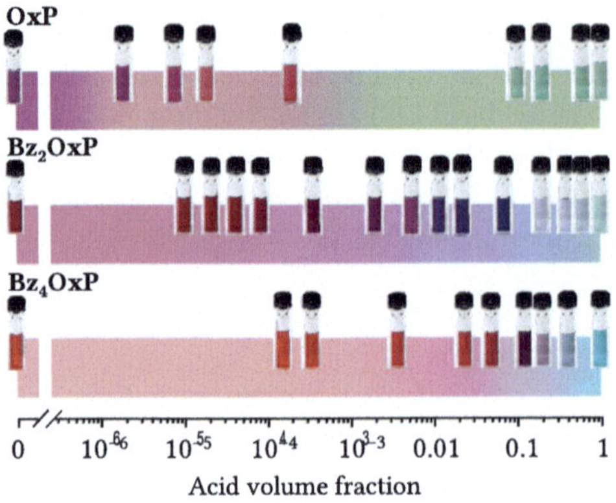

Fig. 6.6 Colorimetric response of **OxP** (with TFA), **Bz$_2$OxP**, and **Bz$_4$OxP** (with DFA) over the entire concentration range of the acids. Colors shown were obtained by interpolation of RGB values (see Appendix C.3) obtained from the actual measured spectra, and the color scales are overlaid with photographs of the sample cuvettes (from the titration experiments) at the appropriate vol.f. Reproduced from [1] with permission of Wiley-VCH Verlag

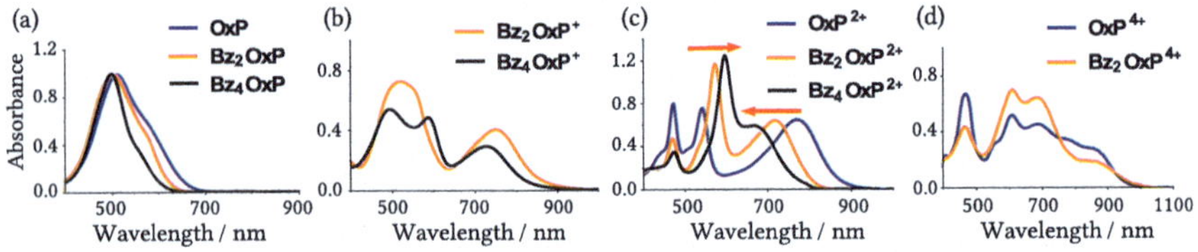

Fig. 6.7 Assignment and spectra of neutral and protonated species of **OxP**, **Bz$_2$OxP** and **Bz$_4$OxP** in the **a** neutral state and in the **b** monoprotonated, **c** diprotonated and **d** tetraprotonated forms. Spectra of neutral species were normalized to unit absorbance at maximum, protonated species are to scale with the neutral ones. Adapted from [1] with permission of Wiley-VCH Verlag

Bz$_4$OxP lacks further counterion stabilization, therefore it does not show any strongly cooperative protonation. In addition, its steric crowding at the core obstructs redistribution of charge, therefore **Bz$_4$OxP** is not completely protonated even in neat acid.

In Figs. 6.3e, 6.4g and 6.5g we have assigned the formal placement of charge at the central nitrogen atoms of the protonated structures. These charges may be shared over all electronegative atoms, albeit with preference for the central pyrrole nitrogens. The hypothesis of charged tetrapyrrole center is supported by significant shifts of pyrrole β-protons resonances in NMR titrations to higher ppm discussed in the next section (green squares in Fig. 6.12).

Based on the formal conjugated structure, charge redistribution can be depicted. Figure 6.8a shows tautomerization of **OxP^{2+}**, Fig. 6.8b clarifies transfer of charge to the molecular center of **Bz$_2$OxP$^+$** upon protonation and Fig. 6.8c illustrates variants of charge placement at the pyrrole nitrogens of **Bz$_2$OxP^{2+}**. The protonation sites

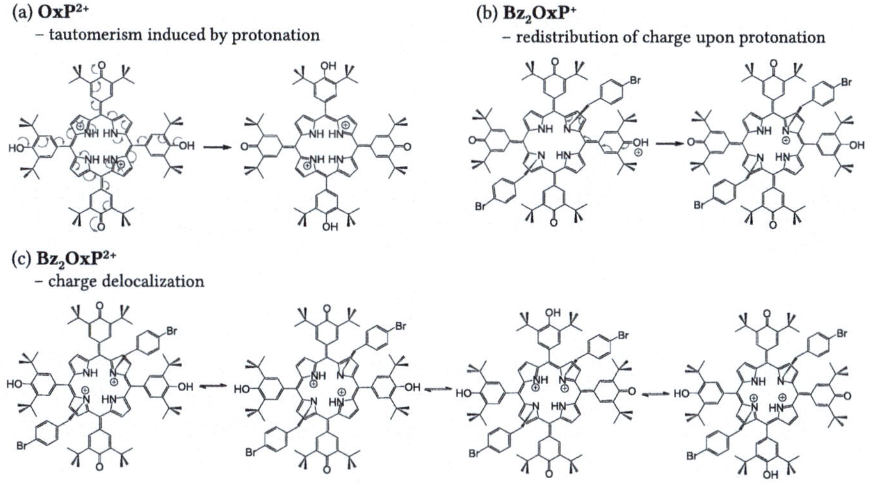

Fig. 6.8 Redistribution of charge upon protonation for **a OxP** and (**b**, **c**) **Bz$_2$OxP**. Reproduced from [1] with permission of Wiley-VCH Verlag

dynamically change by a prototropic tautomerization process, which is described in detail in the next chapter in Sects. 7.1.1 and 7.2.3.

6.5 NMR Analysis of Host-Guest Binding

NMR titration experiments are another tool for assessing binding properties of oxoporphyrinogens. Their interaction with DFA as well as with CSA was investigated in deuterated chloroform. DFA (liquid) is miscible with $CDCl_3$ in arbitrary ratio, however, CSA (powder) has limited solubility (Table C.1). Therefore, high degree of protonation can be achieved only at excess of DFA, not in the presence of CSA. Tetramethylsilan (TMS) was used as internal standard. Addition of maximal soluble amount of CSA into solution of **Bz$_4$OxP** did not result in any spectral change, therefore, these experiments are not shown. Experimental details are given in Appendix C.4, data analysis procedure is described in Appendix C.5. Experimental details about the measurements are given in Appendix C.4.

Assignment of ^{1}H spectra of the investigated **OxP** derivatives is shown in Fig. 6.9. Signals due to *tert*-butyl-H are denoted by star (★), due to *ortho*-H by circle (●), due to β-H by square (■) and signals of the bromobenzyl groups by triangles (▲). Assignment of **OxP** and **Bz$_2$OxP** resonances was published in [12, 13], respectively, assignment of **Bz$_4$OxP** was conducted with help of HMBC measurement, see Appendix E.18. Despite high number of hydrogen atoms, the resulting spectra contain low number of signals due to symmetry. Spectra of pure acids can be found in Fig. E.1 in Appendix E.

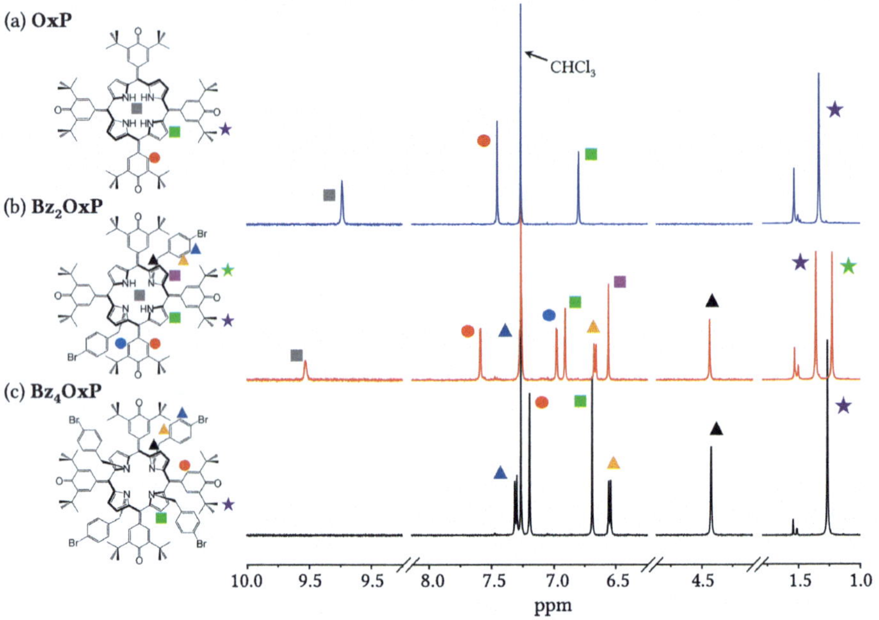

Fig. 6.9 Assignment of ^{1}H NMR signals of investigated oxoporphyrinogens without the presence of acid. Reproduced from [1] with permission of Wiley-VCH Verlag

NMR chemical shifts of the investigated oxoporphyrinogens are influenced by protonation with consequent charge transfer, by acid anion binding and also by the change of solvent polarity at high acid vol.f. [14]. In the case of **OxP**, new isolated peaks of bound/protonated species arise upon acid addition (both DFA and CSA) due to slow exchange regime. Also the host OH signals from protonation can be observed as shown in Fig. 6.10a (top). Relative area of this resonance, shown in Fig. 6.10a (bottom), is saturated at the value of 2, which is a direct evidence for two-fold protonation. The assignment to host OH resonance was confirmed by addition of heavy water after which the resonance disappeared, see Appendix E.17. Full titration spectra of **OxP** with DFA or CSA are shown in Figs. E.2 and E.3 in Appendix E, respectively.

On the other hand, $\mathbf{Bz_2OxP}$ does not show new separate signals upon protonation due to intermediate/fast exchange of free and bound species. Only the NH signals are an exception and signals due to free and bound species are separated, see Fig. 7.12a in the next chapter. We found a candidate OH signal for direct observation of the protonation, see Appendix E.17.

Titration spectra of $\mathbf{Bz_4OxP}$ do not show any formation of separate peaks. The only spectral changes are shifts of *tert*-butyl-H and β-H signals. The significant shift of β-H resonance (■) shown in Fig. 6.10b indicates charge transfer from carbonyl groups to the porphyrinogen center induced by protonation.

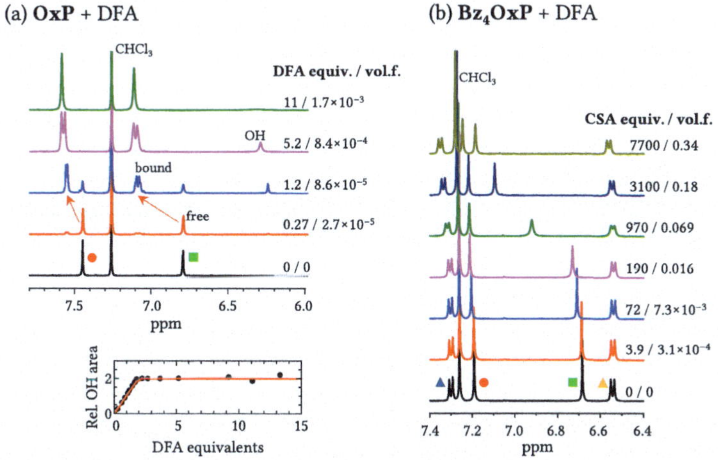

Fig. 6.10 Illustration of spectral changes during titration experiments. **a** Titration of **OxP** with DFA. The spectra (top) show separate peaks of free and bound forms due to *ortho*-H (●) and β-H (■) spins as well as the host OH resonances of protonated host. Relative area of the OH resonances (bottom) normalized to the total area of *tert*-butyl-H peaks, fitted with the function $([\mathbf{HG}] + 2[\mathbf{HG_2}]/[\mathbf{H}]_t$ and yielding parameters $K_1 = 4 \times 10^6$, $K_2 = 8.4 \times 10^6$, compatible with results in Table 6.1. **b** Titration of **Bz4OxP** with DFA showing significant shift of the β-H resonance (■)

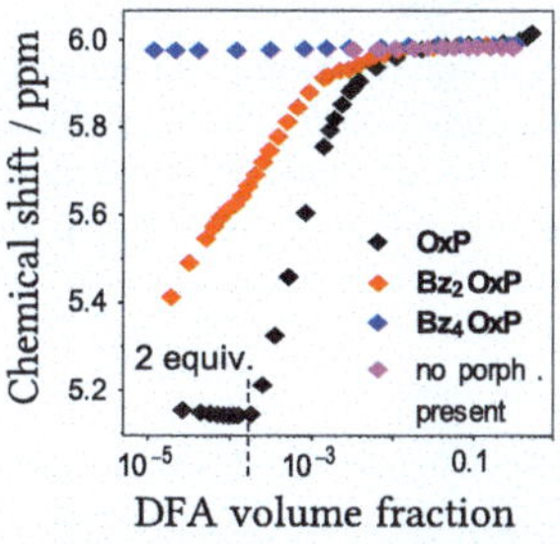

Fig. 6.11 Chemical shift of the DFA non-exchanging proton signal (central frequency of the triplet) in the presence and absence of the oxoporphyrinogens investigated for different vol. frac. of DFA. Reproduced from [1] with permission of Wiley-VCH Verlag

Chemical shift of the center of DFA triplet is plotted in Fig. 6.11 as a function of DFA concentration. The frequency shift reflects acid anion binding rather than protonation of the oxoporphyrinogens. The DFA anions are in fast exchange between their free and bound states, thus, only the averaged triplet signal is observed. During titration of **OxP** all acid anions are bound to the oxoporphyrinogen up to 2 equiv. due to the strong binding affinity (◆). For $\mathbf{Bz_2OxP}$, anion binding is not so strong, so the corresponding chemical shift is not constant (◆). In the case of $\mathbf{Bz_4OxP}$, chemical shift of its acid triplet (◆) is the same as if no oxoporphyrinogen was present (◆). It confirms the lack of acid anion binding to $\mathbf{Bz_4OxP}$.

For each titration experiment, binding isotherms were constructed from integrated peak areas (slow exchange) and peak shifts (fast exchange) as described in Sect. 3.4.2. All binding isotherms were simultaneously fitted with host-guest binding model of

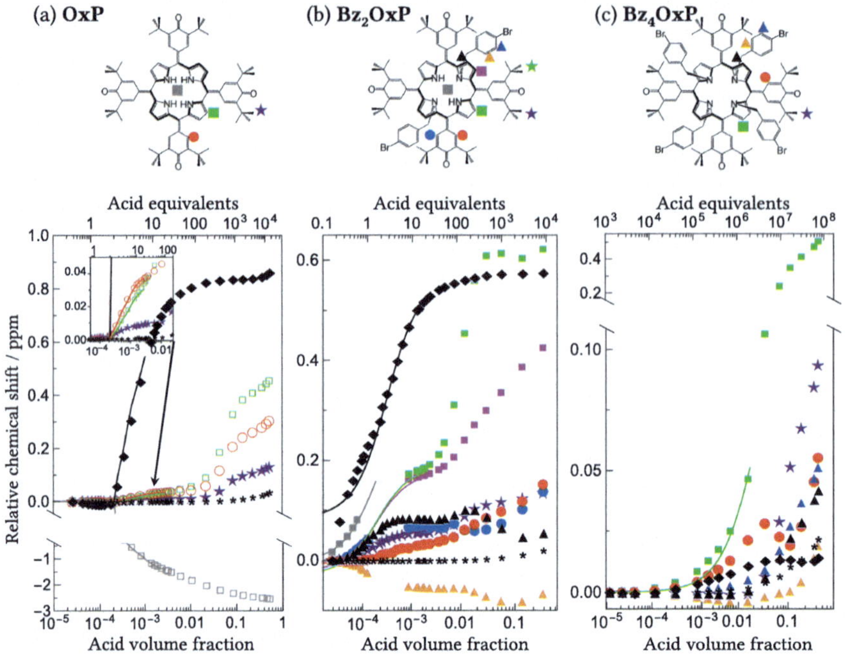

Fig. 6.12 Relative chemical shifts of peaks of the investigated oxoporphyrinogens during titrations with DFA. The value of chemical shift of particular resonance at 0 equiv. of DFA is subtracted from corresponding resonance during the titration process. For two distinct tert-butyl peaks or any doublets central chemical shift is displayed as they shift simultaneously. Non-exchangeable DFA resonance (triplet) is denoted by the diamond symbol (◆), $CHCl_3$ is denoted by asterisk (∗). Solid lines stand for fits, the corresponding binding constants are in agreement with Table 6.1. Reproduced from [1] with permission of Wiley-VCH Verlag

appropriate stoichiometry (1:4 for **OxP**, 1:1 for **Bz₂OxP** and **Bz₄OxP**), the resulting binding constants are listed in Table 6.1. The binding models were not applicable at acid vol. frac. $\varphi > 0.001$ since the change of solvent polarity, caused by addition of acid, influenced the peak shifts [14]. Moreover, for the same reason, activity cannot be approximated by concentration at high acid concentrations, see Sect. 2.1.1. Relative chemical shifts (i.e. with subtracted initial value) of non-overlapping signals of the oxoporphyrinogens as function of DFA vol.f. are shown in Fig. 6.12. Some of the peak shifts were used as binding isotherms in the fitting procedure, the fitting curves are also shown (solid lines). Of the oxoporphyrinogen peaks, the most significant spectral shifts belong to the pyrrole β-H and NH resonances, supporting the hypothesis of charge localization at the pyrrole nitrogen atoms. Note that even residual solvent peak (black asterisk ∗ in Fig. 6.12) is shifted relative to TMS at high acid. Details about the current data analysis, including peak integration using lineshape fitting, fitting of binding isotherms and the implementation for **OxP** + (R)-CSA system, are given in Appendix C.5.

6.6 Solvatochromism of OxP

To examine solvatochromic properties (see Sect. 4.5) of **OxP** and its derivatives, we investigated their behavior in solvents of different polarity, namely in dimethylformamide/chloroform binary mixtures, see Appendix C.2 for experimental details. Increase in concentration of the polar dimethylformamide (DMF) resulted in solvatochromic red shift caused by hydrogen bonding. In the case of **OxP**, unexpected rearrangement of its conjugated system was observed for particular DMF concentration range. Hydrogen bonding between oxoporphyrinogens and solvent molecules has been reported, including carbonyl-containing solvents [15] (UV/vis spectra, X-ray crystallographic structure of $\mathbf{Bz_2OxP}\cdot$ acetone) and trace water in tetrahydrofuran [9] (UV/vis titration with water, DFT calculations of $\mathbf{Bz_2OxP}\cdot\, n\mathrm{H_2O}$).

As shown in Fig. 6.13a–c, titration of **OxP** with DMF resulted in three distinct spectral changes. The first one involves a positive solvatochromic shift (the original electronic structure does not change), identified as hydrogen bonding with either DMF or water, which is also present in the sample (amount of water increased during titration, see NMR spectra in Fig. E.16). Hydrogen bonding of water to **OxP** induces similar spectral change [9] shown in Fig. 6.14a. Other two spectral changes imply formation of new species with different electronic structure since new absorption bands arise.

In our publication [1] we identified the spectra of absorbing components as endpoints of the three spectral changes (implying four species) and consequently performed DSIS analysis. However, in this work we revisited these results and provide more precise analysis based on combination coefficients. As shown in the SVD analysis in Appendix D.5, the first spectral change, although solvatochromic, can be described as linear combination of two spectra of individual species, thus, for the whole experiment it holds $\mathbf{A} = \mathbf{Z}\mathbf{P}^{\mathrm{T}}$ (see Sect. 5.2.1). Singular values indicate the presence of three components, also the first three combination coefficients lie on a 2D plane in their 3D space as confirmed by principal component analysis (PCA, explained in Sect. 5.3), see Fig. D.8 (N_{spc} components form an $(N_{\mathrm{spc}} - 1)$-dimensional object in the combination coefficients space, although small overlap with other dimensions occurs due to the noise). These are sufficient arguments to prove the presence of exactly three absorbing species. Mutual plot of combination coefficients PCs (principal components) in Fig. 6.13f shows that the line defined by the first spectral change intersects the last points of the third spectral change (first and third spectral changes are denoted by green lines). Hence, before the hydrogen bonding in the first spectral change could be saturated, there is a "digression" in the combination coefficients plot identified clearly as formation of a third absorbing component, which looks very distinct from both **OxP** and **OxP H-bond**. Coordinates of **OxP** and this third species were deduced directly from the measured spectra (bold lines in Fig. 6.13a, b) and coordinates of **OxP H-bond** were determined as intersection of the green lines in combination coefficients PCA plot. After obtaining the coordinates, individual spectra were calculated (using Eq. (5.21)) in

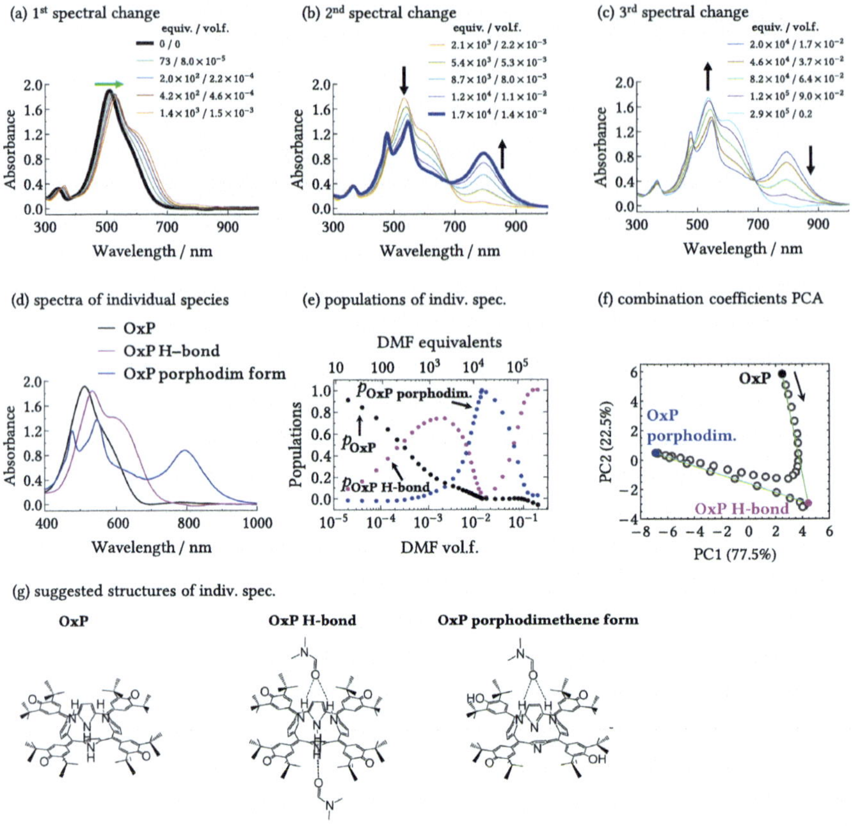

Fig. 6.13 Measurement of **OxP** (1.7×10^{-5} M, spectra rescaled to initial dye concentration to compensate for dilution during experiment, $CDCl_3$) with DMF and corresponding analyses. (**a**, **b**, **c**) Three sets of spectral changes occurring during titration. Starting point of the first change and endpoint of the second change are identified as individual absorbing species. **d** Individual spectra **Z** and **e** their populations $\mathbf{P}_{\text{exp}}$ obtained from coordinates **T** in combination coefficients space in frame of 3D SVD approximation. **f** PCA of combination coefficients, colored points denote coordinates of absorbing components, intersection of green lines helped to identify the coordinate of **OxP H-bond**. **g** Suggested structures of the absorbing components. Adapted from [1] with permission of Wiley-VCH Verlag

frame of 3D SVD approximation (capturing 99.98% of the signal) and together with their populations plotted in Fig. 6.13d, e. Details of this analysis are given in Appendix D.5.

Electronic structure of the third absorbing species implies big reorganization of the conjugated system.[3] Therefore its structure was assigned to **OxP porphodimethene form** shown in Fig. 6.13g (cf. Fig. 1.1), which is based on significant similarity to the

[3] Precautions were taken to exclude acid formation in the solvents, potentially causing such spectral change by protonation, see Appendix C.2.

spectrum of **OxP^{2+}** (cf. Fig. 6.3c). Structures of **OxP porphodimethene form** and **OxP^{2+}** share similar conjugation at the porphyrinogen center. Stabilization of **OxP porphodimethene form** depends on solvent polarity (and perhaps on the strength of hydrogen bonding) as it is prevalent only in DMF vol.f. range of $\varphi \in [0.01, 0.05]$.

NMR titration of **OxP** by DMF shown in Fig. E.16 reveals no conclusive results, however, it indicates interaction of **OxP** with DMF as their peaks are slightly shifted even at low DMF concentration.

Figure 6.15 provides overview of the colorimetric response of all three oxoporphyrinogens. The vol.f. range of stabilized **OxP porphodimethene form** is clearly visible with naked eye. Clearly, there is no such effect in the case of the N-alkylated oxoporphyrinogen derivatives. Note that the color change in **OxP** caused by DMF was saturated at $\varphi(\text{DMF}) < 0.2$, while for **Bz$_2$OxP** and **Bz$_4$OxP** the color change was not saturated even in pure DMF.

6.7 Solvatochromism of Bz$_2$OxP

Measurements of **Bz$_2$OxP** in DMF/$CDCl_3$ at different vol.f. shows two different solvatochromic changes, see Fig. 6.16a, the first one involving only small shift of the original absorbing maxima (green arrow), the second including variation of the original band intensities and formation of another absorbing band at ca. 750 nm (black arrows). SVD analysis indicates presence of three absorbing components (99.99% of signal, see Appendix D.6), which were identified from the PCs of combination coefficients in Fig. 6.16d. The first component **Bz$_2$OxP** was measured in isolation as the first measured spectrum (in pure chloroform), the second was identified as intersection of the orange lines in Fig. 6.16d delineating first and second spectral change. However, placement of third component on the orange line is not unique as the second spectral change is probably not saturated even at $\varphi(\text{DMF}) = 1$. Therefore, we identified range of possible coordinates as a line segment bounded by the last measured spectrum at one side and the position of last reasonable spectral shape (transparent green line segment in Fig. 6.16d). The green point in the same figure denotes coordinates of a representative spectral shape in the middle of the allowed range. Figure 6.16b, c show the absorbing spectra and the their populations during the measurements. Green solid line (or green points) correspond to the green point in (d), green shaded area corresponds to the uncertainty denoted by the transparent line segment in (d). Negative values of the individual spectrum at ca. 700 nm and negative populations at ca. $\varphi(\text{DMF}) = 0.4$ are caused by non-linear effects in the UV/vis data. Details of this analysis are given in Appendix D.6.

The current spectral changes highly resemble those of hydrogen bonding of **Bz$_2$OxP** with water in tetrahydrofuran reported in [9], see Fig. 6.14b. In the same article, DFT calculations were conducted, revealing different effects of simple hydrogen bonding at single carbonyl (**Bz$_2$OxP**·H_2O) compared to multiple carbonyl hydrogen bonding or bonding at the central NH group (**Bz$_2$OxP**·$4H_2O$, **Bz$_2$OxP**·$5H_2O$), see Fig. 6.17. Thus, we assign the second absorbing component to **Bz$_2$OxP** interacting

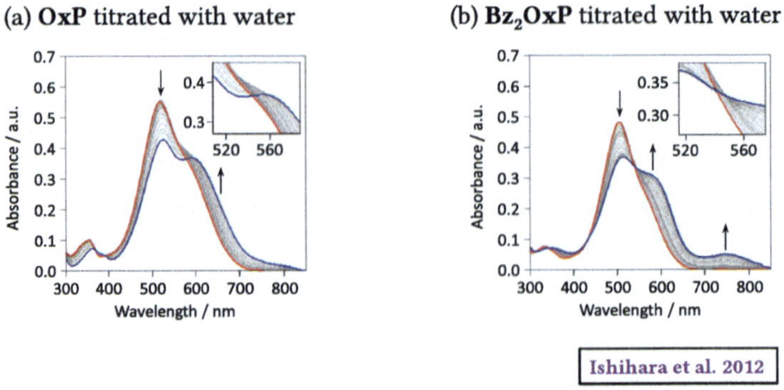

Fig. 6.14 UV/vis titration of **a OxP** and **b Bz_2OxP** with water in tetrahydrofuran solvent. Reproduced from [9] with permission from the Royal Society of Chemistry

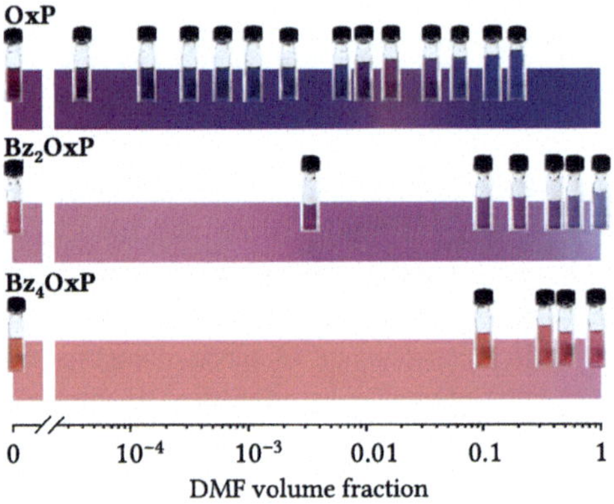

Fig. 6.15 Colorimetric response of **OxP**, **Bz_2OxP**, and **Bz_4OxP** over the entire concentration range of DMF/$CDCl_3$ mixture. Colors shown were obtained by interpolation of RGB values (see Appendix C.3) obtained from the actual measured spectra, and the color scales are overlaid with photographs of the sample cuvettes (from the titration experiments) at the appropriate vol.f. Reproduced from [1] with permission of Wiley-VCH Verlag

with DMF in unspecified way (possibly hydrogen bonding at the host carbonyl sites or only by increasing solvent polarity) and denote it as **Bz_2OxP form 1**. The third absorbing component is assumed to bind DMF at the porphyrinogen center based on the comparison with Fig. 6.14b (from [9]) and is denoted as **Bz_2OxP form 2**. The suggested structures are shown in Fig. 6.16e. Colorimetric response of **Bz_2OxP** to the solvent is shown in Fig. 6.15. The color variation is shifted to higher DMF concentration compared to **OxP**.

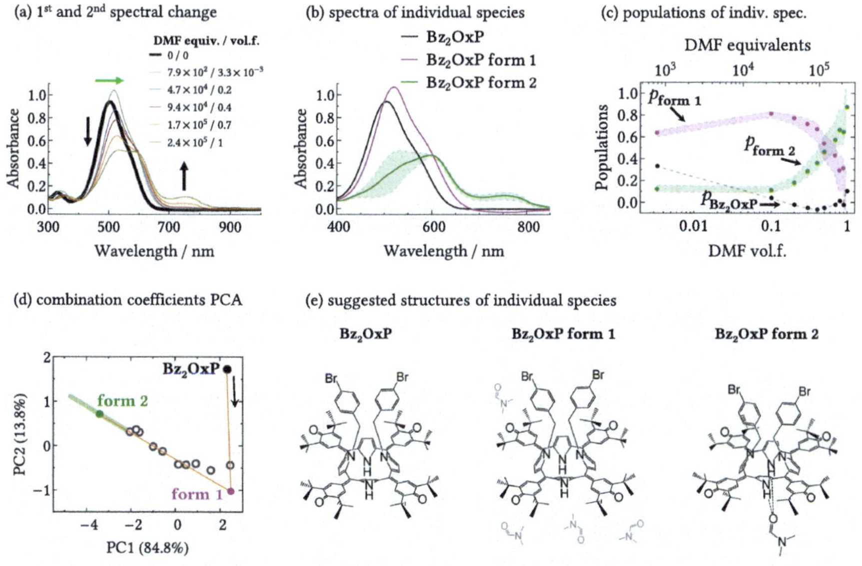

Fig. 6.16 UV/vis measurements of **Bz_2OxP** (5.5×10^{-5} M) in DMF/$CDCl_3$ mixture and corresponding analyses. **a** Two sets of spectral changes occurring at different DMF vol.f. Starting point of the first change is identified as individual absorbing species (bold line). **b** Individual spectra **Z** and **c** their populations $\mathbf{P}_{exp}$ obtained from coordinates **T** in combination coefficients space in frame of 3D SVD approximation. **d** PCA of combination coefficients, colored points denote coordinates of absorbing components. Intersection of orange lines helped to identify the coordinate of **Bz_2OxP form 1**. **e** Suggested structures of absorbing components. Adapted from [1] with permission of Wiley-VCH Verlag

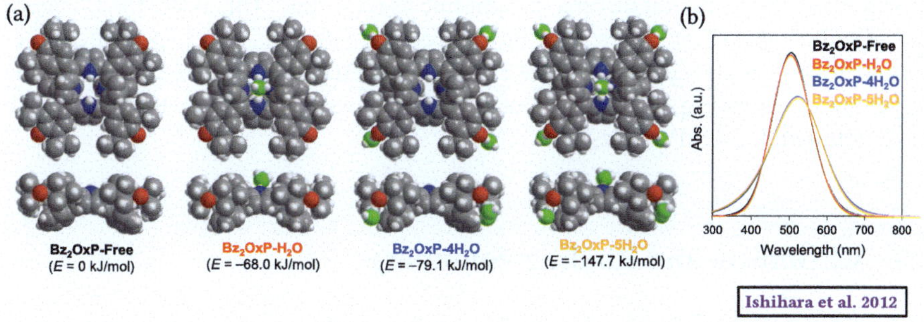

Fig. 6.17 DFT calculations of **Bz_2OxP**. Bromine atoms at benzyl substituents were replaced with hydrogen atoms to simplify calculations. **a** Top and side views of calculated hydrogen bonded structures of **Bz_2OxP** (M06-L/6-31G(d,p), 0 K in vacuum). Total binding energies of water molecules were are shown. **b** Calculated absorption spectra (TD-M06-2x/6-31++G(d,p)). Reproduced from [9] with permission from the Royal Society of Chemistry

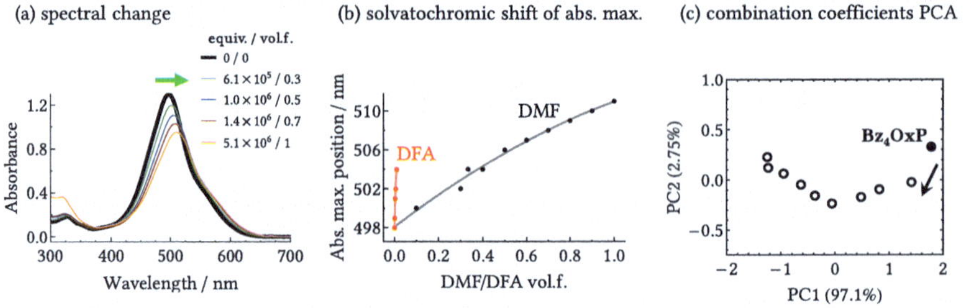

Fig. 6.18 UV/vis measurements of **Bz_4OxP** (6.3×10^{-5} M) in DMF/$CDCl_3$ mixture and corresponding analyses. **a** The only spectral change occurring at different DMF vol.f. The starting point (in pure chloroform) is identified as an individual absorbing species (bold line). **b** Solvatochromic shift of absorbance maxima induced by DMF (black) or DFA (red, replicated from Fig. 6.5e), solid lines are eyeguides. **c** PCA of combination coefficients. Adapted from [1] with permission of Wiley-VCH Verlag

6.8 Solvatochromism of Bz_4OxP

Similarly to solvatochromism in DFA, **Bz_4OxP** shows small solvatochromic shift in DMF, see Fig. 6.18a. Compared to DFA, the red shift caused by DMF requires much higher concentration of the added polar/hydrogen bonding substance as shown in Fig. 6.18b. Due to the small magnitude of the solvatochromic spectral change ($\approx$ 10 nm), SVD cannot unequivocally determine if the spectral change is caused by formation of a new absorbing species. 2D SVD approximation covers 99.994% of the signal suggesting transition between two species, however, combination coefficients shown in Fig. 6.18c do not lie on a straight line as required.[4] This behavior, if caused by individual absorbing species, requires three components. This hypothetical scenario is further analyzed in Appendix D.7. Colorimetric response of **Bz_4OxP** to the solvent is shown in Fig. 6.15. The color variation is weak and shifted to higher DMF vol.f. compared to **OxP** or **Bz_2OxP**.

6.9 Chromism and Binding–Summary

The investigated oxoporphyrinogens are halochromic dyes operating as colorimetric indicators over broad range of acid contents in nonpolar media (φ(DFA) $\in [10^{-6}, 1]$). This was permitted by up to fourfold protonation of the **OxP** derivatives. The maximum protonation level was tuned by N-alkylation, which gradually shifted the sensitivity, i.e. colorimetric response, to higher acid contents, as protonation of the molecules was increasingly obstructed in the series **OxP**, **Bz_2OxP** and **Bz_4OxP**. The

[4] Such deviation might be caused by variation in total dye concentration. However, in this experiment, stock solution with DMF contained the same concentration of **Bz_4OxP** as the measured solution, which should prevent any concentration fluctuations.

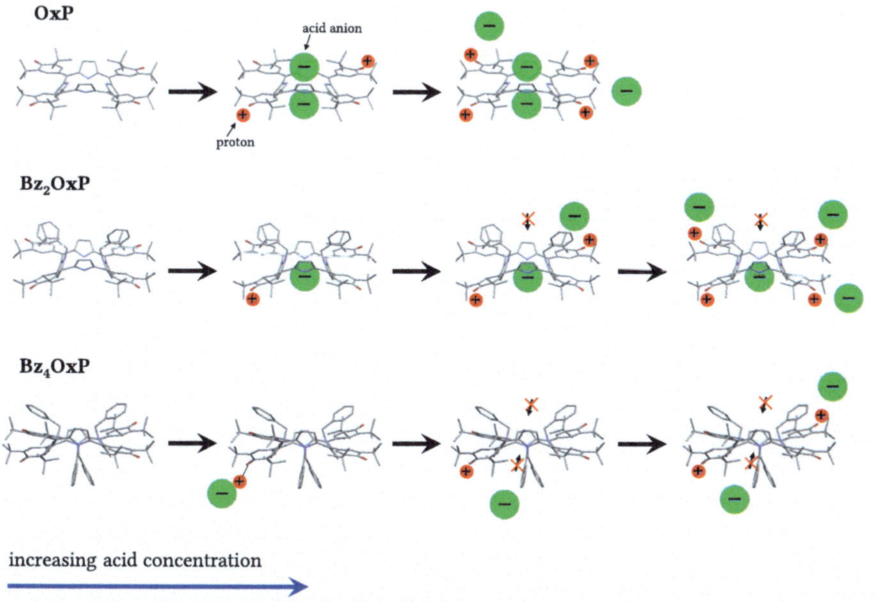

Fig. 6.19 Binding site blocking effect of the oxpoporphyrinogen molecules on their interaction with acid. Crossed arrows denote blocked sites due to N-alkylation

mechanism of operation involves stabilization by counteranion binding and charge delocalization. This stabilization induces strong cooperativity (see Sect. 2.2.4) in the binding step **HG**→**HG**$_2$ (described by the interaction parameter α_{12}). Surprisingly, also the binding step **HG**$_3$ →**HG**$_4$ (α_{34}) is strongly cooperative in **OxP** and **Bz**$_2$**OxP** due to unknown reason. For **Bz**$_4$**OxP**, stabilization by a counteranion is not permitted, thus, its response is limited to higher acid contents (φ(DFA) > 0.1). Schematic overview of the protonated species and the effect of binding site blocking is shown in Fig. 6.19. Note the difference between weak hydrogen bonding of acid to **Bz**$_4$**OxP** at low acid contents and protonation of **Bz**$_4$**OxP** when more acid is added. UV/vis and NMR titration experiments allowed fitting of binding isotherms (at φ(acid) < 0.001) to obtain the corresponding binding constants. At higher acid vol.f., the binding constants could not be obtained because of the the unknown activity coefficients of the species in solution (see Sect. 2.1.2) and peak shifts due to increased solvent polarity in NMR measurements. Also, full protonation could not be achieved in CSA due to its limited solubility in chloroform (see Appendix C.1), hence, CSA induces only monoprotonation[5] in **Bz**$_2$**OxP** and even no protonation in **Bz**$_4$**OxP**. The results are presented in Table 6.1 including the calculated values of interaction parameters for

[5] Peak shifts in NMR titration above ca. 10 equiv. of CSA are induced by the increase of solvent polarity, although they might also be attributed to formation of **HG**$_2$ complex. However, UV/vis titration of **Bz**$_2$**OxP** with CSA proved only the presence of monoprotonation, see Appendix D.1.

OxP protonation.[6] In the case of **OxP**, only the range of the binding constants could be determined due to strong cooperativity. Note that the parameters obtained from NMR titrations are less reliable compared to those from UV/vis because shift of NMR peaks is affected by variation of solvent polarity at high acid contents. Since the values of binding constants were determined with uncertainty over several orders, their decadic logarithms are given.

Solvatochromic effects in DMF/chloroform mixture were observed. Besides other interactions, central NH binding sites allow hydrogen bonding with DMF carbonyl group. All three oxoporphyrinogens show red shifts of absorption bands (positive solvatochromism, see Sect. 4.5) upon addition of DMF attributed to the increase of polarity or to hydrogen bonding (the unavoidable presence of water might have small influence on the spectra too). Furthermore, **OxP** showed transition to another form (assigned to porphodimethene form) stable only at particular DMF concentration range. Analogously to halochromism, the solvatochromic colorimetric response is also shifted to higher DMF concentration with increasing degree of N-alkylation.

UV/vis titration experiments proved as a powerful tool to investigate proton transfer and formation of hydrogen bonds in the oxoporphyrinogen conjugated systems. Together with SVD and DSIS, they enable determination of the number of individual absorbing species including their spectra and populations. Especially, mutual plot of SVD combination coefficients is a very effective instrument to assess the present spectral changes. NMR measurements provide only supporting evidence (except for direct observation of protonation in **OxP** and possibly in **Bz_2OxP** (see Appendix E.17) since at higher acid contents, all peaks undergo fast chemical exchange and thus providing only limited information from shifts of their resonant frequencies. On the contrary, NMR is an excellent tool to investigate intra- and intermolecular dynamic processes as presented in the next chapter.

References

1. Březina V, Ishihara S, Lang J, Hanyková L, Ariga K, Hill JP, Labuta J (2018) Structural modulation of chromic response: effects of binding-site blocking in a conjugated calix [4] pyrrole chromophore. ChemistryOpen 7:323–335
2. Březina V (2014) Spectroscopic and theoretical study of supramolecular complexes of symmetric porphyrins with chiral guests. MA thesis, Univerzita Karlova, Matematicko-fyzikální fakulta May
3. Hill JP, Hewitt IJ, Anson CE, Powell AK, McCarty AL, Karr PA, Zandler ME, D'Souza F (2004) Highly nonplanar, electron deficient, N-substituted tetra-oxocyclohexadienylidene porphyrinogens: structural, computational, and electrochemical investigations. J Org Chem 69:5861–5869
4. Shundo A, Ishihara S, Labuta J, Onuma Y, Sakai H, Abe M, Ariga K, Hill JP (2013) Colorimetric visualization of acid-base equilibria in non-polar solvent. Chem Commun 49:6870–6872

[6] Here, we revised the values of α_{12} and α_{34} published in our article [1] and used the correct prefactor 8/3.

5. Labuta J, Futera Z, Ishihara S, Kouřilová H, Tateyama Y, Ariga K, Hill JP (2014) Chiral guest binding as a probe of macrocycle dynamics and tautomerism in a conjugated tetrapyrrole. J Am Chem Soc 136:2112–2118
6. Labuta J, Ishihara S, Shundo A, Arai S, Takeoka S, Ariga K, Hill JP (2011) Chirality sensing by nonchiral porphines. Chem Eur J 17:3558–3561
7. Medforth CJ, Berget PE, Fettinger JC, Smith KM, Shelnutt JA (2016) Determination of the activation energies for ND tautomerism and anion exchange in a porphyrin monocation. J Porphyr Phthalocyanines 20:307–317
8. Thyagarajan S, Leiding T, Årsköld SP, Cheprakov AV, Vinogradov SA (2010) Highly non-planar dendritic porphyrin for pH sensing: observation of porphyrin monocation. Inorg Chem 49:9909–9920
9. Ishihara S, Labuta J, Šikorský T, Burda JV, Okamoto N, Abe H, Ariga K, Hill JP (2012) Colorimetric detection of trace water in tetrahydrofuran using N, N′-substituted oxoporphyrinogens. Chem Commun 48:3933–3935
10. Reichardt C (1992) Solvatochromism, thermochromism, piezochromism, halochromism, and chiro-solvatochromism of pyridinium N-phenoxide betaine dyes. Chem Soc Rev 21:147–153
11. Machado VG, Stock RI, Reichardt C (2014) Pyridinium N-phenolate betaine dyes. Chem Rev 114:10429–10475
12. Shundo A, Labuta J, Hill JP, Ishihara S, Ariga K (2009) Nuclear magnetic resonance signaling of molecular chiral information using an achiral reagent. J Am Chem Soc 131:9494–9495
13. Labuta J, Ishihara S, Šikorský T, Futera Z, Shundo A, Hanyková L, Burda JV, Ariga K, Hill JP (2013) NMR spectroscopic detection of chirality and enantiopurity in referenced systems without formation of diastereomers. Nat Commun 4:2188
14. Reichardt C, Welton T (2011) Solvents and solvent effects in organic chemistry. John Wiley & Sons
15. Hill JP, Schumacher AL, D'Souza F, Labuta J, Redshaw C, Elsegood MRJ, Aoyagi M, Nakanishi T, Ariga K (2006) Chromogenic indicator for anion reporting based on an N-substituted oxoporphyrinogen. Inorg Chem 45:8288–8296

Chapter 7
Dynamic Processes and Chemical Exchange in OxP Derivatives

Systems of oxoporphyrinogen hosts in the presence of organic acids (chiral CSA or achiral DFA) show not only dynamic host-guest binding (described in the previous chapter) but also other processes such as rotation of bulky side groups. In this chapter, these processes are described qualitatively and some of them even quantitatively using appropriate models.

The dynamic behavior is investigated using NMR lineshape analysis covered in Sect. 3.3. NMR signal is sensitive to the kinetics of a particular nuclear spin serving as a probe of molecular processes. Although in many cases there is no direct one-to-one correspondence between chemical kinetics (involving chemical species present in the sample) and the observed spin kinetics (involving states in different environments of a reference nuclear spin), comparison of chemical and spin kinetics schemes provides the relationship between reaction and transition rate coefficients.

Two types of ^{1}H NMR experiments were used to assess the dynamic properties. The first type consists of titration of host with guest at constant temperature (i.e. measurement of spectra at different guest concentrations when the guest is gradually added into the host solution), which was already used to determine binding properties of oxoporphyrinogens in the previous chapter. Here, these NMR spectra are analyzed in more detail. The second type of experiments consists of variable temperature measurements at constant concentrations of host and guest in the sample. Experimental details about the measurements are given in Appendix C.4. Concentration and temperature dependence of measured transition rate coefficients is different for each process. Full description requires a specific kinetic model, which connects the transition rate coefficients to reaction rate coefficients, see Sect. 3.4.1. The latter are independent of concentration and their temperature dependence follows the Eyring equation, see Sect. 2.1.3 and 2.1.4.

Results in Sect. 7.2.3 mainly follow our publication [1] including parts of the text, figures and schemes. These were reproduced with permission from Springer Nature. Other results are still unpublished in a scientific journal (as of February 2024). Several NMR spectra analyzed here and some preliminary results were presented in the master's thesis [2].

V. Březina, *Supramolecular Complexes of Oxoporphyrinogens with Organic Molecules*, Springer Theses, https://doi.org/10.1007/978-3-031-68646-7_7

7.1 Dynamics of OxP

7.1.1 Prototropic Tautomerization

When **OxP** is double protonated by an acid (while forming a host-guest complex with two acid anions), the protonation is located at the opposite oxoporphyrinogen C = O sites because protonation of adjacent sites is energetically unfavorable [3]. Hence, two different tautomers denoted as (+)-**OxP**$^{2+}$ and (−)-**OxP**$^{2+}$ are formed, see Fig. 7.1. Double protonation reduces the **OxP** symmetry from D_{2d} to D_2, thus, the tautomers are formally mutual mirror images (although the actual shape also depends on the acid structure), i.e. enantiomers since the D_2 symmetry group lacks improper rotation. These tautomeric forms can be interconverted by so-called *prototropic tautomerization* process characterized by the reaction rate coefficient κ_{taut}. Tautomerization[1] is a special type of isomerization process (structural change preserving atomic composition of the molecule), where an atom or a group of atoms is repositioned within a molecule. In **OxP**$^{2+}$, the repositioned atoms are protons (H^+) originating from the acid. Figure 7.1 shows that a reference spin of *tert*-butyl-H (denoted by blue circle) changes its environment between states A_1 and A_2, and thus undergoes chemical exchange characterized by transition rate coefficient k_{taut} (which is related to κ_{taut}).

Besides protonation, the resulting geometry of the **OxP**·acid host-guest complex is shaped by the two bound acid anions (see Sect. 6.9), which create so-called 'chiral field' by their thermal movement around the central binding sites,[2] which is averaged on the NMR timescale [3, 4]. For our purposes, only the symmetry of chiral field is essential, nevertheless, its exact shape can be rigorously obtained from molecular dynamics modeling for example as angular probability function of a chosen axis in the acid anion [3–5]. Symmetry of the chiral field is influenced by symmetry of the protonated **OxP**$^{2+}$ host and by symmetry of the acid anion guest. Molecule of DFA

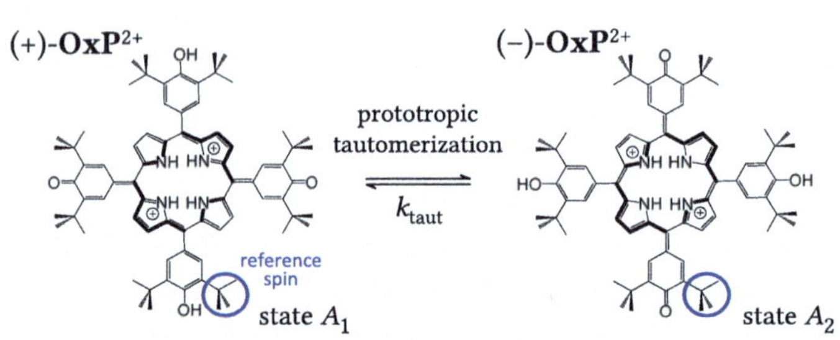

Fig. 7.1 Prototropic tautomerization process. The tautomers are different due to position of protonation relative to spatial orientation of central pyrroles (bold bonds are oriented towards the reader). Influence of acid anion is not shown

[1] Tautomerization https://goldbook.iupac.org/terms/view/T06253.

[2] The anion is held close to the charged host because of hydrophobic interaction with the nonpolar solvent. The counterions do not dissociate as they would in a polar solvent.

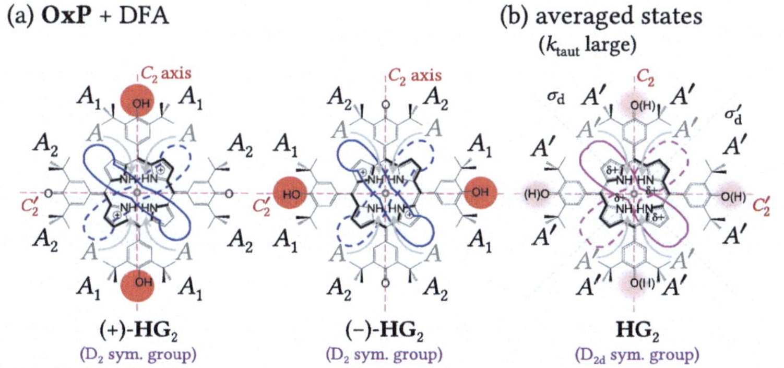

Fig. 7.2 Structures of **OxP** + DFA 1:2 host-guest complex, corresponding states of *tert*-butyl-H spins (black labels; also valid for *ortho*-H and β-H spins) and states of NH spins (gray labels). **a** Spin states of two tautomeric forms (enantiomers) (+)-$\mathbf{HG}_2$ and (−)-$\mathbf{HG}_2$. **b** Spin states of the structure averaged due to fast tautomerization (combination of (+)-$\mathbf{HG}_2$ and (−)-$\mathbf{HG}_2$ structures). Solid and dashed geometric shapes with proper symmetry represent chiral field in front of and behind the host, respectively. Dashed straight lines denote symmetry elements C_2 (two-fold rotation axis) and σ_d (diagonal mirror plane). Averaging of states implies $\delta_{A'} = \frac{1}{2}(\delta_{A_1} + \delta_{A_2})$, state of NH spins does not change due to the fast exchange

possesses mirror symmetry, and thus chiral field of the DFA anion does not change the D_2 symmetry of the host-guest complex as shown in Fig. 7.2a. Here, the chiral field is represented as geometric shapes with proper symmetry.

Note that the tautomers (+)-$\mathbf{HG}_2$ and (−)-$\mathbf{HG}_2$ are mutual mirror images and thus true enantiomers. Number of states of the host nuclei are determined by symmetry of the host-guest complex. Figure 7.2a reveals two distinct states A_1 (closer to protonated site) and A_2 (further from protonated site) for *tert*-butyl-H nuclei. Note that (**i**) the same notation can be used for *ortho*-H or β-H signals in the vicinity of the corresponding *tert*-butyl-H (black state labels in Fig. 7.2a), and (**ii**) another independent notation is used for central NH signals (gray state labels in Fig. 7.2a) since they lie exactly in the mirror symmetry planes of **OxP**. The situation is more complicated when we consider complexation with a chiral acid lacking mirror symmetry such as CSA, where tautomers (+)-$\mathbf{HG}_2$ and (−)-$\mathbf{HG}_2$ are not mutual mirror images [3]. Then the host-guest complex can be formed either with two (*R*)-**G**, two (*S*)-**G** or one (*R*)-**G** and one (*S*)-**G** as shown in [4].

When the tautomerization process is fast on the NMR timescale, the chiral field is averaged and D_{2d} symmetry of free **OxP** is restored, see Fig. 7.2b. This averaging is represented as superposition of geometric shapes representing the chiral field in Fig. 7.2a. The combined geometric shapes properly represent symmetry of the averaged chiral field. Fast exchange also induces new averaged spin states denoted by prime (i.e. A') in Fig. 7.2b, cf. replacement states in Sect. 3.3.8.

NMR spectra of titration of **OxP** with DFA are shown in Fig. 7.3a. At low acid concentrations, the reference *tert*-butyl-H spin assumes three states: equally populated A_1 and A_2 in the $\mathbf{HG}_2$ complex ([**HG**] is negligible due to strong cooperativity

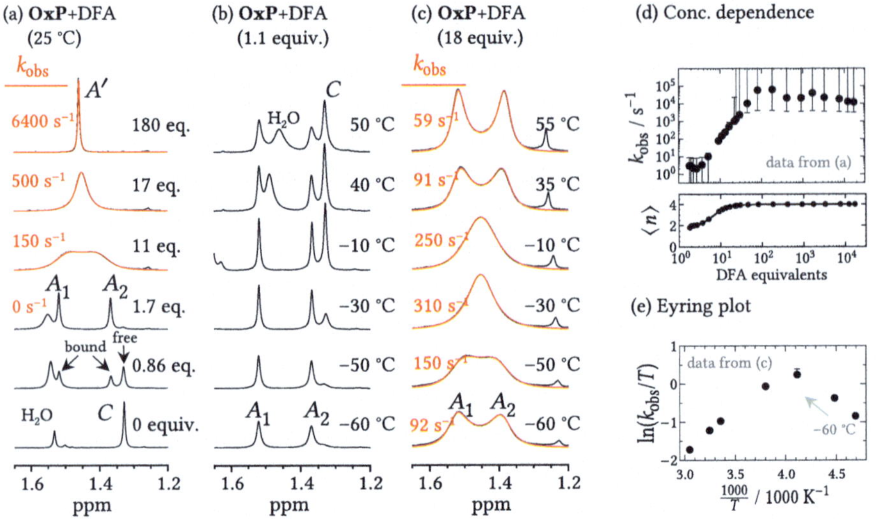

Fig. 7.3 ^{1}H NMR spectral manifestation of prototropic tautomerization at *tert*-butyl signals in **OxP**. **a** Titration of **OxP** with DFA (full spectra in Fig. E.2). States A_1, A_2 and their average A' refer to bound **OxP** (see Fig. 7.2), state C refers to free **OxP**. **b** Variable temperature measurements at 1.1 equiv. of DFA (full spectra in Fig. E.8). **c** Variable temperature measurements at 18 equiv. of DFA (full spectra in Fig. E.9). **d** Analysis of titration measurement from (a), dependence of k_{obs} on DFA concentration (top). Mean protonation number $\langle n \rangle = ([\mathbf{HG}] + 2[\mathbf{HG_2}] + 3[\mathbf{HG_3}]) + 4[\mathbf{HG_4}]/[\mathbf{H}]_t$ calculated using UV/vis binding constants (although for TFA) $K_1 = 6 \times 10^4$, $K_2 = 7 \times 10^8$, $K_3 = 0.02$, $K_4 = 8 \times 10^5\ M^{-1}$ from Table 6.1 (bottom). **e** Analysis of variable temperature measurement from (b), the Eyring plot is nonlinear. Values of k_{obs} in (a,c) were obtained from symmetric two-state exchange fit of states A_1 and A_2 (fits are denoted by red line). Fitting procedure and meaning of error bars are explained in Appendix C.6

α_{12}, see Sect. 6.9) and C in the free **OxP**. Assignment of states A_1 and A_2 is not confirmed[3] by 2D NMR experiments, however, this is not important for the analyses. At more than 2 equiv. of DFA, free host is absent and the exchange can be modeled and fitted with symmetric two-state exchange with observed transition rate coefficient k_{obs}, see red lines in Fig. 7.3a. Details about the fitting procedure are given in Appendix C.6. Fitting results in Fig. 7.3d (top) show increase of k_{obs} with acid concentration (similarly for CSA in Fig. C.5a). The shape of calculated mean protonation number $\langle n \rangle$ in Fig. 7.3d (bottom) indicates strong correlation between k_{obs} and $\langle n \rangle$. However, $\langle n \rangle$ was calculated using binding constants of stronger TFA rather than DFA, and the actual $\langle n \rangle$ might be shifted to higher DFA equivalents. Hence, k_{obs} describes tautomerization of diprotonated **OxP** (k_{taut}) at lower acid concentrations and possibly tautomerization of tetraprotonated **OxP** ($\tilde{k}_{taut}$) at higher acid concentrations, see the illustration in Table 7.4, Sect. 7.3.

[3] The presence of protonation close to *tert*-butyl-H spins in state A_1 is expected to alter the Larmor frequency of free **OxP** state C by greater amount then in state A_2, thus we assume $|\delta_{A_1} - \delta_C| > |\delta_{A_2} - \delta_C|$ in Fig. 7.3a.

Variable temperature measurements at 1.1 equiv. of DFA provide little information since the exchange is in slow regime, see Fig. 7.3b. At 18 equiv. of DFA causing at least twofold protonation of the host, chemical exchange is in intermediate regime and the lineshape can be fitted as shown in Fig. 7.3c. Fitting results in Fig. 7.3e show that in the range from −60 °C to −30 °C the transition rate coefficient k_{obs} increases but decreases at higher temperatures. Henceforth, this process cannot be described by single energy barrier with constant $\Delta H^{\ddagger}$ and $\Delta S^{\ddagger}$, see discussion at the end of Sect. 2.1.4.

As known from the literature, transfer of protons between (+)-$\mathbf{HG}_2$ and (−)-$\mathbf{HG}_2$ is realized in three different ways, see Fig. 7 in [3]. The first way is proton transfer by free acid molecules, manifested by the increase of k_{obs} with total acid concentration as described above. The second way is mediation by **OxP**-**OxP** collisions since the transition rate coefficient increases with $[\mathbf{OxP}]_t$. The transition rate coefficient also increases with water concentration, thus, the process is mediated by water (carrying the protons in the form of H_3O^+).

7.1.2 Anion Binding at the Porphyrinogen Center

Behavior of the acid anion is governed by several processes. The first one is *host-guest binding*, transition between free host and the host-guest complex. This process consists of two steps (indiscernible in NMR), protonation of a host carbonyl group and subsequent binding of guest anion at the porphyrinogen center, where the positive charge is located. In addition, intermediate species **HG** is neglected due to strong cooperativity (high α_{12}). Host-guest association is characterized by k_{ass} and dissociation by k_{diss} as shown in Scheme 7.1. Both transition rate coefficients are connected by the expression

$$\frac{k_{ass}}{k_{diss}} = \frac{p_{bound}}{p_{free}}, \tag{7.1}$$

where p_{free} and p_{bound} ($p_{free} + p_{bound} = 1$) are relative populations of free and bound host, respectively. **OxP** in the presence of either (*R*)-CSA or DFA shows separate peaks (due to *tert*-butyl-H, β-H and *ortho*-H) due to free and bound species, see Figs. E.2, E.8 and E.3, therefore, the process of host-guest binding is slow on the NMR timescale. Half-symmetric three-state lineshape fitting (Eq. (3.37)) of *tert*-butyl signals of **OxP** in the presence of 1 equiv. of DFA[4] according to Scheme 7.1 yields k_{ass}, k_{diss}, $k_{taut} < 5\ \mathrm{s}^{-1}$.

Another process is *anion exchange* at the porphyrinogen center characterized by k_{ex}, when an incoming anion from free acid replaces the original bound anion in the host-guest complex. Anion exchange of enantiopure or achiral acid guests has

[4] The fitted spectrum was measured during the titration of **OxP** with DFA, fitting curve is not shown. Parameters δ_j were obtained at the peak positions, parameter R_2 was obtained from the spectrum at 0 guest equivalents. Equation (7.1) was used and the free/bound populations were obtained from peak integration.

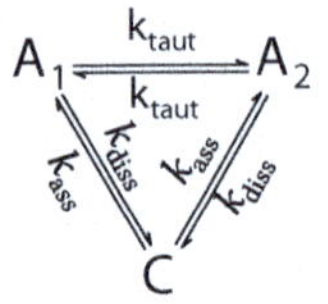

Scheme 7.1 Exchange scheme for **OxP** + DFA (or enantiopure CSA) 1:2 host-guest complex applicable to *tert*-butyl-H, β-H and *ortho*-H signals. Intermediate species **HG** is neglected due to high cooperativity α_{12}

no effect on the host spin states. However, the presence of both enantiomers of a chiral guest enables anion exchange (R)-$\mathbf{G}^-$ $\leftrightarrow$ (S)-$\mathbf{G}^-$, which affects the spin states by changing the corresponding chiral field, see details in [4, 6]. This process must be fast on the NMR timescale to enable determination of enantiomeric excess (e.e.) as explained in Sect. 3.4.3. In **OxP** + (R)-CSA system, the e.e.-induced resonance splitting ($\Delta\delta$) is small but visible at the *ortho*-H resonance in the whole acid concentration range, see Fig. E.3 (cf. significant splitting due to mandelic acid in Fig. 11 in [4]). Analogously to k_{taut}, the anion exchange process is different (occurs in different protonation states) at lower acid concentrations, where the acid anion binds at the porphyrinogen center (k_{ex}) and at higher acid concentrations without acid anion binding ($\tilde{k}_{\text{ex}}$), see the illustration in Table 7.4.

7.1.3 Macrocyclic Inversion

Another dynamic process previously found in **OxP** is *macrocyclic inversion* of the porphyrinogen macrocycle characterized by k_{flip}. It consists in collective flipping of all pyrrole subunits by 90° and interconverts the tautomeric forms (+)-$\mathbf{HG}_2$ and (−)-$\mathbf{HG}_2$. This process was reported in [3] for **OxP** + mandelic acid system (in $CDCl_3$), its progress as suggested in the same article is reproduced in Fig. 7.4a–c. Free **OxP** (a) is unable of the flipping due to steric hindrance at the porphyrinogen center. In the second step (b), $\mathbf{HG}_2$ complex is formed, yet not enabling the inversion. Finally (c), each of the two acid anions removes a proton from the central NH groups rendering the host neutral, which makes the porphyrinogen center less crowded and activates the flipping process. The transition rate coefficient of flipping (not exceeding $300\,\text{s}^{-1}$ at 50 °C for **OxP** in the presence of mandelic acid [3]) decreases with the acid concentration since strongly bound acid anion prevents flipping. Water content in the sample increases the rate probably because water can expel the bound acid molecules (while keeping the double protonation in place), which enables the ring flip process and preserves its detectability, see Fig. 7 in [3].

States of **OxP** +(R)-CSA (or another enantiopure acid, e.g. (R)-mandelic acid) are shown in Fig. 7.5. Each diastereomer (+)-$\mathbf{HG}_2$ and (−)-$\mathbf{HG}_2$ induces spin states with different chemical shift as shown in (a). Averaging of these states occurs in three possible ways as shown in (b): (**i**) Fast prototropic tautomerization (i.e. large k_{taut} and

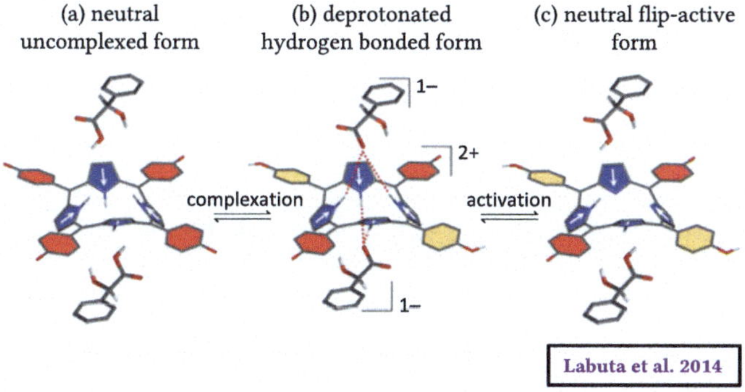

Fig. 7.4 Macrocyclic inversion process in **OxP** + mandelic acid system. Adapted with permission from [3]. Copyright 2014 American Chemical Society

small k_{flip}), which does not change D_2 symmetry of the diastereomers, and thus does not change the number of observed peaks. (**ii**) Fast macrocyclic inversion (i.e. large k_{flip} and small k_{taut}), which again does not change the symmetry and the number of observed peaks. (**iii**) Both prototropic tautomerization and macrocyclic inversion are fast (i.e. large k_{taut} and k_{flip}), which averages *tert*-butyl-H (as well as *ortho*-H and β-H) and NH resonances to a single peak. Figure 7.5c shows that the reference *ortho*-H spin (denoted by blue circle) undergoes four-state chemical exchange governed by k_{taut} and k_{flip}. In the case of mandelic acid in [3], all four states were observed as separate peaks (see Fig. 5 in [3]) for *ortho*-H and β-H nuclear spins. Macrocyclic inversion has little effect on distant *tert*-butyl-H spins, it influences mainly the porphyrinogen center.

Unlike mandelic acid, the presence of (*R*)-CSA induces a doublet in *ortho*-H in the **HG**$_2$ complex, see Fig. E.3. We suggest that splitting due to *tert*-butyl protonation is averaged out by prototropic tautomerization above 2 equiv. of guest (i.e. k_{taut} is large compared to the corresponding *ortho*-H splitting, although small compared to splitting of *tert*-butyl-H). The observable splitting is ascribed to the chiral field. This situation implies the absence/slow regime of macrocycle inversion process, which would otherwise cause averaging of the *ortho*-H resonance to a singlet. This conclusion is also supported by the splitting of bound guest methyl signal at 0.57 ppm below 2 guest equiv. (each signal belongs to different diastereomer), which contradicts the presence of fast/intermediate macrocyclic inversion or prototropic tautomerization. The structure in Fig. 7.5b (left) implying two different *ortho*-H states A'_{I} and A'_{II} are relevant for this system at high (*R*)-CSA concentration. The reason for the absence of macrocyclic inversion is probably the greater binding strength of (*R*)-CSA compared to mandelic acid ($K_1K_2 > 10^8$ M^{-2} for CSA compared to $K_1K_2 < 10^7$ M^{-2} for mandelic acid, see Sect. 6.9 and [3]) preventing the activation step in Fig. 7.4c.

In **OxP** + DFA system, macrocyclic inversion does not alter the spin states shown in Fig. 7.2a. Hence, this process cannot be detected by NMR in this system. Macrocyclic inversion is not possible in **Bz$_2$OxP** and **Bz$_4$OxP** due to steric hindrance.

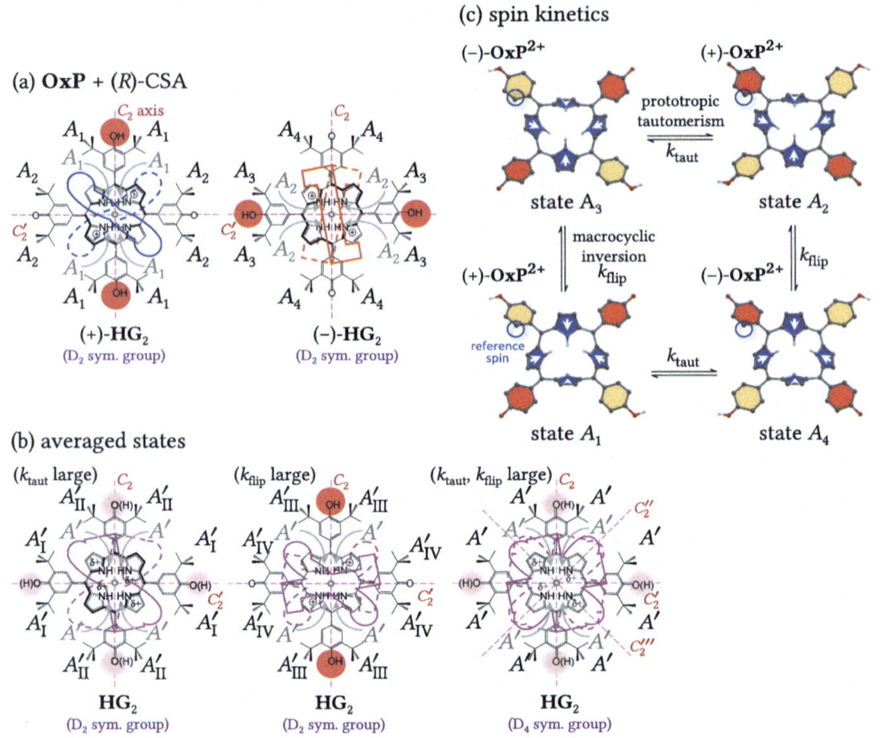

Fig. 7.5 Structures of **OxP** + (*R*)-CSA 1:2 host-guest complex, corresponding states of *tert*-butyl-H spins (black labels; also valid for *ortho*-H and *β*-H spins) and states of NH spins (gray labels). **a** Spin states in the two tautomeric forms (diastereomers) (+)-$\mathbf{HG_2}$ and (−)-$\mathbf{HG_2}$, **b** Averaged structures (combination of (+)-$\mathbf{HG_2}$ and (−)-$\mathbf{HG_2}$ structures) due to fast prototropic tautomerization (left), fast macrocyclic inversion (middle) and due to the combined effect of these two (right). Solid and dashed geometric shapes with proper symmetry represent chiral field in front of and behind the host, respectively. Dashed straight lines denote symmetry elements C_2 (two-fold rotation axis). Fast exchange implies (Eq. (3.62)) in *tert*-butyl-H peaks: $\delta_{A'_{I}} = p_+\delta_{A_2} + p_-\delta_{A_3}$, $\delta_{A'_{II}} = p_+\delta_{A_1} + p_-\delta_{A_4}$, $\delta_{A'_{III}} = p_+\delta_{A_1} + p_-\delta_{A_3}$, $\delta_{A'_{IV}} = p_+\delta_{A_2} + p_-\delta_{A_4}$, $\delta_{A'} = \frac{p_+}{2}(\delta_{A_1} + \delta_{A_2}) + \frac{p_-}{2}(\delta_{A_3} + \delta_{A_4})$, where p_+ and p_- are relative populations of the species (+)-**H**(*R*)-$\mathbf{G_2}$ and (−)-**H**(*R*)-$\mathbf{G_2}$, respectively. Fast exchange implies in NH peaks: $\delta_{A'} = p_+\delta_{A_1} + p_-\delta_{A_2}$. **c** Processes of macrocyclic inversion (k_{flip}) and prototropic tautomerization (k_{taut}) in **OxP** in the presence of enantiopure guest. The reference *ortho*-H spin (denoted by blue circle) changes its position among states A_1, A_2, A_3 and A_4 (corresponding to *a*', *b*', *a*" and *b*" in [3], respectively). *Tert*-butyl groups and some hydrogen atoms are removed for clarity. Yellow colored hemiquinonoid groups are subject to protonation

7.2 Dynamics of Bz$_2$OxP

7.2.1 Topomerization at Hemiquinonoid Groups

Topomerization[5] is a process converting a chemical species into itself, typically rotation of a side group. In protonated **Bz$_2$OxP**, double bond at the hemiquinonoid group reduces to single bond enabling rotation of the whole group about its *meso*-position bond as illustrated in Fig. 7.6. Although the chemical structure does not change, the reference *tert*-butyl nuclear spin (denoted by circle) changes its environment between states A and B. Consequently, **Bz$_2$OxP** hemiquinone topomerization is manifested as symmetric two-state exchange in the NMR spectrum (*ortho*-H signals are sensitive to this process as well, but they do not coalesce due to large peak separation). In fact, each of states A and B consists of multiple substates, which are averaged due to other processes, as will be shown later. Due to this averaging, all four hemiquinonoid groups appear as if they rotate simultaneously on the NMR timescale even when the mean protonation number $\langle n \rangle < 4$. Although hemiquinonoid topomerization almost certainly takes place also in **OxP**, it is undetectable because the spin environments are identical (the rotation axis coincides with the C_2 symmetry axis, which also lies in the mirror symmetry plane). However, **Bz$_2$OxP** possesses intrinsic asymmetry due to N-alkylation revealing the topomerization process.

As expected, the observed transition rate coefficient k_{obs} (determined by lineshape fitting, see details in Appendix C.6) indicating the hemiquinone topomerization[6] depends on acid concentration, see Fig. 7.7a. Initially, increasing the mean protonation number $\langle n \rangle$ on average reduces double bonds of the hemiquinonoid groups to single bonds and k_{obs} increases, see Fig. 7.7f ($\langle n \rangle$ is calculated only up to 20 equiv. because only the value of K_1 is available, see Sect. 6.9). Above ca. 200 equiv. of DFA, rate of the process decreases with acid concentration. The absence of chemical exchange in NMR spectra without any acid (even at high temperatures) proves the necessity of protonation to activate the rotation, see Fig. 7.7b. Concentration dependence at room temperature was observed only in the presence of DFA not CSA (see

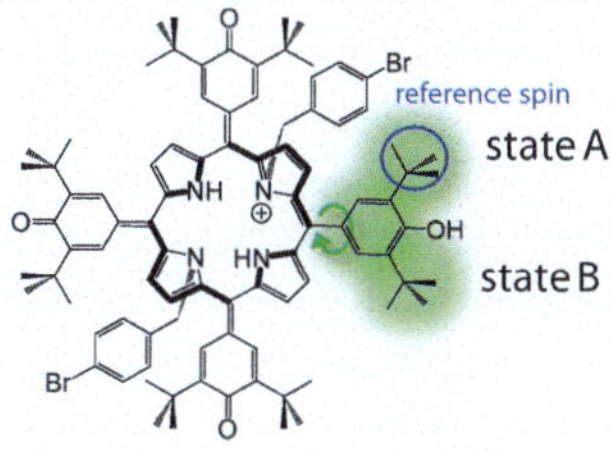

Fig. 7.6 Rotation of hemiquinonoid group (topomerization) in protonated **Bz$_2$OxP**

[5] Topomerization https://goldbook.iupac.org/terms/view/T06396.

[6] Relationship between k_{obs} and the "true" reaction/transition rate coefficient $\kappa_{\mathrm{topo}} = k_{\mathrm{topo}}$ is non-trivial as discussed at the end of this section.

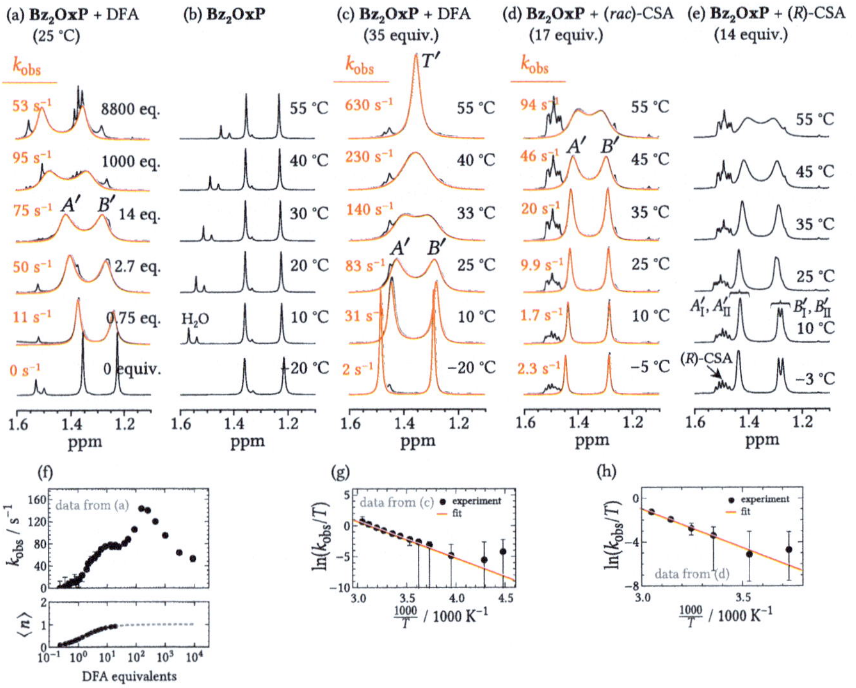

Fig. 7.7 Analysis of NMR *tert*-butyl signals of $\mathbf{Bz_2OxP}$ in various conditions reflecting rotation of the hemiquinonoid groups. **a** Titration of $\mathbf{Bz_2OxP}$ with DFA (full spectra in Fig. E.4). **b** Variable temperature measurement of $\mathbf{Bz_2OxP}$. There is no chemical exchange in the absence of acid. **c** Variable temperature measurements of $\mathbf{Bz_2OxP}$ with 35 equiv. of DFA (full spectra in Fig. E.11). **d** Variable temperature measurements of $\mathbf{Bz_2OxP}$ with 17 equiv. of (*rac*)-CSA (full spectra in Fig. E.15). **e** Variable temperature measurements of $\mathbf{Bz_2OxP}$ with 14 equiv. of (*R*)-CSA (full spectra in Fig. E.13). **f** Concentration dependence of k_{obs} for titration of $\mathbf{Bz_2OxP}$ with DFA obtained from lineshape fitting of spectra in panel (a) (top) and the corresponding mean protonation number $\langle n \rangle = [\mathbf{HG}]/[\mathbf{H}]_{\mathrm{t}}$ calculated using $K_1 = 1300\ \mathrm{M^{-1}}$ from Table 6.1 (bottom). **g** Eyring plot of $\mathbf{Bz_2OxP}$ + DFA system obtained from lineshape fitting of spectra in panel (c). **h** Eyring plot of $\mathbf{Bz_2OxP}$ + (*rac*)-CSA system obtained from lineshape fitting of spectra in panel (d). Details about meaning of error bars, lineshape fitting procedure and Eyring plot fitting are given in Appendix C.6. Structures corresponding to states in $\mathbf{Bz_2OxP}$ in the presence of DFA or (*R*)-CSA are given in Fig. C.7 in Appendix C.6 or in Fig. 7.8, respectively

spectrum in Fig. E.5), energy barrier in the latter case is too high and the effect of topomerization is only observed at higher temperatures as discussed below.

Variable temperature measurements in the presence of constant amount of DFA or CSA (enough to induce $\langle n \rangle \geq 1$ so that free host contribution to the spectral lineshape is negligible) in Fig. 7.7c, d show increase in k_{obs}. In the system of $\mathbf{Bz_2OxP}$ + DFA (Fig. 7.7c), the topomerization is fast enough to induce an averaged state T' formed from A' and B', see the corresponding structures in Fig. C.7 in Appendix C.6. Temperature dependence of k_{obs} is linear in Eyring plot, see Fig. 7.7g, h, thus, the hemiquinone topomerization process can be described by a Gibbs energy barrier

Table 7.1 Results of Eyring plot fitting for topomerization of hemiquinonoid group of **Bz$_2$OxP** in the presence of acid. The fitting procedure is described in Appendix C.6

Acid	$\Delta H^{\ddagger}$/kJ.mol^{-1}	$\Delta S^{\ddagger}$/J.mol^{-1}.K^{-1}	$\Delta G^{\ddagger}$ (25 °C)/kJ.mol^{-1}	$-T\Delta S^{\ddagger}$ (25 °C)/kJ.mol^{-1}
DFA	47.4 ± 0.9	-49.1 ± 2.8	62.0 ± 1.2	14.6 ± 0.8
CSA	58.7 ± 2.6	-29.0 ± 8.0	67.3 ± 3.5	8.6 ± 2.4

$\Delta G^{\ddagger} = \Delta H^{\ddagger} - T\Delta S^{\ddagger}$ (Sect. 2.1.4). Weighted linear fit of the Eyring plots quantifies the Gibbs energy barrier as shown in Table 7.1. Room temperature (298 K) entropic contribution to the Gibbs energy barrier ($-T\Delta S^{\ddagger}$) is smaller than the enthalpic contribution ($\Delta H^{\ddagger}$), however, it is not negligible as would be expected for rotation processes with no bond breaking [7]. Hence, the structure of protonated complex is rearranged during the topomerization process. Relevance of using k_{obs} in the Eyring plot (instead of reaction rate coefficient κ_{topo}) is discussed at the end of this section.

Comparison of **Bz$_2$OxP** spectra with racemic CSA and its (*R*)-enantiomer in Fig. 7.7d, e reveals an important difference, the presence of (*R*)-CSA induces four *tert*-butyl states, while (*rac*)-CSA only two. To understand this effect caused by chirality of CSA, it is necessary to analyze all possible *tert*-butyl spin states of the monoprotonated **Bz$_2$OxP** host (states of higher protonation are negligible for **Bz$_2$OxP** + CSA system, see Sect. 6.9). Figure 7.8a shows host structures with four different protonation sites in **Bz$_2$OxP** + (*R*)-CSA system. They comprise of two chemically distinct tautomers (+)-**HG** and (−)-**HG**. The labels (+) and (−) have been assigned arbitrarily (without the influence of guest anion, protonated species (+) − **H**$^+$ and (−) − **H**$^+$ are mutual mirror images, i.e. enantiomers). Asymmetric acid anions create different chiral field at the porphyrinogen center of (+)-**HG** and (−)-**HG**, which is schematically denoted by blue and red asymmetric shapes, thus, NMR spectra of (+)-**HG** (*tert*-butyl states $A_1, \ldots, A_4$) and (−)-**HG** (*tert*-butyl states $A_5, \ldots, A_8$) tautomers are different.[7] First and third structures in Fig. 7.8a (i.e. (+)-**H**(*R*)-**G**) are chemically identical, but a reference *tert*-butyl nuclear spin in both structures is located in different environments, similarly for (-)-**H**(*R*)-**G**. Note that B_j states are always in the vicinity of N-bromobenzyl groups contrary to A_j states. Figure 7.8b shows four different protonation states in **Bz$_2$OxP** + (*S*)-CSA system. The structure of (-)-**H**(*S*)-**G** is a mirror image of (+)-**H**(*R*)-**G** and the structure of (+)-**H**(*S*)-**G** is a mirror image of (-)-**H**(*R*)-**G** as schematically denoted by the shape and orientation of the chiral field. This implies that, for example, spin state A_1 is located at (+)-**H**(*R*)-**G** as well as (-)-**H**(*S*)-**G** and similarly for other states.

Besides hemiquinone topomerization, other processes change the environment of a reference *tert*-butyl spin as described in next two sections. These processes are characterized by k_{taut}, k^*_{taut} and k_{ex}, the corresponding chemical exchange is fast on the NMR scale above −20 °C, thus, the NMR signals of spin states A_j, B_j are averaged, see Figs. E.4, E.5 and E.6. In the case of **Bz$_2$OxP** in the presence of (*R*)-CSA, all

[7] As in the case of **OxP**, the chiral field is formed since the charged species interact strongly together due to hydrophobicity of the $CDCl_3$ solvent.

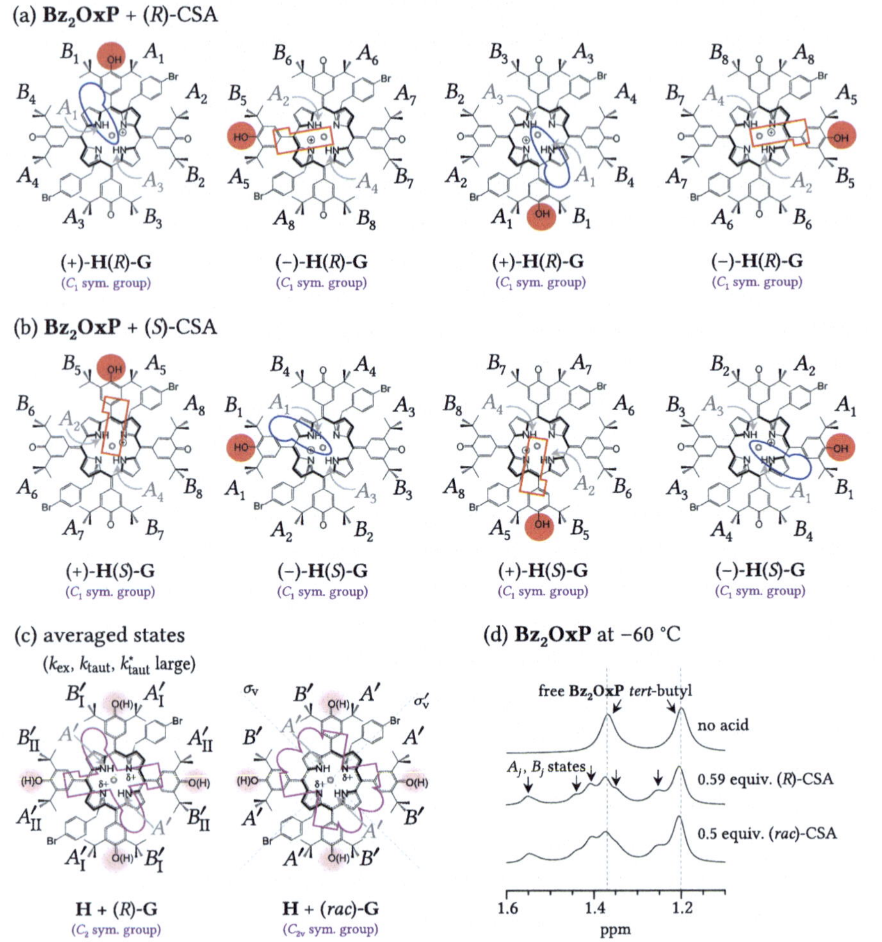

Fig. 7.8 Structures of **Bz₂OxP** + CSA complex, corresponding states of *tert*-butyl-H spins (black labels; also valid for *ortho*-H and β-H spins) and states of NH spins (gray labels). **a** Spin states in **Bz₂OxP**$^+$·(*R*)-CSA$^-$ complex, **b** spin states in **Bz₂OxP**$^+$·(*S*)-CSA$^-$ complex. **c** Fast prototropic tautomerization processes cause averaging of chiral field in **Bz₂OxP** + (*R*)-CSA system (consists of averaged states from (a)) and **Bz₂OxP** + (*rac*)-CSA system (consists of averaged states from (a) and (b), also requires fast anion exchange process). Geometric shapes with proper symmetry represent chiral field created by the acid anion. Dashed straight lines denote symmetry elements σ_v (vertical mirror plane). Fast exchange implies (Eq. (3.62)) in *tert*-butyl-H peaks: $\delta_{A'_{\mathrm{I}}} = \frac{p_{+,R}}{2}(\delta_{A_1} + \delta_{A_3}) + \frac{p_{-,R}}{2}(\delta_{A_6} + \delta_{A_8})$, $\delta_{A'_{\mathrm{II}}} = \frac{p_{+,R}}{2}(\delta_{A_2} + \delta_{A_4}) + \frac{p_{-,R}}{2}(\delta_{A_5} + \delta_{A_7})$, $\delta_{B'_{\mathrm{I}}} = \frac{p_{+,R}}{2}(\delta_{B_1} + \delta_{B_3}) + \frac{p_{-,R}}{2}(\delta_{B_6} + \delta_{B_8})$, $\delta_{B'_{\mathrm{II}}} = \frac{p_{+,R}}{2}(\delta_{B_2} + \delta_{B_4}) + \frac{p_{-,R}}{2}(\delta_{B_5} + \delta_{B_7})$, $\delta_{A'} = \frac{1}{2}(\delta_{A'_{\mathrm{I}}} + \delta_{A'_{\mathrm{II}}}) = \frac{p_{+,R}}{4}(\delta_{A_1} + \delta_{A_2} + \delta_{A_3} + \delta_{A_4}) + \frac{p_{-,R}}{4}(\delta_{A_5} + \delta_{A_6} + \delta_{A_7} + \delta_{A_8})$ and $\delta_{B'} = \frac{1}{2}(\delta_{B'_{\mathrm{I}}} + \delta_{B'_{\mathrm{II}}}) = \frac{p_{+,R}}{4}(\delta_{B_1} + \delta_{B_2} + \delta_{B_3} + \delta_{B_4}) + \frac{p_{-,R}}{4}(\delta_{B_5} + \delta_{B_6} + \delta_{B_7} + \delta_{B_8})$, where $p_{+,R}$, $p_{-,R}$, $p_{+,S}$ and $p_{-,S}$ are relative populations of the species (+)-**H**(*R*)-**G**, (−)-**H**(*R*)-**G**, (+)-**H**(*S*)-**G** or (−)-**H**(*S*)-**G**, respectively. Since some species are exact mirror images of others, and thus possess the same energy, it follows that $p_{+,R} = p_{-,S}$ and $p_{-,R} = p_{+,S}$. Fast exchange implies in NH peaks: $\delta_{A'} = \frac{p_{+,R}}{2}(\delta_{A_1} + \delta_{A_3}) + \frac{p_{-,R}}{2}(\delta_{A_2} + \delta_{A_4})$. (**d**) Some of the states from (a) and (b) are detected in low-temperature NMR spectra (full spectra in Figs. E.11, E.12 and E.14)

four structures in Fig. 7.8a are combined resulting to averaged chiral field inducing C_2 symmetry of the host-guest complex, see Fig. 7.8c (left). Averaged states A'_{I} and A'_{II} are formed from substates $A_1, \ldots, A_8$, averaged states B'_{I} and B'_{II} are formed from substates $B_1, \ldots, B_8$. These averaged states A'_{I}, A'_{II}, B'_{I} and B'_{II} are observed in Fig. 7.7e for **Bz₂OxP** in the presence of (*R*)-CSA (two resonances overlap at 1.44 ppm). Hemiquinone topomerization causes further averaging of these states, in particular $A'_{\mathrm{I}} \leftrightarrow B'_{\mathrm{I}}$ and $A'_{\mathrm{II}} \leftrightarrow B'_{\mathrm{II}}$.

The presence of racemate combines eight states from Fig. 7.8a, b resulting to averaged chiral field inducing C_{2v} symmetry of the host-guest complex, see Fig. 7.8c (right). This yields only two states A and B and explains the spectra in Fig. 7.7d for **Bz_2OxP** in the presence of (*rac*)-CSA. Note that the aforementioned sensitivity to chirality follows the mechanism of determination of enantiomeric excess by prochiral solvating agents as described in Sect. 3.4.3. Spectra of **Bz_2OxP** in the presence of DFA in Fig. 7.7c include only two *tert*-butyl states A' and B' because the **Bz_2OxP**·DFA complex has C_{2v} symmetry (induced by the mirror symmetry of DFA), see the corresponding structures in Fig. C.7 in Appendix C.6. Hemiquinone topomerization causes further averaging of states A' and B' manifested as symmetric two-state chemical exchange.

At −60 °C and low number of acid equivalents, all k_{topo}, k_{ex}, k_{taut} and k^*_{taut} are small enough that the exchange enters intermediate regime and some of the states $A_1, \ldots, A_8$ and $B_1, \ldots, B_8$ are directly visible in NMR although being broadened, see Fig. 7.8d. Note that chirality does not play a role and low-temperature spectra in the presence of (*R*)-CSA or (*rac*)-CSA are identical.

Determination of energy barrier of topomerization process requires knowledge of κ_{topo} temperature dependence. Our previous analysis in Table 7.1 made a simple assumption $\kappa_{\mathrm{topo}} = k_{\mathrm{obs}}$, however, according to Fig. 7.7d k_{obs} depends on acid concentration, and thus it is not equal to a reaction rate coefficient in general. We suggest a simple model, which naturally takes the concentration dependence into account,

$$k_{\mathrm{obs}} = \langle n \rangle \kappa_{\mathrm{topo}} \,. \tag{7.2}$$

This model predicts linear dependence (with zero intercept) of k_{obs} on $\langle n \rangle$ at $\langle n \rangle \leq 1$, which is approximately confirmed experimentally in Fig. 7.9. However, the presence of free host affects the spectral lineshapes and introduces error to k_{obs} determined by two-state fitting as discussed in Appendix C.6. Thus, the observed dependence is not perfectly linear. In the presence of 35 equiv. of DFA or 17 equiv. of CSA $\langle n \rangle \approx 1$, thus, our simple assumption $\kappa_{\mathrm{topo}} = k_{\mathrm{obs}}$ used in Table 7.1 is valid. At more than 200 equiv. of DFA, k_{obs} decreases with concentration. This effect is probably related to increasing steric hindrance due to the high abundance of guest molecules around carbonyl groups.

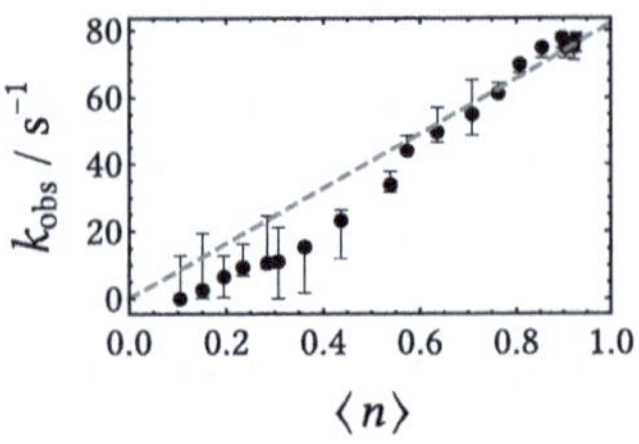

Fig. 7.9 Linear correlation between $\langle n \rangle$ and k_{obs} in titration of $\mathbf{Bz_2OxP}$ with DFA. Dashed line stands for weighted linear fit. Plotted data originate from Fig. 7.7f

7.2.2 Anion Binding at the Porphyrinogen Center

Processes in $\mathbf{Bz_2OxP}$ + acid systems involving transfer of the acid anion are analogous to those described for **OxP** in Sect. 7.1.2. The first process is the host-guest binding (accompanied by host protonation), where the association and dissociation is characterized by k_{ass} and k_{diss} (or by concentration-independent κ_{ass} and κ_{diss}). Unlike for **OxP**, there are no separate peaks of free and bound host except for NH signals with large separation $|\delta_{bound} - \delta_{free}|$. Other host signals are significantly broadened in the presence of low acid concentration implying intermediate exchange regime of k_{ass} and k_{diss}. Values of k_{ass} and k_{diss} are determined in the next section by lineshape fitting of NH signals.

Another process is the exchange of the bound acid anion characterized by k_{ex} (described in detail for **OxP** in Sect. 7.1.2). This process can also be accompanied by the prototropic tautomerization, see the next section and Appendix C.9. Exchange of achiral anions or identical enantiomers does not alter spin state of a reference spin. However, the spin state is altered when, for example, (*R*)-CSA anion exchanges for (*S*)-CSA anion, thus enabling determination of e.e. [2] in the fast exchange regime of k_{ass}, k_{taut} and k_{ex} (exchange broadening must not be present). In theory, all hydrogen signals are split due to e.e. except NH since they lie exactly in the prochiral plane, see Sect. 3.4.3 for details. Actually, only β-protons at alkylated pyrroles (□ in Fig. E.13) provide well resolved e.e.-induced splitting[8] $\Delta\delta$ at room temperature [2, 5], compare Figs. E.13 and E.15. Splitting of *tert*-butyl resonances due to e.e. in the $\mathbf{Bz_2OxP}$ + (*R*)-CSA system is observable below 10 °C but obscured by exchange broadening at higher temperatures, see Fig. 7.7e.

At high acid concentrations, host-guest stoichiometries higher than 1:1 appear implying guest binding at the C=O sites at the edge of the host, which is characterized by another set of transition rate coefficients $\tilde{k}_{ass}$, $\tilde{k}_{diss}$ and $\tilde{k}_{ex}$, see Table 7.5.

[8] Interestingly, β-protons at non-alkylated pyrroles (□ in Fig. E.13) are J-coupled to each other and also to the NH hydrogen, which results to involved spectral lineshape with badly detectable e.e.-induced splitting [5].

7.2.3 Prototropic Tautomerization

Prototropic tautomerization, i.e. change of the protonation site in the host-guest complex, is observed at *tert*-butyl resonances in the case of **OxP** + acid systems. However, **Bz₂OxP** + acid systems show *tert*-butyl peak broadening rather than splitting upon protonation due to their small peak separation. Thus, **Bz₂OxP** *tert*-butyl peaks cannot be used as a probe to the tautomerization process. However, the protonation induces partial charge in the porphyrinogen center (see Sect. 6.4), and thus affects state of the central NH spins possessing significant peak separation upon protonation. Hence, NH groups serve as probes of the tautomerization process. In this section, we describe the prototropic tautomerization in the **Bz₂OxP** + (*R*)-CSA system in detail.

The central NH protons of the **Bz₂OxP** host in the presence of (*R*)-CSA and unavoidable presence of water are subject to chemical exchange between six environments, which is experimentally manifested in its NMR spectrum as three resonances (some of the resonances are already merged due to fast exchange, see below) in the titration spectra at low $[\mathbf{G}]_t$, see Fig. 7.12a (full spectra in Fig. E.5). At low $[\mathbf{G}]_t$, the three observable resonances of the reference spin (denoted by green arrow in Fig. 7.10a) are well separated and the exchange between them is in a slow regime. The peak at 9.3 ppm, denoted as state C', vanishes at 1 guest equivalent (equivalents defined as $[\mathbf{G}]_t/[\mathbf{H}]_t$). The state C' actually comprises substates C and D averaged due to fast chemical exchange, see Fig. 7.10b. These substates correspond to free host and its complex with water, respectively (details are discussed below). The two resonances at 13.1 and 11.8 ppm correspond to two equally populated (due to the σ_v' symmetry of the host) states denoted as A_I' and A_{II}', respectively. The state A_I' comprises substates A_1 and A_2, the state A_{II}' comprises substates A_3 and A_4, each two substates are averaged due to the fast exchange. States $A_1, \ldots, A_4$ of the green-labeled reference NH spin in Fig. 7.10c correspond to protonated host with protonation at four different C = O sites. Hence, states A_1 and A_3 correspond to (+)-**HG** tautomer and states A_2 and A_4 to (−)-**HG** tautomer (NH spin states of the host in the presence of (*R*)- or (*S*)-form of the guest are also shown in Fig. 7.8a, b including chiral field induced by the acid anion). Considering the mirror symmetry plane σ_v' of the host (shown in Fig. 7.10a), the reference spin is in state A_1 or A_2 (forming A_I') when the protonation is at the same side of the molecule as the reference spin and in state A_3 or A_4 (forming A_{II}') when the protonation is on the other side.

Transitions between substates $A_1, \ldots, A_4$ are accomplished by the prototropic tautomerization process, mediated by another incoming guest molecule (details are discussed below). For simplicity, we characterize this process by transition rate coefficient $k_{A'}$ (corresponding to reaction rate coefficient κ_{taut}), when the protonation changes across the σ_v' plane (i.e. transitions $A_1 \leftrightarrow A_3$, $A_1 \leftrightarrow A_4$, $A_2 \leftrightarrow A_3$ or $A_2 \leftrightarrow A_4$); or $k_{A'}^*$ (corresponding to κ_{taut}^*) otherwise (i.e. transitions $A_1 \leftrightarrow A_2$ or $A_3 \leftrightarrow A_4$). Large value of $k_{A'}^*$ causes averaging of the substates to A_I' and A_{II}' as shown in detail in Fig. 7.11a including the corresponding chiral field (cf. Fig. 7.8a). The presence of (*S*)-CSA induces identical A_I' and A_{II}' states in the host since the NH groups lie in the prochiral plane σ_v, see Fig. 7.11b (cf. Fig. 7.8b). Therefore,

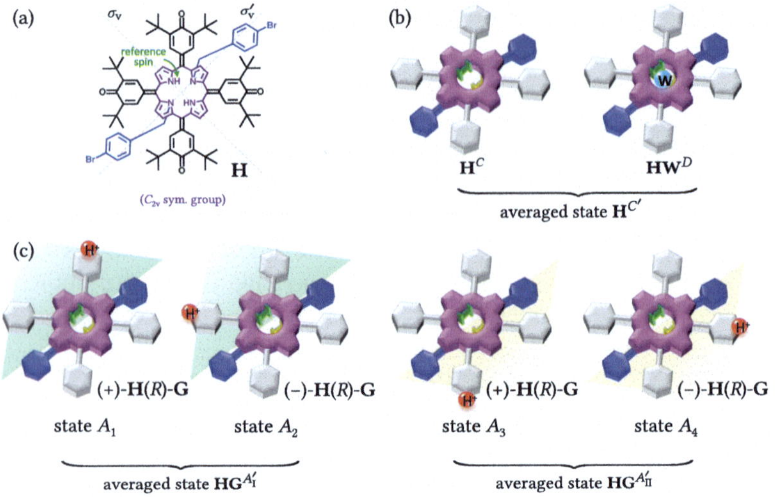

Fig. 7.10 Spin states of central NH protons of **Bz$_2$OxP** host molecule (**H**) in the presence of CSA (**G**). **a** Structure of **H**. Bromobenzyl groups are situated behind the molecule. The spin states are described with respect to reference proton denoted by green arrow. **b** Schematic representation of free host $\mathbf{H}^C$ and host-water complex $\mathbf{HW}^D$ inducing the spin states denoted in superscript. **c** Protonated host-guest complex in two distinct tautomeric forms (+)-**H**(*R*)-**G** and (−)-**H**(*R*)-**G** differing in the site of protonation. NH spins of (+)-**HG** tautomer can be present in two spin states $\mathbf{HG}^{A_1}$ or $\mathbf{HG}^{A_3}$ (similarly NH spins of (−)-**HG** can be in states $\mathbf{HG}^{A_2}$ or $\mathbf{HG}^{A_4}$) depending on the carbonyl protonation proximity to the green-labeled reference proton. Different sites of protonation within the green or yellow zones, form the averaged states A_I' and A_{II}', respectively. The superscripts (e.g. $\mathbf{H}^C$, $\mathbf{HG}^{A_1}$) denote the spin state with respect to the green-labeled reference NH proton. Adapted from [1] with permission from Springer Nature

also the presence of (*rac*)-CSA induces identical states A_I' and A_{II}' and chemical shift of these averaged states does not depend on enantiomeric excess of guest. Note that averaged states of other hydrogen spins not located in the prochiral plane are in different states in the presence of (*rac*)-CSA compared to (*R*)-CSA. These states are denoted by tilde in Fig. 7.11c and also require fast anion exchange k_{ex} in order to induce single resonance. At more than 1 guest equiv., the resonance in state C' vanishes and the resonances in states A_I' and A_{II}' start to coalesce and enter intermediate then fast exchange regimes. The presence of only one resonance implies that fast prototropic tautomerization process restores the C_2 symmetry even in protonated host, see the structures in Fig. 7.8c. Furthermore, titration with (*rac*)-CSA produced the same spectral behavior of the NH resonances (see Fig. E.6). Note that in the **Bz$_2$OxP** + DFA system, the tautomerization process is fast, and thus only one NH resonance is observed in the whole DFA concentration range, the corresponding structures are in Fig. C.7 in Appendix C.6.

All states A_1, ..., A_4 are directly observed at low temperature as shown in Fig. 7.11d. Since the chiral field induced by the acid anion is different in (+)-**HG** and (−)-**HG** (due to the absence of mirror symmetry of the guest), Larmor frequency

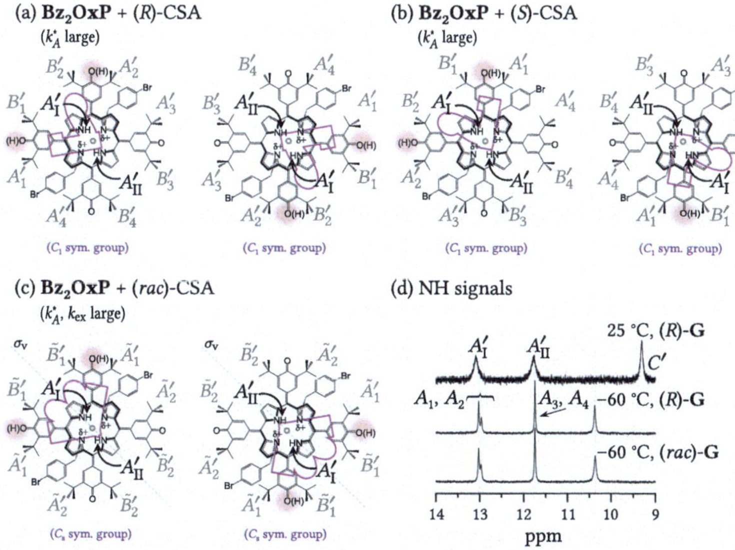

Fig. 7.11 **a** Averaged structures induced by fast exchange for the system of **Bz₂OxP** + (*R*)-CSA (large $k^*_{A'}$), **b** **Bz₂OxP** + (*S*)-CSA (large $k^*_{A'}$) and **c** **Bz₂OxP** + (*rac*)-CSA (large $k^*_{A'}$ and k_{ex}). Corresponding states of NH spins (black labels) and states of *tert*-butyl-H spins (gray labels; also valid for *ortho*-H and β-H spins) are formed as combination of spin states in Fig. 7.8a, b. Fast exchange implies (Eq. (3.62)) in NH peaks: $\delta_{A'_{\mathrm{I}}} = p_{+,R}\delta_{A_1} + p_{-,R}\delta_{A_2}$ and $\delta_{A'_{\mathrm{II}}} = p_{+,R}\delta_{A_3} + p_{-,R}\delta_{A_4}$. Fast exchange implies in *tert*-butyl-H peaks: $\delta_{A'_1} = p_{+,R}\delta_{A_4} + p_{-,R}\delta_{A_5}$, $\delta_{A'_2} = p_{+,R}\delta_{A_1} + p_{-,R}\delta_{A_6}$, $\delta_{A'_3} = p_{+,R}\delta_{A_2} + p_{-,R}\delta_{A_7}$, $\delta_{A'_4} = p_{+,R}\delta_{A_3} + p_{-,R}\delta_{A_8}$, $\delta_{B'_1} = p_{+,R}\delta_{B_4} + p_{-,R}\delta_{B_5}$, $\delta_{B'_2} = p_{+,R}\delta_{B_1} + p_{-,R}\delta_{B_6}$, $\delta_{B'_3} = p_{+,R}\delta_{B_2} + p_{-,R}\delta_{B_7}$, $\delta_{B'_4} = p_{+,R}\delta_{B_3} + p_{-,R}\delta_{B_8}$, $\delta_{\tilde{A}'_1} = \frac{p_{+,R}}{2}(\delta_{A_1} + \delta_{A_4}) + \frac{p_{-,R}}{2}(\delta_{A_5} + \delta_{A_6})$, $\delta_{\tilde{A}'_2} = \frac{p_{+,R}}{2}(\delta_{A_2} + \delta_{A_3}) + \frac{p_{-,R}}{2}(\delta_{A_7} + \delta_{A_8})$, $\delta_{\tilde{B}'_1} = \frac{p_{+,R}}{2}(\delta_{B_1} + \delta_{B_4}) + \frac{p_{-,R}}{2}(\delta_{B_5} + \delta_{B_6})$, $\delta_{\tilde{B}'_2} = \frac{p_{+,R}}{2}(\delta_{B_2} + \delta_{B_3}) + \frac{p_{-,R}}{2}(\delta_{B_7} + \delta_{B_8})$, where $p_{+,R}$, $p_{-,R}$, $p_{+,S}$ and $p_{-,S}$ are relative populations of the species (+)-**H**(*R*)-**G**, (−)-**H**(*R*)-**G**, (+)-**H**(*S*)-**G** and (−)-**H**(*S*)-**G**, respectively. Since some species are exact mirror images of others, and thus possess the same energy, it follows that $p_{+,R} = p_{-,S}$ and $p_{-,R} = p_{+,S}$. (**d**) ^{1}H NMR spectra of NH resonances of host (8.4×10^{-4} M^{-1}, $CDCl_3$) with 0.59 equiv. (*R*)-**G** at −60 °C and 25 °C, host with 0.5 equiv. (*rac*)-**G** at 25 °C. The low-temperature spectra show the presence of four states A_1, ..., A_4 (intensity ratio of the peaks at 13.03 and 12.96 ppm is 74:26), the room-temperature spectrum shows only two averaged states A'_{I} and A'_{II}

of state A_1 differs from that of A_2, although Larmor frequencies of A_3 and A_4 are coincidentally similar. The presence of (*rac*)-CSA does not alter the corresponding chemical shifts compared to (*R*)-CSA. At room temperature, only averaged states A'_{I} and A'_{II} are detected due to fast exchange. We were not able to unequivocally assign states A_1, ..., A_4 (corresponding to structures in Fig. 7.10c) to particular resonances in Fig. 7.11d. However, this information is not essential for analyzing the system's kinetics. Therefore, we assumed that Larmor frequency of the reference NH spin (denoted by green color in Fig. 7.10c) is likely similar when the protonation is in its vicinity (i.e. states A_1 and A_2) contrary to protonation across the symmetry plane (i.e. states A_3 and A_4), which results in assignment shown in Fig. 7.11d.

The observed spectral behavior can be described as the half-symmetric three-state exchange with states A'_{I}, A'_{II} and C' in terms of reduced equivalent scheme from Sect. 3.3.8 (Table 3.2b). It is characterized by two independent transition rate coefficients $k_{A'}$ and $k_{A'C'}$ ($k_{C'A'} = k_{A'C'}p_{C'}/p_{A'_{\mathrm{I}}}$ is not independent as follows from Eq. (3.37)). The corresponding populations are $p_{A'_{\mathrm{I}}} = p_{A_1} + p_{A_2}$, $p_{A'_{\mathrm{II}}} = p_{A_3} + p_{A_4}$ and $p_{C'} = p_C + p_D$. Resonance positions $\delta_{A'_{\mathrm{I}}}$ and $\delta_{A'_{\mathrm{II}}}$ shift to lower ppm values at higher acid concentrations due to the increase in polarity of the medium upon addition of acid, see arrow in Fig. 7.12b. The resonance position $\delta_{C'}$ shifts to larger ppm values during the titration (while $\delta_{A'_{\mathrm{I}}}$ and $\delta_{A'_{\mathrm{II}}}$ still remain constant), see blue arrow in Fig. 7.12b. This is a direct evidence that state C' consists of two substates C and D because the three-state solution in Eq. (3.52) does not allow shifts of peak maxima in the slow exchange regime. In the current model, the peak due to state C' can shift when populations p_C and p_D change ($\delta_{C'} = p_C\delta_C + p_D\delta_D$). It is already known that the host molecule can bind water with a binding constant $K_{\mathbf{HW}} = 240 \pm 35\ \mathrm{M}^{-1}$ [5]. During experiments it is not reasonably possible to prepare $CDCl_3$ solutions of **H** containing no residual water. Moreover, small quantities of water are added together with the hydrophilic guest during the titration process. Hence, the spin state D most likely corresponds to a host-water complex (**HW**). Competitive binding of guest and water is assumed [5], i.e. complexation with one ligand precludes binding of the other (see Sect. 2.2.3). Because of the high mobility of water (small molecule) and low value of $K_{\mathbf{HW}}$ we assume infinitely fast exchange between states C and D. To summarize, the chemical kinetics description of this system includes competitive host-guest and host-water binding. The observed spin exchange kinetics (for host NH signals) can be described using the half-symmetric three-state model. Fast exchange effectively reduces the number of states to three, enabling the use of the reduced equivalent scheme.

Scheme 7.2 shows three different levels of description of chemical kinetics in this system, the underlying equations are given in Appendix C.8. Scheme 7.2a shows the simplified chemical kinetics of the processes involved assuming the presence of host molecule as three distinct chemical species, **H**, **HG** and **HW**. The simplified chemical kinetics scheme is sufficient to describe the competitive host-ligand binding characterized by equilibrium constants $K_{\mathbf{HG}}$ and $K_{\mathbf{HW}}$, the corresponding equations are shown in Sect. 2.2.3. The constant $K_{\mathbf{HG}}$ accounts for the protonation of any of the four sites. The low-temperature spectrum of $\mathbf{Bz_2OxP}$ + (*R*)-CSA in Fig. 7.11d suggests that the sites are not equivalent since states A_1 and A_2 are not equally populated with intensity ratio 74:26. The populations cannot be determined at room temperature. However, we can treat the sites as equivalent (with equal populations) at room temperature with reasonable accuracy since the populations tend to equalize with increasing temperature. Hence, the reaction rate coefficient $4\kappa_{\mathrm{ass}}$ in Scheme 7.2a contains an integer prefactor to account for the four equivalent protonation sites. Protonation of one particular site is characterized by $K_{\mathbf{HG}}^{\mathrm{micro}} = K_{\mathbf{HG}}/4$ (cf. Table 2.2 in Sect. 2.2.4). The microscopic equilibrium constant $K_{\mathbf{HG}}^{\mathrm{micro}}$ is also equal to the ratio of reaction rate coefficients for the molecular processes of protonation and deprotonation, i.e. $K_{\mathbf{HG}}^{\mathrm{micro}} = \kappa_{\mathrm{ass}}/\kappa_{\mathrm{diss}}$.

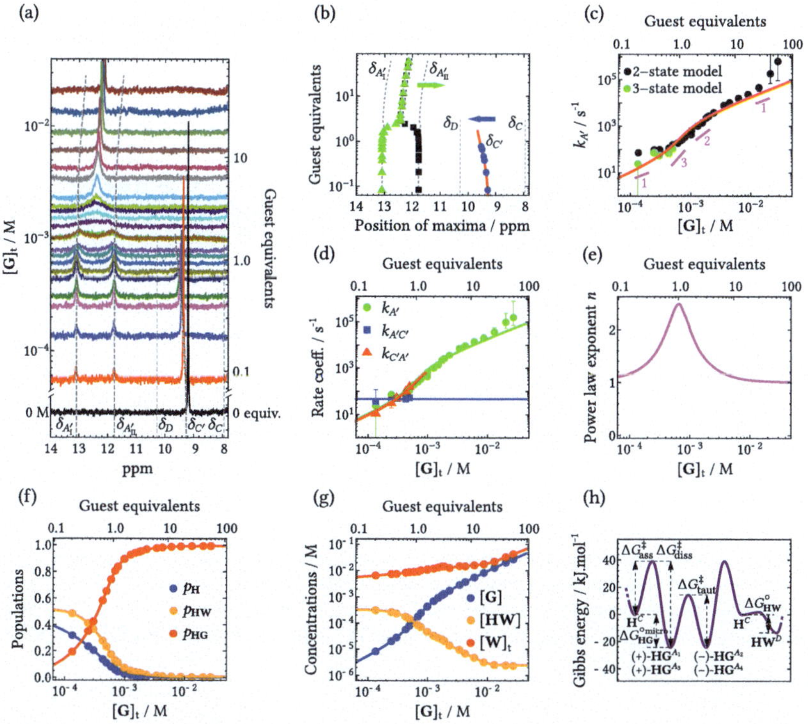

Fig. 7.12 Experimental results for the host-guest system of **Bz_2OxP** + (*R*)-CSA. **a** NH portion of ^{1}H NMR spectra of host **H** (initial concentration 6×10^{-4} M, $CDCl_3$) during the titration with guest **G** (y-scaling of spectra is adjusted for clarity). Concentration of guest **G** corresponds to the value where the spectrum meets the y-axis. **b** Apparent positions of peak maxima during the titration, green and blue arrows denote the shift of maxima due to solvent polarity increase and fast exchange between states C (free **H**) and D (host-water complex **HW**), respectively. Red line is fit of the frequency $\omega_{C'} = p_C\omega_C + p_D\omega_D$ rescaled to ppm. **c** Concentration dependence of the transition rate coefficient $k_{A'}$ obtained using two-state lineshape fitting on states A'_{I} and A'_{II} (black circles) and using three-state lineshape fitting on states A'_{I}, A'_{II} and C' (green circles). The red line is the best fit of the $k_{A'}$ concentration dependence within three-state model (two-state model is used above 1 equiv. due to the disappearance of the C state from NMR spectra) using Eqs. (7.3a) and (A.3a). Magenta lines denote the slope of the red fitting curve. **d** Other transition rate coefficients describing the half-symmetric three-state exchange. The parameter $k_{A'C'}$ was fitted and $k_{C'A'}$ was calculated ($k_{C'A'} = k_{A'C'}p_{C'}/p_{A'_{\mathrm{I}}}$). **e** Concentration dependence of the power law exponent. **f** Concentration dependence of populations of host-related species. **g** Concentration dependence of free guest, host-water complex (both determined from the model in Sect. 2.2.3) and total concentration of water (as determined from water peak integration, red solid line is interpolation). **h** Gibbs energy profile of all chemical species as calculated from the Eyring equation, Eq. (2.20), from the reaction rate coefficients at $T = 298$ K and setting the transition probability $\eta = 1$. Barriers between states C and D were not determined since the corresponding exchange process was too fast. The Larmor frequencies of states A'_{I} and A'_{II} are not unequivocally assigned with respect to the structures in Fig. 7.10c. Error bars in (**c**) and (**d**) denote maximum errors, see discussion in Sect. C.7. Adapted from [1] with permission from Springer Nature

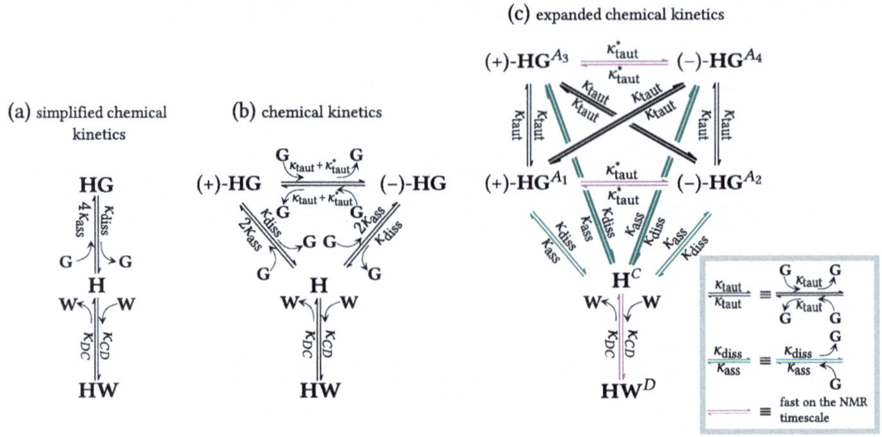

Scheme 7.2 Chemical kinetics schemes for the multi-state system of **Bz$_2$OxP** (host **H**) in the presence of two ligands, (*R*)-CSA (ligand **G**) and water (ligand **W**). **a** Simplified chemical kinetics scheme corresponding to 1:1 **H**:**G** binding with competitive 1:1 **H**:**W** binding (see Sect. 2.2.3 for details). **b** Chemical kinetics scheme describing interconversion of all distinguishable chemical species. **c** Expanded chemical kinetics scheme, equal to the full spin kinetics scheme. All relevant molecular processes and their reaction rate coefficients are shown. Processes denoted by magenta arrows are fast on the NMR timescale. Adapted from [1] with permission from Springer Nature

In fact, the protonation forms two different tautomeric species (+)-**HG** and (−)-**HG** as discussed above, see Fig. 7.10c. This situation is captured by the chemical kinetics Scheme 7.2b. Both tautomers can be formed in two different ways, hence the integer prefactor in $2\kappa_{\text{ass}}$. As discussed above, chemical species (+)-**HG** and (−)-**HG** can be interconverted by the prototropic tautomerization processes characterized by κ_{taut} or κ^*_{taut}. This interconversion between the two tautomeric forms can occur in two ways implying overall reaction rate coefficient $\kappa_{\text{taut}} + \kappa^*_{\text{taut}}$ (Scheme 7.2b). It is a second order reaction since the guest molecule has to collide with the host-guest complex. Details of these tautomerization processes, including structure of transition states, are given in Appendix C.9.

The chemical kinetics scheme can be further expanded to remove degeneracy from all microstates, see in Scheme 7.2c. Here, all chemically distinct species in all spin states (as listed in Fig. 7.10b, c) are shown as separate entities, so that no prefactors or sums of reaction rate coefficients are present for the reaction rate coefficients. The spin states are denoted in superscripts at the chemical species, e.g. $\mathbf{H}^C$, $\mathbf{HG}^{A_1}$. The processes characterized by κ^*_A (denoted by magenta arrows in Scheme 7.2c) have low energy barriers and therefore are fast at room temperature. This makes κ^*_A indeterminable by lineshape analysis. The expanded chemical kinetics scheme is equal to the full spin kinetics scheme and forms a basis for the connection between chemical kinetics and observed spin kinetics as explained in the next paragraph.

Due to fast exchange (large κ^*_{taut}, κ_{CD} and κ_{DC}), the kinetics scheme in Scheme 7.2c must be contracted from six to three states, which results in the spin kinetics

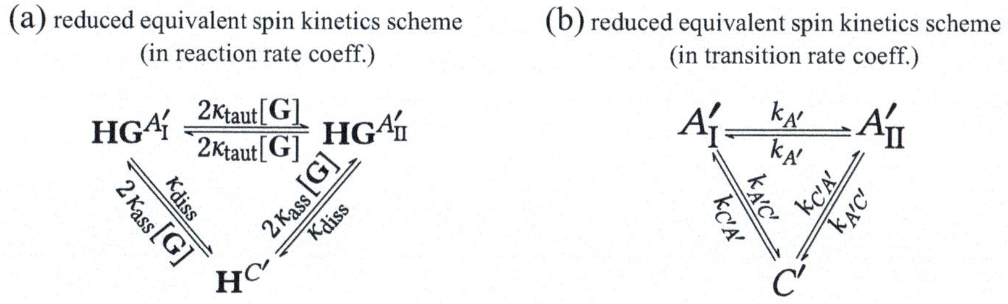

Scheme 7.3 Reduced equivalent spin kinetics schemes for the multi-state system of **Bz$_2$OxP** (host **H**) in the presence of two ligands, (*R*)-CSA (ligand **G**) and water (ligand **W**). Schemes refer to the central NH protons of the host molecule. **a** Spin kinetics in terms of reaction rate coefficients as obtained from contraction of the Scheme 7.2c. **b** Corresponding spin kinetics in terms of transition rate coefficients. It has the form of half-symmetric three-state exchange. Comparison with (**a**) gives the relationship between transition and reaction rate coefficients in Eq. (7.3). This scheme represents experimentally observed spin kinetics. It was used for lineshape fitting, see Appendix C.7 for details. Adapted from [1] with permission from Springer Nature

Scheme 7.3a. Then this scheme expressed in reaction rate coefficients can be directly compared to the corresponding scheme expressed in transition rate coefficients in Scheme 7.3b, and subsequently, the relations between reaction and transition rate coefficients can be established. In addition, the concentration dependence of the transition rate coefficients is obtained. Details of this procedure are described in Appendix C.8. The resulting equations are

$$k_{A'} = 2\kappa_{\text{taut}}[\mathbf{G}]\,, \tag{7.3a}$$

$$k_{A'C'} = \kappa_{\text{diss}}\,, \tag{7.3b}$$

$$k_{C'A'} = 2\kappa_{\text{ass}}[\mathbf{G}]\,. \tag{7.3c}$$

The transition rate coefficient $k_{A'C'}$, which corresponds to decay of **HG** complex, is independent of concentration. On the other hand, formation of a complex is a bimolecular reaction, therefore $k_{C'A'}$ is proportional to [**G**]. Also, the guest-mediated prototropic tautomerization is a bimolecular reaction (based on the suggested reaction scheme in Appendix C.9), implying $k_{A'}$ is proportional to [**G**].

The use of three-state reduced equivalent spin kinetics scheme enables us to describe the observed spectra with only two transition rate coefficients $k_{A'}$ and $k_{A'C'}$. At guest concentrations higher than 1 equiv. the two-state model with transition rate coefficient $k_{A'}$ is sufficient because the state C' is unpopulated. During the fitting procedure, the parameter $\delta_{C'}$, which describes the averaged frequency of the state C', is fixed exactly at the corresponding peak position in the spectra, frequencies of the states A'_{I} and A'_{II} are also fixed, for details see Appendix C.8. To show the actual usefulness of the three-state model, we also fitted the states A'_{I} and A'_{II} with the two-state model below 1 guest equiv. (fitting parameter $k_{A'}$). From Fig. 7.12c it is obvious that the three-state model systematically shifts $k_{A'}$ to lower values than obtained using the two-state model (presence of the third state C' causes broadening of the peaks due to states A'_{I} and A'_{II}). Examples of actual fitted spectra are shown in Fig. C.8 in Appendix C.7.

After the values of $k_{A'}$ and $k_{A'C'}$ were obtained from the raw spectra (green and blue points in Fig. 7.12d), fitting of the concentration dependence of $k_{A'}$ (using Eq. (7.3a)) was conducted simultaneously with other changes in spectra. This is shown together with other technical details of the fitting procedure in Appendix C.7. The resulting fitted curve for $k_{A'}$ (red line in Fig. 7.12c) describes the experimental values from the three-state exchange model with good accuracy and confirms the assumption of proportionality to [**G**] for the guest-mediated prototropic tautomerization, expressed in Eq. (7.3a). The value of equilibrium constant $K_{\mathbf{HG}} = (7.4 \pm 3.0) \times 10^4\ \mathrm{M}^{-1}$ was also obtained. This value is comparable with the results of our previous analysis from Sect. 6.9 $K_{\mathbf{HG}} = (8.0 \pm 5.0) \times 10^4\ \mathrm{M}^{-1}$ (obtained from NMR binding isotherms) or $K_{\mathbf{HG}} = (5.0 \pm 1.0) \times 10^4\ \mathrm{M}^{-1}$ (obtained from UV-vis), which did not take into account the competitive binding of water and the concentration dependence of $k_{A'}$. The concentration dependence of $k_{C'A'}$ was determined using the formula $k_{C'A'} = 2\kappa_{\mathrm{ass}}[\mathbf{G}] = K_{\mathbf{HG}} k_{A'C'}[\mathbf{G}]/2$, see Fig. 7.12d. The value of $k_{C'A'}$ is calculated directly from fitted values of $k_{A'C'}$ up to 1 equiv. (red points in Fig. 7.12d). For higher guest concentrations, the population of state C' is negligible, and values of $k_{A'C'}$ could not be determined by three-state lineshape fitting. The red line in Fig. 7.12d is an extrapolation from the guest concentration dependence using the mean value of $k_{A'C'}$ (blue line). The fitting procedure also provides concentrations of all species present in the sample, see Fig. 7.12f, g. Values of equilibrium constants and reaction rate coefficients are listed in Table 7.2 and an overview of all parameters used during the fitting procedure is given in Table C.3.

The concentration dependence of $k_{A'}$ can also be expressed in the form of a power law as $k_{A'} \propto [\mathbf{G}]_{\mathrm{t}}^n$ (for constant $[\mathbf{H}]_{\mathrm{t}}$). The exponent n can easily be extracted from the log-log plot in Fig. 7.12c as the gradient (first derivative) of the red curve. The concentration dependence of the power law exponent is shown in Fig. 7.12e. It can be seen that for low and high guest concentrations $n = 1$. However, between these limiting cases, the power law exponent reaches values over $n = 3$. It is interesting to mention that the $k_{A'}$ dependence on free guest concentration [**G**] has a simple linear relationship (see Eq. (7.3a)) while its dependence on total guest concentration $[\mathbf{G}]_{\mathrm{t}}$ has a nonlinear form with the largest deviation from linearity around 1 equiv. of total guest concentration (Fig. 7.12e).

The chemical kinetics, according to Scheme 7.2b, can also be viewed in terms of the Gibbs energy landscape as shown in Fig. 7.12h. The energy barriers are calculated from the corresponding reaction rate coefficients using the Eyring equation, Eq. (2.20), at $T = 298$ K with the transition probability $\eta = 1$. Standard reaction Gibbs energies for both **HG** and **HW** complexes were calculated from the equilibrium constants $K_{\mathbf{HG}}^{\mathrm{micro}}$ and $K_{\mathbf{HW}}$, respectively. All parameters of the Gibbs energy profile are listed in Table 7.2. The barrier between states C and D was not determined since it is very low (i.e. fast exchange regime).

Table 7.2 Parameters of Gibbs energy profile at $T = 298$ K. In the Eyring equation (Eq. (2.20)), the assumption for transition probability $\eta = 1$ was used

Equilibrium constants		Std. Gibbs energy of reaction		Mutual relationship
$K_{\mathbf{HG}}$	$(7.4 \pm 3.0) \times 10^4\ \mathrm{M}^{-1}$	$\Delta G^\circ_{\mathbf{HG}}$	$-27.8 \pm 1.0\ \mathrm{kJ.mol}^{-1}$	$\Delta G^\circ_{\mathbf{HG}} = -RT \ln K_{\mathbf{HG}}$
$K^{\mathrm{micro}}_{\mathbf{HG}}$	$(1.9 \pm 0.8) \times 10^4\ \mathrm{M}^{-1}$	$\Delta G^{\circ\mathrm{micro}}_{\mathbf{HG}}$	$-24.3 \pm 1.0\ \mathrm{kJ.mol}^{-1}$	$\Delta G^{\circ\mathrm{micro}}_{\mathbf{HG}} = -RT \ln K^{\mathrm{micro}}_{\mathbf{HG}}$
$K_{\mathbf{HW}}$	$240 \pm 35\ \mathrm{M}^{-1}$	$\Delta G^\circ_{\mathbf{HW}}$	$-13.6 \pm 0.4\ \mathrm{kJ.mol}^{-1}$	$\Delta G^\circ_{\mathbf{HW}} = -RT \ln K_{\mathbf{HW}}$
Reaction rate coefficients		Gibbs energy of barrier		Mutual relationship
κ_{diss}	$47 \pm 6\ \mathrm{s}^{-1}$	$\Delta G^\ddagger_{\mathrm{diss}}$	$63.4 \pm 0.3\ \mathrm{kJ.mol}^{-1}$	$\Delta G^\ddagger_{\mathrm{diss}} = -RT \ln \frac{h\kappa_{\mathrm{diss}}}{k_B T}$
κ_{ass}	$(9 \pm 4) \times 10^5\ \mathrm{M}^{-1}\mathrm{s}^{-1}$	$\Delta G^\ddagger_{\mathrm{ass}}$	$39.0 \pm 1.1\ \mathrm{kJ.mol}^{-1}$	$\Delta G^\ddagger_{\mathrm{ass}} = -RT \ln \frac{h\kappa_{\mathrm{ass}}}{k_B T}$
κ_{taut}	$(10 \pm 1) \times 10^5\ \mathrm{M}^{-1}\mathrm{s}^{-1}$	$\Delta G^\ddagger_{\mathrm{taut}}$	$38.7 \pm 0.2\ \mathrm{kJ.mol}^{-1}$	$\Delta G^\ddagger_{\mathrm{taut}} = -RT \ln \frac{h\kappa_{\mathrm{taut}}}{k_B T}$

7.3 Dynamic Processes Summary

Multiple processes have been identified in **OxP** and **Bz$_2$OxP**, which is summarized and schematically illustrated in Tables 7.4 and 7.5, respectively. NMR spectra of **Bz$_4$OxP** in the presence of DFA do not show any dynamic behavior since the tautomeric processes are fast (OH signal due to protonated **Bz$_4$OxP** is broadened and thus not observable), anion binding at the center is prevented by N-alkylation, and observation of topomerization is impossible due to high symmetry of **Bz$_4$OxP**. Our analyses quantified some of these processes in terms of transition rate coefficients summarized in Table 7.3, reaction rate coefficients and energy barrier heights. These results were obtained in two steps: **(i)** lineshape fitting of NMR spectra to obtain transition rate coefficients and **(ii)** fitting of binding and exchange models to obtain reaction rate coefficients, binding constants and other parameters included in binding

Table 7.3 Overview of transition rate coefficients. Host-guest binding was characterized at low guest concentration, where the free host signal can be observed. Question mark stands for the inability to obtain the corresponding parameter from current experiments

	OxP		Bz$_2$OxP	
Process	DFA	CSA	DFA	CSA
Host-guest bind	$k_{\mathrm{ass}}, k_{\mathrm{diss}} < 5\ \mathrm{s}^{-1}$	Slow	Intermediate	$k_{\mathrm{diss}} = 47\ \mathrm{s}^{-1}$, $k_{\mathrm{ass}} < 200\ \mathrm{s}^{-1}$
Anion exchange	Undetectable	?	Undetectable	Fast at acid excess
Prototrop. tautomer	$k_{\mathrm{taut}} < 10^5\ \mathrm{s}^{-1}$	$k_{\mathrm{taut}} < 2000\ \mathrm{s}^{-1}$	?	$k_{\mathrm{taut}} < 2 \times 10^5\ \mathrm{s}^{-1}$
Macrocyc. inv	Undetectable	Not occurring	Impossible	Impossible
Hemiquin. topomer	Undetectable	Undetectable	$k_{\mathrm{topo}} < 150\ \mathrm{s}^{-1}$	Slow

Table 7.4 Schematic overview of all processes in **OxP**. When the processes can be realized by multiple ways, only one instance is shown as illustration. Transition states are not shown

process	illustration	notes
host-guest binding	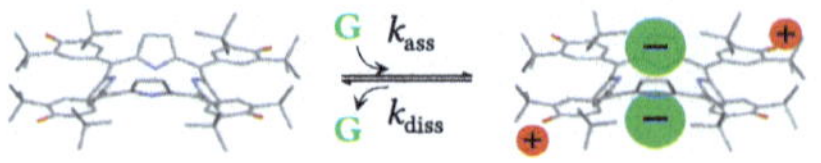	in the binding step **H→HG$_2$**
host-guest binding at acid excess	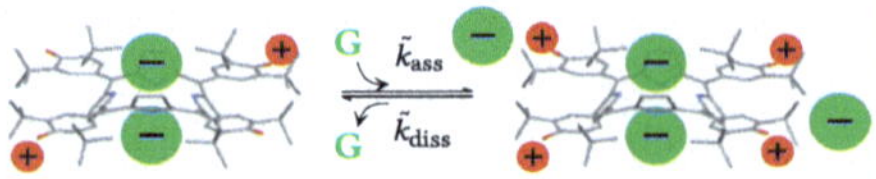	in the binding step **HG$_2$→HG$_4$**
anion exchange	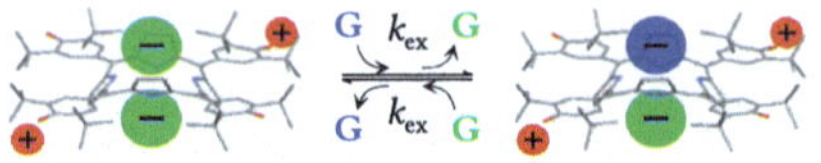	shares transition states with prototropic tautomerization; in **HG$_2$**
anion exchange at acid excess	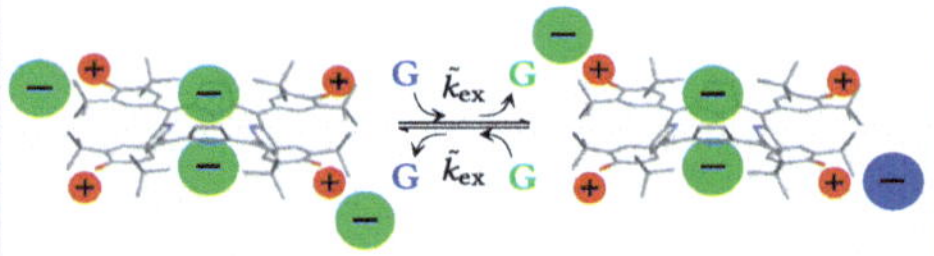	either the center anion or the guest at the edge can be exchanged; in **HG$_4$**
prototropic tautomerization	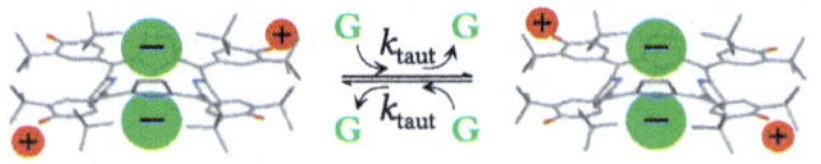	mediated by either (i) free guest, (ii) water or (iii) host-host collisions; involved temperature dependence; in **HG$_2$**
prototropic tautomerization at acid excess	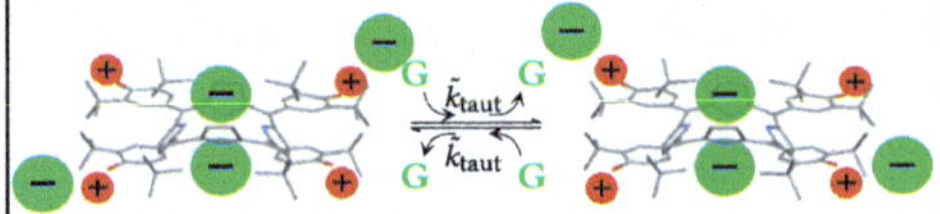	in **HG$_4$**
macrocyclic inversion	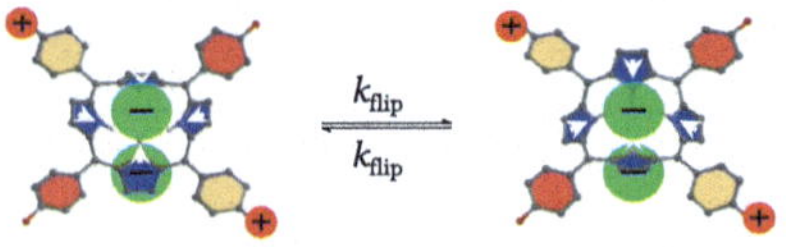	activated by the acid binding; in **HG$_2$**

isotherms (see Sect. 3.4.2). These steps are described in detail in Appendix C.7 for the analysis of NH signals in **Bz$_2$OxP** + (*R*)-CSA titration.

The first process in Table 7.3 is *dynamic host-guest binding*. This process could not be accurately quantified in **OxP** due to the slow exchange regime, but the corresponding transition rate coefficients were determined in **Bz$_2$OxP** from NH resonances (see Sect. 7.2.3). Dissociation of the complex is unimolecular reaction, thus, k_{diss} does not depend on concentration, while k_{ass} representing a bimolecular reaction is proportional to [**G**]. The *anion exchange process* k_{ex} changes the host spin states only when chiral and not enantiopure guest is present in the sample. This condition was fulfilled for **Bz$_2$OxP** + CSA systems, which revealed fast anion exchange at acid excess (see Sect. 7.2.2 or [2]). *Prototropic tautomerization* is mediated by free guest molecules,

Table 7.5 Schematic overview of all processes in **Bz$_2$OxP**. When the processes can be realized by multiple ways, only one instance is shown as illustration. Proposed transition states of prototropic tautomerization and anion exchange are shown in Sect. C.9

process	illustration	notes
host-guest binding	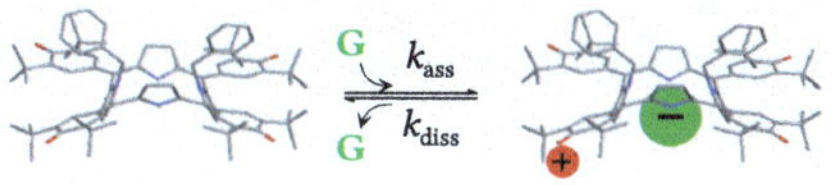	in the binding step **H→HG**
host-guest binding at acid excess	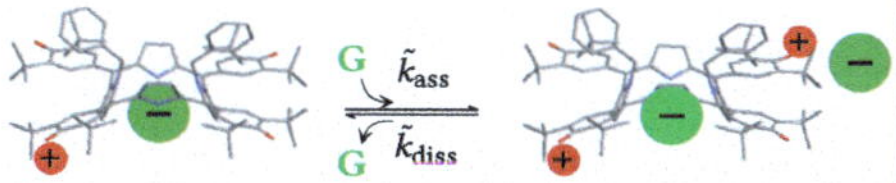	in the binding step **HG→HG$_2$** or **HG$_2$→HG$_4$**
anion exchange	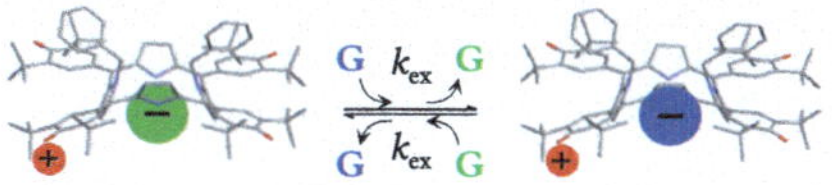	shares transition states with prototropic tautomerization; in **HG**
anion exchange at acid excess	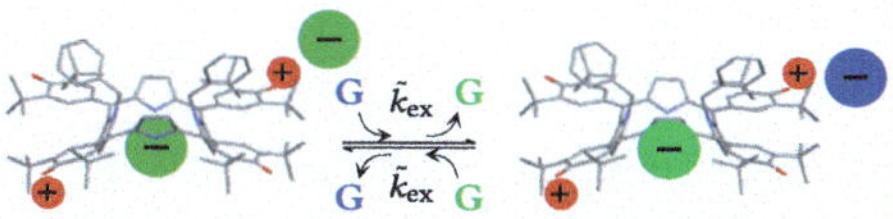	either center anion or the guest at the edge can be exchanged; in **HG$_2$** or **HG$_4$**
prototropic tautomerization	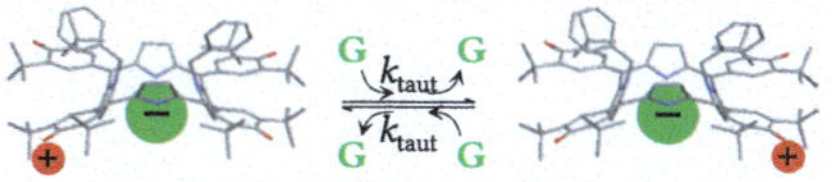	in **HG**
prototropic tautomerization at acid excess	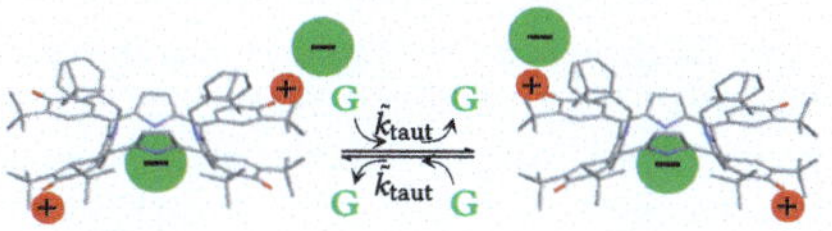	in **HG$_2$** or **HG$_4$**
hemiquinonoid topomerization	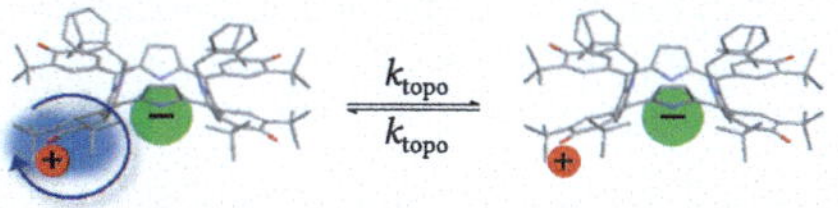	rotation around single bond; activated by protonation; in **HG**, **HG$_2$** or **HG$_4$**

the concentration dependence was measured in both **OxP** and **Bz$_2$OxP**. In **Bz$_2$OxP** + (*R*)-CSA system, the concentration dependence was successfully described by a kinetics model (see Sect. 7.2.3 or our article [1]). *Macrocyclic inversion* is an unexpected dynamic process taking part in **OxP** in the presence of mandelic acid reported in [3]. However, this process was not observed in the **OxP** + CSA system probably due to strong host-guest binding, which prevented formation of the neutral flip-active form. In the presence of DFA it cannot be decided whether this process takes place because the host spin states are not changed due to mirror symmetry of DFA (see Sect. 7.1.3). Finally, rotation of the bulky side groups referred to as *hemiquinonoid topomerization* was detected in **Bz$_2$OxP**. This process is activated by protonation, and thus shows complicated dependence on concentration when host-guest complex with higher than 1:1 stoichiometry is formed (see Sect. 7.2.1).

Noteworthy, high symmetry of **OxP** in combination with mirror symmetry of DFA hinders the observability of some processes because the corresponding nuclear spins undergo chemical exchange between identical chemical environments (first column of Table 7.3). On the other hand, N-alkylation lowers the symmetry of **Bz_2OxP**, which in combination with asymmetric (*R*)-CSA allows, in theory, observation of all processes (last column of Table 7.3).

References

1. Březina V, Hanyková L, Velychkivska N, Hill JP, Labuta J (2022) NMR lineshape analysis using analytical solutions of multi-state chemical exchange with applications to kinetics of host-guest systems. Sci Rep 12:17369
2. Březina V (2014) Spectroscopic and theoretical study of supramolecular complexes of symmetric porphyrins with chiral guests. MA thesis, Univerzita Karlova, Matematicko-fyzikální fakulta May
3. Labuta J, Futera Z, Ishihara S, Kouřilová H, Tateyama Y, Ariga K, Hill JP (2014) Chiral guest binding as a probe of macrocycle dynamics and tautomerism in a conjugated tetrapyrrole. J Am Chem Soc 136:2112–2118
4. Labuta J, Ishihara S, Ariga K, Hill JP (2014) Dynamic processes in prochiral solvating agents (pro-CSAs) studied by NMR spectroscopy. Symmetry 6:345–367
5. Labuta J, Ishihara S, Šikorský T, Futera Z, Shundo A, Hanyková L, Burda JV, Ariga K, Hill JP (2013) NMR spectroscopic detection of chirality and enantiopurity in referenced systems without formation of diastereomers. Nat Commun 4:2188
6. Shundo A, Labuta J, Hill JP, Ishihara S, Ariga K (2009) Nuclear magnetic resonance signaling of molecular chiral information using an achiral reagent. J Am Chem Soc 131:9494–9495
7. Casarini D, Lunazzi L, Mazzanti A (2010) Recent advances in stereodynamics and conformational analysis by dynamic NMR and theoretical calculations. Eur J Org Chem 2010:2035–2056

Chapter 8
Conclusion

OxP, **Bz_2OxP** and **Bz_4OxP** serve as colorimetric sensors for acids in nonpolar media in wide range of acid volume fractions. The halochromic response of **OxP** (φ(DFA) $\in [10^{-6}, 10^{-3}]$) is very sensitive to small amounts of acid (detects ca. 1 ppm of DFA), however, it also saturates at relatively low acid contents. The color response is shifted to higher acid contents in **Bz_2OxP** (φ(DFA) $\in [10^{-4}, 1]$) and **Bz_4OxP** (φ(DFA) $\in [0.01, 0.6]$) due to N-alkylation (Fig. 6.6). The color changes were permitted by protonation and subsequent anion binding (i.e. host-guest binding), which was studied by UV/vis and NMR spectroscopy in Chap. 6. UV/vis spectra were analyzed using SVD to determine the number of absorbing components and using DSIS to obtain the components' populations. Up to fourfold protonation in **OxP** and **Bz_2OxP**, and up to twofold protonation in **Bz_4OxP** was observed, all UV/vis spectra of the individual protonated species were obtained (Figs. 6.7 and 6.19). Experimental binding isotherms were constructed from both UV/vis and NMR spectroscopic data and fitted with appropriate host-guest binding models (applicable at φ(DFA) < 0.001) to obtain binding constants (Table 6.1). It turns out that N-alkylation in **Bz_2OxP** and **Bz_4OxP** reduces the binding affinity in two ways: (**i**) it blocks the binding sites for acid anions at the porphyrinogen center and (**ii**) prevents redistribution of charge, which would stabilize the protonated species. Upon host protonation, the acid anions are located near the charged complex because the nonpolar solvent does not allow full dissociation.

Solvatochromic effects were studied in DMF/chloroform mixture. High contents of DMF increase the solvent polarity and induce hydrogen bonding of DMF to the oxoporphyrinogen host, causing positive solvatochromic shifts in UV/vis spectra of all investigated oxoporphyrinogens. Analogously to halochromism, solvatochromic response threshold is shifted to higher DMF concentrations with increasing degree of N-alkylation (Fig. 6.15). The spectral changes in **OxP** and **Bz_2OxP** could be modeled as interconversion of individual absorbing species, which was further confirmed by SVD. Besides solvatochromic shifts, **OxP** showed transition to another form with

V. Březina, *Supramolecular Complexes of Oxoporphyrinogens with Organic Molecules*, Springer Theses, https://doi.org/10.1007/978-3-031-68646-7_8

distinct absorption bands assigned to porphodimethene form, which was stable only at particular DMF concentration range. Detailed summary of chromic and binding properties is given in Sect. 6.9.

Various molecular dynamic processes are manifested as chemical exchange in NMR spectra of protonated **OxP** and **Bz$_2$OxP** as shown in Chap. 7. In both oxoporphyrinogens, we identified the processes of anion exchange and prototropic tautomerization, also the host-guest binding was confirmed to be a dynamic process (Figs. 7.4 and 7.5). Protonation also activates rotation of bulky side groups (i.e. process of hemiquinonoid topomerization) as observed in **Bz$_2$OxP**. This process cannot be detected in **OxP** or **Bz$_4$OxP** due to their high symmetry. NMR spectra of protonated **Bz$_4$OxP** do not show any chemical exchange due to its symmetry and fast rate of the molecular processes.

To enable fitting of three-state chemical exchange lineshapes in NMR spectra (in the absence of J-coupling), we calculated the analytical formula, which has not been published before (Sect. 3.3.7). We also devised so-called 'reduced equivalent schemes' for exchange schemes containing a fast-exchanging state in order to reduce the number of parameters for lineshape fitting (Sect. 3.3.8). Furthermore, we derived an approximative formula to determine the coalescence point in symmetric two-state exchange with nonzero transverse relaxation rate (Eq. (3.20)).

Using these theoretical instruments, temperature and concentration dependence of transition rate coefficients corresponding to the present molecular processes was obtained. Effect of temperature on the topomerization process in protonated **Bz$_2$OxP** obeys Eyring equation (Table 7.1), however, transition rate coefficient of tautomerization in protonated **OxP** shows non-monotonous temperature dependence. Concentration dependence of transition rate coefficients in **Bz$_2$OxP** systems has been described in detail using a kinetic model derived from competitive host-guest binding (Scheme C.2 and Table 7.2). Note that chirality of the acid influences the number of exchanging states in NMR spectra because the bound acid anion determines overall symmetry of the complex through the so-called chiral field. Detailed summary of dynamic processes in the oxoporphyrinogens is given in Sect. 7.3.

OxP and its derivatives have potential practical applications as colorimetric sensors in nonpolar media. N-alkylation modulates the color response so that the full acid volume fraction (φ(DFA) $\in [10^{-6}, 1]$) is covered by either of the compounds. Similar structure of the compounds assures matching solubility, thus, these sensors might be applied as blends analogously to the Universal Indicator.

Our NMR study of dynamic processes in the oxoporphyrinogens contributes to understanding the concentration dependence of transition rate coefficients, which is always system-specific. In this regard, models derived from chemical kinetics proved to be useful. We have also shown the advantage of reduced equivalent schemes, derived from analytical exchange lineshapes. The presented methods can be used to describe molecular kinetics in a wide range of systems with a variety of intra- or intermolecular processes, such as those involving host-guest binding, enzymatic reactions or nontrivial conformational dynamics of proteins or other complex molecules.

We also provide parts of the code in the *Mathematica* software, which were used in our analyses. Appendix A offers an overview of host-guest binding models, Appendix B provides an overview of various NMR spectral lineshapes with or without the presence of chemical exchange including the derivation of multi-state exchange lineshape.

Correction to: Supramolecular Complexes of Oxoporphyrinogens with Organic Molecules

Correction to:
V. Březina, *Supramolecular Complexes of Oxoporphyrinogens with Organic Molecules*, Springer Theses,
https://doi.org/10.1007/978-3-031-68646-7

In the original version of the book, the following corrections have been incorporated: The author name "Václav Bezina" has been changed to "Václav Březina" in the Frontmatter, Backmatter and in Chapter number level. The book has been updated with the changes.

The updated version of this book can be found at
https://doi.org/10.1007/978-3-031-68646-7

V. Březina, *Supramolecular Complexes of Oxoporphyrinogens with Organic Molecules*, Springer Theses, https://doi.org/10.1007/978-3-031-68646-7_9

Appendix A
Overview of Binding Models

In this section, *Mathematica* code for the calculation of binding isotherms is provided.

A.1 Binding Model 1:1

The simple 1:1 stoichiometry enables simple analytical solution. The concentrations are calculated as the following functions

```
(* H:G 1:1 *)
H[K1_,Ht_,Gt_] := (-1 - Gt K1 + Ht K1 + Sqrt[4 Ht K1 + (-1 - Gt K1 + Ht K1)^2])/(2 K1);
HG[K1_,Ht_,Gt_] := (1 + Gt K1 + Ht K1 - Sqrt[4 Ht K1 + (-1 - Gt K1 + Ht K1)^2])/(2 K1);
G[K1_,Ht_,Gt_] := (-1 + Gt K1 - Ht K1 + Sqrt[4 Ht K1 + (-1 - Gt K1 + Ht K1)^2])/(2 K1);
```

Using artificial data, numeric values can be obtained as follows

```
(* creating artificial data: *)
Htdata = ConstantArray[.001, 100];
Gtdata = Subdivide[0, .01, 99];
K1 = 5000;
(* calculating the concentrations *)
Gdata=G[K1, Htdata, Gtdata];
Hdata=H[K1, Htdata, Gtdata];
HGdata=HG[K1, Htdata, Gtdata];
```

A.2 Binding Model 1:2

In principle, this model can be solved analytically using the formula for cubic equation. However, numerical approach from [1] can be applied universally as explained in Sect. 2.2.2. The derivation and implementation is shown here.

V. Březina, *Supramolecular Complexes of Oxoporphyrinogens with Organic Molecules*, Springer Theses, https://doi.org/10.1007/978-3-031-68646-7

First, the host-related species (i.e. [**H**] and [$\mathbf{HG}_j$]) and $[\mathbf{G}]_t$ are expressed as functions of [**G**]. Then the free guest concentration is calculated numerically from a polynomial equation in the form $[\mathbf{G}]_t = [\mathbf{G}]_t([\mathbf{G}], [\mathbf{H}]_t, K_j)$. The equations for stepwise binding constants and total mass balance are

```
(* H:G 1:2 *)
equations = {K1 == HG/(H G), K2 == HG2/(HG G), Ht == H + HG + HG2, Gt == G + HG + 2 HG2};
solutionRule = Solve[And @@ equations, {H, HG, HG2, Gt}][[1]]
```

This solution provides formulae for host-related species

```
H[K1_, K2_, Ht_, G_] := Ht/(1 + G K1 + G^2 K1 K2);
HG[K1_, K2_, Ht_, G_] := G Ht K1/(1 + G K1 + G^2 K1 K2);
HG2[K1_, K2_, Ht_, G_] := G^2 Ht K1 K2/(1 + G K1 + G^2 K1 K2);
```

The concentration [**G**] can be calculated from a polynomial equation, which can be obtained from the solution for $[\mathbf{G}]_t$ by the following procedure

```
polynomEq = (Gt == (Gt /. solutionRule));
polynomEq = MultiplySides[polynomEq, Denominator[(Gt /. solutionRule)]][[1, 1, 1]];
polynomEq = SubtractSides[polynomEq, polynomEq[[2]]];
polynomEq = Collect[polynomEq, G]  (* polynomial equation for [G] *)
```

which yields the `polynomEq` in the form

```
Gt + G (-1 + Gt K1 - Ht K1) - G^3 K1 K2 + G^2 (-K1 + Gt K1 K2 - 2 Ht K1 K2) == 0
```

The code for calculation of [**G**] is then obtained by

```
G[K1_?NumericQ, K2_?NumericQ, Ht_, Gt_] :=
G /. FindRoot[{Gt + G (-1 + Gt K1 - Ht K1) - G^3 K1 K2 + G^2 (-K1 + Gt K1 K2 - 2 Ht K1 K2)
== 0}, {G, Gt}];
SetAttributes[G, Listable];
```

The command `SetAttributes[G, Listable]` allows accepting values of `Ht` and `Gt` as lists of numbers. For the use in fitting procedure, the check for numeric value `?NumericQ` must be added to input arguments.

Using artificial data, numeric values can be obtained as follows

```
(* creating artificial data: *)
Htdata = ConstantArray[.001, 100];
Gtdata = Subdivide[0, .01, 99];
K1 = 5000; K2 = 1000;
(* calculating the concentrations *)
Gdata=G[K1, K2, Htdata, Gtdata];
Hdata=H[K1, K2, Htdata, Gdata];
HGdata=HG[K1, K2, Htdata, Gdata];
HG2data=HG2[K1, K2, Htdata, Gdata];
```

A.3 Binding Model 1:3

The equations describing the model are

```
(* H:G 1:3 *)
equations = {K1 == HG/(H G), K2 == HG2/(HG G), K3 == HG3/(HG2 G),
   Ht == H + HG + HG2 + HG3, Gt == G + HG + 2 HG2 + 3 HG3};
solutionRule = Solve[And @@ equations, {H, HG, HG2, HG3, Gt}][[1]]
```

Polynomial equation for [**G**] is derived analogously to the previous section, the resulting code is

```
H[K1_, K2_, K3_, Ht_, G_] := Ht/(1 + G K1 + G^2 K1 K2 + G^3 K1 K2 K3);
HG[K1_, K2_, K3_, Ht_, G_] := G Ht K1/(1 + G K1 + G^2 K1 K2 + G^3 K1 K2 K3);
HG2[K1_, K2_, K3_, Ht_, G_] := G^2 Ht K1 K2/(1 + G K1 + G^2 K1 K2 + G^3 K1 K2 K3);
HG3[K1_, K2_, K3_, Ht_, G_] := G^3 Ht K1 K2 K3/(1 + G K1 + G^2 K1 K2 + G^3 K1 K2 K3);
G[K1_?NumericQ, K2_?NumericQ, K3_?NumericQ, Ht_, Gt_] :=
  G /. FindRoot[{Gt + G (-1 + Gt K1 - Ht K1) + G^2 (-K1 + Gt K1 K2 - 2 Ht K1 K2)
  - G^4 K1 K2 K3 + G^3 (-K1 K2 + Gt K1 K2 K3 - 3 Ht K1 K2 K3) == 0}, {G, Gt}];
SetAttributes[G, Listable];
```

Using artificial data, numeric values can be obtained as follows

```
(* creating artificial data: *)
Htdata = ConstantArray[.001, 100];
Gtdata = Subdivide[0, .01, 99];
K1 = 5000; K2 = 1000; K3 = 200;
(* calculating the concentrations *)
Gdata=G[K1, K2, K3, Htdata, Gtdata];
Hdata=H[K1, K2, K3, Htdata, Gdata];
HGdata=HG[K1, K2, K3, Htdata, Gdata];
HG2data=HG2[K1, K2, K3, Htdata, Gdata];
HG3data=HG3[K1, K2, K3, Htdata, Gdata];
```

A.4 Binding Model 1:4

The equations describing the model are

```
(* H:G 1:4 *)
equations = {K1 == HG/(H G), K2 == HG2/(HG G), K3 == HG3/(HG2 G), K4 == HG4/(HG3 G),
 Ht == H + HG + HG2 + HG3 + HG4, Gt == G + HG + 2 HG2 + 3 HG3 + 4 HG4};
solutionRule = Solve[And @@ equations, {H, HG, HG2, HG3, HG4, Gt}][[1]]
```

Polynomial equation for [**G**] is derived analogously to previous sections, the resulting code is

```
H[K1_, K2_, K3_, K4_, Ht_, G_] :=
  Ht/(1 + G K1 + G^2 K1 K2 + G^3 K1 K2 K3 + G^4 K1 K2 K3 K4);
HG[K1_, K2_, K3_, K4_, Ht_, G_] :=
  G Ht K1/(1 + G K1 + G^2 K1 K2 + G^3 K1 K2 K3 + G^4 K1 K2 K3 K4);
HG2[K1_, K2_, K3_, K4_, Ht_, G_] :=
  G^2 Ht K1 K2/(1 + G K1 + G^2 K1 K2 + G^3 K1 K2 K3 + G^4 K1 K2 K3 K4);
HG3[K1_, K2_, K3_, K4_, Ht_, G_] :=
  G^3 Ht K1 K2 K3/(1 + G K1 + G^2 K1 K2 + G^3 K1 K2 K3 + G^4 K1 K2 K3 K4);
```

```
HG4[K1_, K2_, K3_, K4_, Ht_, G_] :=
  G^4 Ht K1 K2 K3 K4/(1 + G K1 + G^2 K1 K2 + G^3 K1 K2 K3 + G^4 K1 K2 K3 K4);
G[K1_?NumericQ, K2_?NumericQ, K3_?NumericQ, K4_?NumericQ, Ht_, Gt_] :=
  G /. FindRoot[{Gt + G (-1 + Gt K1 - Ht K1) + G^2 (-K1 + Gt K1 K2 - 2 Ht K1 K2) +
    G^3 (-K1 K2 + Gt K1 K2 K3 - 3 Ht K1 K2 K3) - G^5 K1 K2 K3 K4 +
      G^4 (-K1 K2 K3 + Gt K1 K2 K3 K4 - 4 Ht K1 K2 K3 K4) == 0}, {G, Gt}];
SetAttributes[G, Listable];
```

Using artificial data, numeric values can be obtained as follows

```
(* creating artificial data: *)
Htdata = ConstantArray[.001, 100];
Gtdata = Subdivide[0, .01, 99];
K1 = 5000; K2 = 1000; K3 = 200; K4 = 300;
(* calculating the concentrations *)
Gdata=G[K1, K2, K3, K4, Htdata, Gtdata];
Hdata=H[K1, K2, K3, K4, Htdata, Gdata];
HGdata=HG[K1, K2, K3, K4, Htdata, Gdata];
HG2data=HG2[K1, K2, K3, K4, Htdata, Gdata];
HG3data=HG3[K1, K2, K3, K4, Htdata, Gdata];
HG4data=HG4[K1, K2, K3, K4, Htdata, Gdata];
```

A.5 Competitive Binding 1:1 Host-Ligand Binding

The competitive binding of a host **H** with two different ligands **G** and **W** is described by the following scheme with the corresponding governing equations

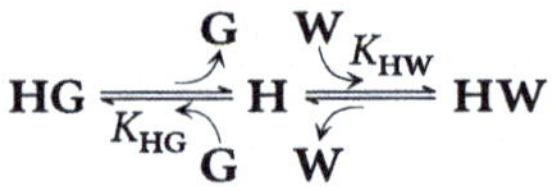

$$K_{\mathbf{HG}} = \frac{[\mathbf{HG}]}{[\mathbf{H}][\mathbf{G}]}, \tag{A.1a}$$

$$K_{\mathbf{HW}} = \frac{[\mathbf{HW}]}{[\mathbf{H}][\mathbf{W}]}, \tag{A.1b}$$

$$[\mathbf{H}]_t = [\mathbf{H}] + [\mathbf{HG}] + [\mathbf{HW}], \tag{A.1c}$$

$$[\mathbf{G}]_t = [\mathbf{G}] + [\mathbf{HG}], \tag{A.1d}$$

$$[\mathbf{W}]_t = [\mathbf{W}] + [\mathbf{HW}]. \tag{A.1e}$$

A.5.1 First Solution

First, the concentrations of host-related species are expressed as functions of [**G**] and [**W**] (instead of $[\mathbf{G}]_t$ and $[\mathbf{W}]_t$) from the governing Eq. (A.1), yielding

$$[\mathbf{H}] = \frac{[\mathbf{H}]_t}{1 + K_{\mathbf{HG}}[\mathbf{G}] + K_{\mathbf{HW}}[\mathbf{W}]}, \tag{A.2a}$$

$$[\mathbf{HG}] = \frac{K_{\mathbf{HG}}[\mathbf{G}][\mathbf{H}]_t}{1 + K_{\mathbf{HG}}[\mathbf{G}] + K_{\mathbf{HW}}[\mathbf{W}]}, \tag{A.2b}$$

$$[\mathbf{HW}] = \frac{K_{\mathbf{HW}}[\mathbf{W}][\mathbf{H}]_t}{1 + K_{\mathbf{HG}}[\mathbf{G}] + K_{\mathbf{HW}}[\mathbf{W}]}. \tag{A.2c}$$

In the next step, substitution of Eq. (A.2) into Eq. (A.1) yields a set of two second-order polynomial equations in [**G**] and [**W**] after rearrangement,

$$0 = [\mathbf{G}] - [\mathbf{G}]_t + [\mathbf{G}]^2 K_{\mathbf{HG}} - [\mathbf{G}][\mathbf{G}]_t K_{\mathbf{HG}} + [\mathbf{G}][\mathbf{H}]_t K_{\mathbf{HG}} + [\mathbf{G}]K_{\mathbf{HW}}[\mathbf{W}] - [\mathbf{G}]_t K_{\mathbf{HW}}[\mathbf{W}], \tag{A.3a}$$

$$0 = [\mathbf{W}] + [\mathbf{G}]K_{\mathbf{HG}}[\mathbf{W}] + [\mathbf{H}]_t K_{\mathbf{HW}}[\mathbf{W}] + K_{\mathbf{HW}}[\mathbf{W}]^2 - [\mathbf{W}]_t - [\mathbf{G}]K_{\mathbf{HG}}[\mathbf{W}]_t - K_{\mathbf{HW}}[\mathbf{W}][\mathbf{W}]_t. \tag{A.3b}$$

This equation is solved numerically in the interval $([\mathbf{G}], [\mathbf{W}]) \in [0, [\mathbf{G}]_t] \times [0, [\mathbf{W}]_t]$.

In *Mathematica*, Eqs. (A.2) and (A.3) can be derived with the following code

```
(* host-ligand binding H:G 1:1, H:W 1:1 *)
equations = {KHG == HG/(H G), KHW == HW/(H W),
  Ht == H + HG + HW, Gt == G + HG, Wt == W + HW};
solutionRule = Solve[And @@ equations, {H, HG, HW, Gt, Wt}][[1]];
polynomEq1 = (Gt == (Gt /. solutionRule));
polynomEq1 =
  MultiplySides[polynomEq1, Denominator[(Gt /. solutionRule)]][[1, 1, 1]];
polynomEq1 = SubtractSides[polynomEq1, polynomEq1[[2]]];
polynomEq1 = Collect[polynomEq1, G] (* 1st polynomial equation for [G] and [W] *)
polynomEq2 = (Wt == (Wt /. solutionRule));
polynomEq2 =
  MultiplySides[polynomEq2, Denominator[(Wt /. solutionRule)]][[1, 1, 1]];
polynomEq2 = SubtractSides[polynomEq2, polynomEq2[[2]]];
polynomEq2 = Collect[polynomEq2, W] (* 2nd polynomial equation for [G] and [W] *)
```

Then the model can be implemented using

```
H[KHG_, KHW_, Ht_, G_, W_] := Ht/(1 + G KHG + KHW W);
HG[KHG_, KHW_, Ht_, G_, W_] := G Ht KHG/(1 + G KHG + KHW W);
HW[KHG_, KHW_, Ht_, G_, W_] := Ht KHW W/(1 + G KHG + KHW W);
G[KHG_?NumericQ, KHW_?NumericQ, Ht_, Gt_, Wt_] :=
  G /. FindRoot[{Gt - G^2 KHG + Gt KHW W + G (-1 + Gt KHG - Ht KHG - KHW W) == 0,
    -KHW W^2 + Wt + G KHG Wt + W (-1 - G KHG - Ht KHW + KHW Wt) == 0},
      {{G, Gt}, {W, Wt}}];
W[KHG_?NumericQ, KHW_?NumericQ, Ht_, Gt_, Wt_] :=
  W /. FindRoot[{Gt - G^2 KHG + Gt KHW W + G (-1 + Gt KHG - Ht KHG - KHW W) == 0,
    -KHW W^2 + Wt + G KHG Wt + W (-1 - G KHG - Ht KHW + KHW Wt) == 0},
      {{G, Gt}, {W, Wt}}];
SetAttributes[G, Listable];
SetAttributes[W, Listable];
```

Using artificial data, numeric values can be obtained as follows

```
(* creating artificial data: *)
Htdata = ConstantArray[.001, 100];
Gtdata = Subdivide[0, .01, 99];
Wtdata = ConstantArray[.05, 100];
KHG = 5000; KHW = 50;
(* calculating the concentrations *)
Gdata=G[KHG, KHW, Htdata, Gtdata, Wtdata];
Wdata=W[KHG, KHW, Htdata, Gtdata, Wtdata];
Hdata=H[KHG, KHW, Htdata, Gdata, Wdata];
HGdata=HG[KHG, KHW, Htdata, Gdata, Wdata];
HWdata=HW[KHG, KHW, Htdata, Gdata, Wdata];
```

A.5.2 Second Solution

Instead of numeric solution of set of two second-order polynomial equations as in the previous section, this method results in numeric solutions of one third-order polynomial equation, which is more numerically stable in some circumstances. Derivation of the equations described in Sect. 2.2.3 is implemented in *Mathematica* in the following code

```
(* host-ligand binding H:G 1:1, H:W 1:1 *)
equations = {KHG == HG/(H G), KHW == HW/(H W), Ht == H + HG + HW,
   Gt == G + HG, Wt == W + HW};
 solutionRule = Solve[And @@ equations, {HG, HW, G, W, Ht}][[1]]
polynomEq = (Ht == (Ht /. solutionRule));
polynomEq =
  MultiplySides[polynomEq, Denominator[(Ht /. solutionRule)]][[1, 1, 1]];
polynomEq = SubtractSides[polynomEq, polynomEq[[2]]];
polynomEq = Collect[polynomEq, H] (* polynomial equation for [H] *)
```

Then the model can be implemented using

```
H[KHG_?NumericQ, KHW_?NumericQ, Ht_, Gt_, Wt_] :=
  H /. FindRoot[{Ht - H^3 KHG KHW + H (-1 - Gt KHG + Ht KHG + Ht KHW - KHW Wt) +
      H^2 (-KHG - KHW - Gt KHG KHW + Ht KHG KHW - KHG KHW Wt) == 0}, {H, Ht}];
HG[KHG_?NumericQ, KHW_?NumericQ, H_, Gt_, Wt_] := Gt H KHG/(1 + H KHG);
HW[KHG_?NumericQ, KHW_?NumericQ, H_, Gt_, Wt_] := H KHW Wt/(1 + H KHW);
G[KHG_?NumericQ, KHW_?NumericQ, H_, Gt_, Wt_] := Gt/(1 + H KHG);
W[KHG_?NumericQ, KHW_?NumericQ, H_, Gt_, Wt_] := Wt/(1 + H KHW);
SetAttributes[H, Listable];
```

Using artificial data, numeric values can be obtained as follows

```
(* creating artificial data: *)
Htdata = ConstantArray[.001, 100];
Gtdata = Subdivide[0, .01, 99];
Wtdata = ConstantArray[.05, 100];
KHG = 5000; KHW = 50;
(* calculating the concentrations *)
Hdata=H[KHG, KHW, Htdata, Gtdata, Wtdata];
HGdata=HG[KHG, KHW, Hdata, Gtdata, Wtdata];
HWdata=HW[KHG, KHW, Hdata, Gtdata, Wtdata];
Gdata=G[KHG, KHW, Hdata, Gtdata, Wtdata];
Wdata=W[KHG, KHW, Hdata, Gtdata, Wtdata];
```

Reference

1. Hargrove AE, Zhong Z, Sessler JL, Anslyn EV (2010) Algorithms for the determination of binding constants and enantiomeric excess in complex host: guest equilibria using optical measurements. New Journal of Chemistry 34:348–354.

Appendix B
Overview of NMR Spectral Lineshapes

B.1 Lineshape Formulae and *Mathematica* Codes

In this appendix, an overview of basic NMR spectral lineshapes is given. For practical application in fitting of experimental spectra, real parts of the lineshapes $\mathrm{Re}[S(\Omega)]$ are implemented in *Mathematica* functions. Implementation in *MATLAB* can be found in [1]. Spectral lineshapes in Chap. 3 are constructed using angular frequency Ω in units of rad.s^{-1}, but experimental NMR spectra are usually measured as chemical shift δ in units of ppm. Therefore, conversion formula from ppm to rad.s^{-1} is present in all *Mathematica* functions calculating the spectral lineshapes in this appendix, and it reads

$$\Delta\Omega\ [\mathrm{s}^{-1}] = 2\pi \times \nu_0\ [\mathrm{MHz}] \times \Delta\delta\ [\mathrm{ppm}]\,, \tag{B.1}$$

where ν_0 is the spectrometer frequency in MHz, see Sect. 3.1.2. Be aware that this formula is true only for differences of angular frequencies $\Delta\Omega$ and differences of chemical shift $\Delta\delta$. The conversion factor `δToΩConvFac` (equal to $\Delta\Omega/\Delta\delta$) is preset to spectrometer frequency 500.13 MHz (corresponding to $B_0 = 11.7$ T, cf. Eq. (3.4)).

In the presence of chemical exchange, state populations must be specified. In the presented *Mathematica* functions, these are calculated internally from analytical solution of $\mathbf{Kp} = 0$, which implies equilibrium or steady-state populations. Note that the R_2 parameters correspond to transverse relaxation including the inhomogeneous broadening effect (see Sect. 3.1.3). Importantly, area below the curve of all presented spectral lineshapes does not depend on peak position Ω_0, transverse relaxation rate R_2, transition rate coefficients k_{ij} or on J-coupling constant J, and it can be obtained as

V. Březina, *Supramolecular Complexes of Oxoporphyrinogens with Organic Molecules*, Springer Theses, https://doi.org/10.1007/978-3-031-68646-7

$$\text{peak area} = \int_{-\infty}^{+\infty} \text{Re}[S(\Omega)]\,d\Omega = \pi M_0 \; \textit{(ang. freq. scale)}, \tag{B.2a}$$

$$\text{peak area} = \int_{-\infty}^{+\infty} \text{Re}[S(\delta)]\,d\delta = \frac{M_0}{2\nu_0\,[\text{MHz}]} \; \textit{(ppm scale)}. \tag{B.2b}$$

Equation (B.2) is obtained from Eq. (B.2) using the substitution $\Omega = 2\pi\nu_0\,[\text{MHz}]\delta$ from Eq. (B.1). Note that M_0 plays a role of arbitrary proportionality constant in measured spectra, it is not equal to the actual magnetization.

B.1.1 Lorentzian Lineshape

Lorentzian lineshape is detected in the absence of chemical exchange and J-coupling. The analytical lineshape formula is

$$S(\Omega) = \frac{M_0}{R_2 - i(\Omega - \Omega_0)} = \frac{M_0}{\alpha}, \tag{B.3}$$

where $\alpha = R_2 - i(\Omega - \Omega_0)$, with the real part

$$\text{Re}[S(\Omega)] = \frac{M_0 R_2}{R_2^2 + (\Omega - \Omega_0)^2}. \tag{B.4}$$

Figure B.1 shows that full width at half-maximum of the Lorentzian lineshape equals to $2R_2$ in the angular frequency scale and 2γ in the chemical shift scale. Thus, R_2 can be calculated from a NMR spectrum (measured in chemical shift scale) using

$$R_2\,[\text{s}^{-1}] = 2\pi \times \nu_0\,[\text{MHz}] \times \gamma\,[\text{ppm}], \tag{B.5}$$

resulting from the conversion formula in Eq. (B.1).

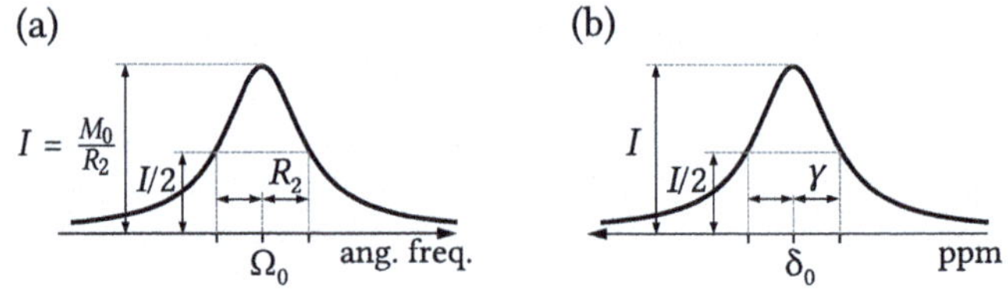

Fig. B.1 Illustration of Lorentzian lineshape with the horizontal axis in (**a**) angular frequency and (**b**) chemical shift scale

Code for implementation in *Mathematica* is

```
(* Lorentzian *)
lorentzianFunc[δ_, δ0_, R2_, M0_] :=
Module[{δToΩConvFac, Ω, Ω0},
 δToΩConvFac = 2π*500.13;
 Ω = δ*δToΩConvFac;
 Ω0 = δ0*δToΩConvFac;
M0 R2/(R2^2 + (Ω - Ω0)^2) ];
(* the use is illustrated in the following lines *)
deltaData = Range[0,2,.005];
absorptionSpectrum = lorentzianFunc[deltaData,(*δ0*)1,(*R2*)150,(*M0*)10];
```

Alternative formulation of the Lorentzian lineshape (real part) is

$$\mathrm{Re}[S(\Omega)] = \frac{I\gamma^2}{\gamma^2 + (\delta - \delta_0)^2}, \tag{B.6a}$$

$$\text{peak area} = \int_{-\infty}^{+\infty} \mathrm{Re}[S(\delta)]\,\mathrm{d}\delta = \pi\gamma I\ (ppm\ scale), \tag{B.6b}$$

where $I = M_0/R_2$. This is convenient for fitting of spectral lineshapes because variation in γ does not influence the peak height (which always equals I).

Mathematica implementation of Eq. (B.6) reads

```
(* Lorentzian - alternative formulation *)
lorentzianAltFunc[δ_, δ0_, γ_, I_] :=
I γ^2/(γ^2 + (δ - δ0)^2);
(* the use is illustrated in the following lines *)
deltaData = Range[0,2,.005];
absorptionSpectrum = lorentzianAltFunc[deltaData,(*δ0*)1,(*γ*)0.04,(*I*)0.25];
```

B.1.2 Two-State Exchange Lineshape

The analytical formula for two-state exchange lineshape is

$$S_{\text{two-state exch.}}(\Omega) = M_0 \frac{p_A\alpha_B + p_B\alpha_A + k_{AB} + k_{BA}}{\alpha_A\alpha_B + k_{AB}\alpha_B + k_{BA}\alpha_A}, \tag{B.7}$$

where $\alpha_j = R_2^j + i(\Omega - \Omega_j)$ for $j = A, B$. Equilibrium populations are calculated as

$$p_A^{\text{eq}} = \frac{k_{BA}}{k_{AB} + k_{BA}}, \tag{B.8a}$$

$$p_B^{\text{eq}} = \frac{k_{AB}}{k_{AB} + k_{BA}}. \tag{B.8b}$$

The *Mathematica* implementation is

```
(* two-state exchange lineshape *)
twoStateExchFunc[δ_, δ0List_, R2List_, kList_, M0_] :=
 Module[{αA, αB, δA, δB, δToΩConvFac, Ω, ΩA, ΩB,
  RA, RB, kAB, kBA, pA, pB, P, Q},
 δToΩConvFac = 2π*500.13;
 {δA, δB} = δ0List;
 Ω = δ*δToΩConvFac;
 ΩA = δA*δToΩConvFac;
 ΩB = δB*δToΩConvFac;
 {kAB, kBA} = kList;
 {RA, RB} = R2List;
 {pA, pB} = {kBA/(kAB + kBA), kAB/(kAB + kBA)};
 αA = RA + I (Ω - ΩA); αB = RB + I (Ω - ΩB);
 P = pA αB + pB αA + kAB + kBA;
 Q = αA αB + kAB αB + kBA αA;
 M0 Re[P/Q] ];
(* the use is illustrated in the following lines *)
deltaData = Range[0,2,.005];
absorptionSpectrum = twoStateExchFunc[deltaData,{(*δA*).7,(*δB*)1.4},
{(*RA*)40,(*RB*)40},{(*kAB*)200,(*kBA*)300},(*M0*)10];
```

Any spectral exchange lineshape can be formally written as

$$S(\Omega) = M_0\frac{\mathcal{A}+i\mathcal{B}}{\mathcal{C}+i\mathcal{D}} = M_0\left(\frac{\mathcal{A}\mathcal{C}+\mathcal{B}\mathcal{D}}{\mathcal{C}^2+\mathcal{D}^2}+i\frac{\mathcal{B}\mathcal{C}-\mathcal{A}\mathcal{D}}{\mathcal{C}^2+\mathcal{D}^2}\right). \tag{B.9}$$

Hence, the real part of asymmetric two-state exchange ($k_{AB} \neq k_{BA}$) can be explicitly written as

$$\begin{aligned}
\mathcal{A} &= k_{AB} + k_{BA} + p_A R_B + p_B R_A\,,\\
\mathcal{B} &= p_A(\Omega - \Omega_B) + p_B(\Omega - \Omega_A)\,,\\
\mathcal{C} &= k_{AB} R_B + k_{BA} R_A + R_A R_B - (\Omega - \Omega_A)(\Omega - \Omega_B)\,,\\
\mathcal{D} &= (k_{AB} + R_A)(\Omega - \Omega_B) + (k_{BA} + R_B)(\Omega - \Omega_A)
\end{aligned} \tag{B.10}$$

with the corresponding code

```
(* two-state exchange lineshape with explicit real part *)
twoStateExchReFunc[δ_, δ0List_, R2List_, kList_, M0_] :=
 Module[{δA, δB, δToΩConvFac, Ω, ΩA, ΩB,
  RA, RB, kAB, kBA, pA, pB, A, B, C, D},
 δToΩConvFac = 2π*500.13;
 {δA, δB} = δ0List;
 Ω = δ*δToΩConvFac;
 ΩA = δA*δToΩConvFac;
 ΩB = δB*δToΩConvFac;
 {kAB, kBA} = kList;
 {RA, RB} = R2List;
 {pA, pB} = {kBA/(kAB + kBA), kAB/(kAB + kBA)};
 A = kAB + kBA + pA RB + pB RA;
 B = pA (Ω - ΩB) + pB (Ω - ΩA);
 C = kAB RB + kBA RA + RA RB - (Ω - ΩA) (Ω - ΩB);
 D = (kAB + RA) (Ω - ΩB) + (kBA + RB) (Ω - ΩA);
 M0 (A C + B D)/(C^2 + D^2) ];
(* the use is illustrated in the following lines *)
deltaData = Range[0,2,.005];
absorptionSpectrum = twoStateExchReFunc[deltaData,{(*δA*).7,(*δB*)1.4},
{(*RA*)40,(*RB*)40},{(*kAB*)200,(*kBA*)300},(*M0*)10];
```

In the case of symmetric two-state exchange ($k_{AB} = k_{BA} = k$), the formula simplifies to

$$\begin{aligned}
\mathcal{A} &= (4k + R_A + R_B)/2\,, \\
\mathcal{B} &= (2\Omega - \Omega_A - \Omega_B)/2\,, \\
\mathcal{C} &= k(R_A + R_B) + R_A R_B - (\Omega - \Omega_A)(\Omega - \Omega_B)\,, \\
\mathcal{D} &= (k + R_A)(\Omega - \Omega_B) + (k + R_B)(\Omega - \Omega_A)\,.
\end{aligned} \tag{B.11}$$

B.1.3 Three-State Exchange Lineshape

The analytical formula for two-state exchange lineshape is

$$S_{\text{three-state exch.}}(\Omega) = M_0 \frac{\mathcal{P}}{\mathcal{Q}}\,, \tag{B.12}$$

where

$$\begin{aligned}
\mathcal{P} &= p_A[\alpha_B\alpha_C + \alpha_B(k_{CA} + k_{CB} + k_{AC}) + \alpha_C(k_{BA} + k_{BC} + k_{AB})] \\
&\quad + p_B[\alpha_A\alpha_C + \alpha_A(k_{CA} + k_{CB} + k_{BC}) + \alpha_C(k_{AB} + k_{AC} + k_{BA})] \\
&\quad + p_C[\alpha_A\alpha_B + \alpha_A(k_{BA} + k_{BC} + k_{CB}) + \alpha_B(k_{AB} + k_{AC} + k_{CA})] \\
&\quad + \pi_A + \pi_B + \pi_C\,, \\
\mathcal{Q} &= \alpha_A\alpha_B\alpha_C + \alpha_A\alpha_B(k_{CA} + k_{CB}) + \alpha_A\alpha_C(k_{BA} + k_{BC}) + \alpha_B\alpha_C(k_{AB} + k_{AC}) \\
&\quad + \alpha_A\pi_A + \alpha_B\pi_B + \alpha_C\pi_C\,, \\
\pi_A &= k_{BA}k_{CA} + k_{BC}k_{CA} + k_{BA}k_{CB}\,, \\
\pi_B &= k_{AB}k_{CA} + k_{AB}k_{CB} + k_{AC}k_{CB}\,, \\
\pi_C &= k_{AC}k_{BA} + k_{AB}k_{BC} + k_{AC}k_{BC}\,.
\end{aligned}$$

Steady-state populations are calculated as

$$p_j^{\text{ss}} = \frac{\pi_j}{\pi_A + \pi_B + \pi_C} \tag{B.13}$$

with $j = A, B, C$. Equilibrium lineshapes are obtained when the input transition rate coefficients fulfill an additional condition $k_{AB} = \frac{k_{AC}k_{BA}k_{CB}}{k_{BC}k_{CA}}$, cf. Sect. 3.3.3.

The *Mathematica* implementation is

```
(* three-state exchange lineshape *)
threeStateExchFunc[δ_, δ0List_, R2List_, kList_, M0_] :=
 Module[{αA, αB, αC, δA, δB, δC, δToΩConvFac, Ω, ΩA, ΩB, ΩC,
  RA, RB, RC, kAB, kBA, kAC, kCA, kBC, kCB, pA, pB, pC, πA, πB, πC, P, Q},
 δToΩConvFac = 2π*500.13;
 {δA, δB, δC} = δ0List;
 Ω = δ*δToΩConvFac;
 ΩA = δA*δToΩConvFac;
 ΩB = δB*δToΩConvFac;
 ΩC = δC*δToΩConvFac;
 {kAB, kBA, kAC, kCA, kBC, kCB} = kList;
 πA = kBA kCA + kBC kCA + kBA kCB;
 πB = kAB kCA + kAB kCB + kAC kCB;
 πC = kAC kBA + kAB kBC + kAC kBC;
 {RA, RB, RC} = R2List;
 {pA, pB, pC} = {πA, πB, πC}/(πA + πB + πC);
 αA = RA + I (Ω - ΩA); αB = RB + I (Ω - ΩB); αC = RC + I (Ω - ΩC);
 P = pA (αB αC + αB (kCA + kCB + kAC) + αC (kBA + kBC + kAB))
   + pB (αA αC + αA (kCA + kCB + kBC) + αC (kAB + kAC + kBA))
   + pC (αA αB + αA (kBA + kBC + kCB) + αB (kAB + kAC + kCA))
   + (πA + πB + πC);
 Q = αA αB αC + αA αB (kCA + kCB) + αA αC (kBA + kBC) + αB αC (kAB + kAC)
   + αA πA + αB πB + αC πC;
 M0 Re[P/Q] ];
(* the use is illustrated in the following lines *)
deltaData = Range[0,2,.005];
absorptionSpectrum = threeStateExchFunc[deltaData,{(*δA*).7,(*δB*)1,(*δC*)1.4},
{(*RA*)40,(*RB*)40,(*RC*)40},{(*kAB*)200,(*kBA*)80,(*kAC*)100,(*kCA*)150,
(*kBC*)70,(*kCB*)80},(*M0*)10];
```

Using Eq. (B.9), the real part of three-state exchange lineshape can be expressed explicitly as

$$\begin{aligned}
\mathcal{A} = {} & p_A[R_2^B R_2^C + R_2^B(k_{CA}+k_{CB}+k_{AC}) + R_2^C(k_{BA}+k_{BC}+k_{AB}) - (\Omega-\Omega_B)(\Omega-\Omega_C)] \\
& + p_B[R_2^A R_2^C + R_2^A(k_{CA}+k_{CB}+k_{BC}) + R_2^C(k_{AB}+k_{AC}+k_{BA}) - (\Omega-\Omega_A)(\Omega-\Omega_C)] \\
& + p_C[R_2^A R_2^B + R_2^A(k_{BA}+k_{BC}+k_{CB}) + R_2^B(k_{AB}+k_{AC}+k_{CA}) - (\Omega-\Omega_A)(\Omega-\Omega_B)] \\
& + \pi_A + \pi_B + \pi_C, \\
\mathcal{B} = {} & p_A[(R_2^C + k_{CA}+k_{CB}+k_{AC})(\Omega-\Omega_B) + (R_2^B + k_{BA}+k_{BC}+k_{AB})(\Omega-\Omega_C)] \\
& + p_B[(R_2^C + k_{CA}+k_{CB}+k_{BC})(\Omega-\Omega_A) + (R_2^A + k_{AB}+k_{AC}+k_{BA})(\Omega-\Omega_C)] \\
& + p_C[(R_2^B + k_{BA}+k_{BC}+k_{CB})(\Omega-\Omega_A) + (R_2^A + k_{AB}+k_{AC}+k_{CA})(\Omega-\Omega_B)], \\
\mathcal{C} = {} & R_2^A R_2^B R_2^C + (k_{CA}+k_{CB})R_2^A R_2^B + (k_{BA}+k_{BC})R_2^A R_2^C + (k_{AB}+k_{AC})R_2^B R_2^C \\
& - (R_2^C + k_{CA}+k_{CB})(\Omega-\Omega_A)(\Omega-\Omega_B) \\
& - (R_2^B + k_{BA}+k_{BC})(\Omega-\Omega_A)(\Omega-\Omega_C) \\
& - (R_2^A + k_{AB}+k_{AC})(\Omega-\Omega_B)(\Omega-\Omega_C) \\
& + R_2^A\pi_A + R_2^B\pi_B + R_2^C\pi_C, \\
\mathcal{D} = {} & (R_2^B R_2^C + (k_{CA}+k_{CB})R_2^B + (k_{BA}+k_{BC})R_2^C + \pi_A)(\Omega-\Omega_A) \\
& + (R_2^A R_2^C + (k_{CA}+k_{CB})R_2^A + (k_{AB}+k_{AC})R_2^C + \pi_B)(\Omega-\Omega_B) \\
& + (R_2^A R_2^B + (k_{BA}+k_{BC})R_2^A + (k_{AB}+k_{AC})R_2^B + \pi_C)(\Omega-\Omega_C) \\
& - (\Omega-\Omega_A)(\Omega-\Omega_B)(\Omega-\Omega_C)
\end{aligned} \tag{B.14}$$

with *Mathematica* code

```
(* three-state exchange lineshape with explicit real part *)
threeStateExchReFunc[δ_, δ0List_, R2List_, kList_, M0_] :=
 Module[{δA, δB, δC, δToΩConvFac, Ω, ΩA, ΩB, ΩC, RA,
  RB, RC, kAB, kBA, kAC, kCA, kBC, kCB, pA, pB, pC, πA, πB, πC, A, B, C, D},
 δToΩConvFac = 2π*500.13;
 {δA, δB, δC} = δ0List;
 Ω = δ*δToΩConvFac;
 ΩA = δA*δToΩConvFac; ΩB = δB*δToΩConvFac; ΩC = δC*δToΩConvFac;
 {kAB, kBA, kAC, kCA, kBC, kCB} = kList;
 πA = kBA kCA + kBC kCA + kBA kCB;
 πB = kAB kCA + kAB kCB + kAC kCB;
 πC = kAC kBA + kAB kBC + kAC kBC;
 {RA, RB, RC} = R2List;
 {pA, pB, pC} = {πA, πB, πC}/(πA + πB + πC);
 A = pA (RB RC + RB (kAC + kCA + kCB) + RC (kBA + kBC + kAB) - (Ω - ΩB) (Ω - ΩC))
   + pB (RA RC + RA (kCA + kCB + kBC) + RC (kAB + kAC + kBA) - (Ω - ΩA) (Ω - ΩC))
   + pC (RA RB + RA (kBA + kBC + kCB) + RB (kAB + kAC + kCA) - (Ω - ΩA) (Ω - ΩB))
   + πA + πB + πC;
 B = pA ((RC + kCA + kCB + kAC) (Ω - ΩB) + (RB + kBA + kBC + kAB) (Ω - ΩC))
   + pB ((RC + kCA + kCB + kBC) (Ω - ΩA) + (RA + kAB + kAC + kBA) (Ω - ΩC))
   + pC ((RB + kBA + kBC + kCB) (Ω - ΩA) + (RA + kAB + kAC + kCA) (Ω - ΩB));
 C = RA RB RC + (kCA + kCB) RA RB + (kBA + kBC) RA RC + (kAB + kAC) RB RC
   - (RC + kCA + kCB) (Ω - ΩA) (Ω - ΩB)
   - (RB + kBA + kBC) (Ω - ΩA) (Ω - ΩC)
   - (RA + kAB + kAC) (Ω - ΩB) (Ω - ΩC)
   + RA πA + RB πB + RC πC;
 D = (RB RC + (kCA + kCB) RB + (kBA + kBC) RC + πA) (Ω - ΩA)
   + (RA RC + (kCA + kCB) RA + (kAB + kAC) RC + πB) (Ω - ΩB)
   + (RA RB + (kBA + kBC) RA + (kAB + kAC) RB + πC) (Ω - ΩC)
   - (Ω - ΩA) (Ω - ΩB) (Ω - ΩC);
 M0 (A C + B D)/(C^2 + D^2) ];
 (* the use is illustrated in the following lines *)
deltaData = Range[0,2,.005];
absorptionSpectrum = threeStateExchReFunc[deltaData,{(*δA*).7,(*δB*)1,(*δC*)1.4},
{(*RA*)40,(*RB*)40,(*RC*)40},{(*kAB*)200,(*kBA*)80,(*kAC*)100,(*kCA*)150,
(*kBC*)70,(*kCB*)80},(*M0*)10];
```

Exchange lineshapes for $N > 3$ can be readily derived in *Mathematica* software, which will be shown in Sect. B.2.

B.1.4 Spectral Lineshape of Two J-coupled Spins

Two J-coupled nuclear spins $I = 1/2$ with offset Larmor frequencies Ω_A and Ω_B produce NMR spectrum with two doublets as shown in Fig. B.2 [2, 3]. If $\Omega_A = \Omega_B$, no splitting is observable. The coupling constant J (in Hz) describes strength of indirect dipole-dipole interaction (Sect. 3.1.4), two strongly interacting spins form so-called 'AB system' and weakly interacting spins form so-called 'AX system'[1] ($|J| \ll |\Omega_A - \Omega_B|$). AB system shows so-called 'roof effect' when the inner peaks are more intensive than the outer peaks ($I_{in} > I_{out}$), see Fig. B.2, while peaks of AX system are equally intensive.

[1] Strength of J-coupling depends on the distance between spins, therefore A and B being close in the alphabet indicate strong interaction, A and X being far indicate weak interaction.

Spectral lineshape for J-coupled spins cannot be derived from Bloch equations, it requires quantum mechanical approach. The solution in the case of homonuclear AB system and $R_2^A = R_2^B = R_2$ can be found in [2] (p. 615) and consists of four Lorentzian peaks,

$$S(\Omega) = \frac{M_{\text{out}}}{\alpha_1} + \frac{M_{\text{in}}}{\alpha_2} + \frac{M_{\text{in}}}{\alpha_3} + \frac{M_{\text{out}}}{\alpha_4}, \tag{B.15}$$

where

$$\begin{aligned}
\alpha_1 &= R_2 - i(\Omega - \Omega_1), \\
\alpha_2 &= R_2 - i(\Omega - \Omega_2), \\
\alpha_3 &= R_2 - i(\Omega - \Omega_3), \\
\alpha_4 &= R_2 - i(\Omega - \Omega_4), \\
\Omega_1 &= \frac{\Omega_A + \Omega_B}{2} + \frac{1}{2}(D + 2\pi J), \\
\Omega_2 &= \frac{\Omega_A + \Omega_B}{2} + \frac{1}{2}(D - 2\pi J), \\
\Omega_3 &= \frac{\Omega_A + \Omega_B}{2} - \frac{1}{2}(D - 2\pi J), \\
\Omega_4 &= \frac{\Omega_A + \Omega_B}{2} - \frac{1}{2}(D + 2\pi J), \\
D &= \sqrt{(\Omega_A - \Omega_B)^2 + (2\pi J)^2}.
\end{aligned} \tag{B.16a}$$

As shown in Fig. B.2, peaks and Ω_1 and Ω_2 as well as Ω_3 and Ω_4 are separated by $2\pi J$ in the angular frequency scale, which corresponds to splitting by $\tilde{J}$ [ppm] $=$ J [Hz]$/\nu_0$ [MHz] in the ppm scale, ν_0 is the spectrometer frequency. Furthermore, $I_{\text{in}} = M_{\text{in}}/R_2$ and $I_{\text{out}} = M_{\text{out}}/R_2$. Intensities of inner and outer peaks satisfy

$$\frac{M_{\text{in}}}{M_{\text{out}}} = \frac{I_{\text{in}}}{I_{\text{out}}} = \frac{D + 2\pi J}{D - 2\pi J}, \tag{B.17a}$$

$$M_{\text{in}} + M_{\text{out}} = M_0, \tag{B.17b}$$

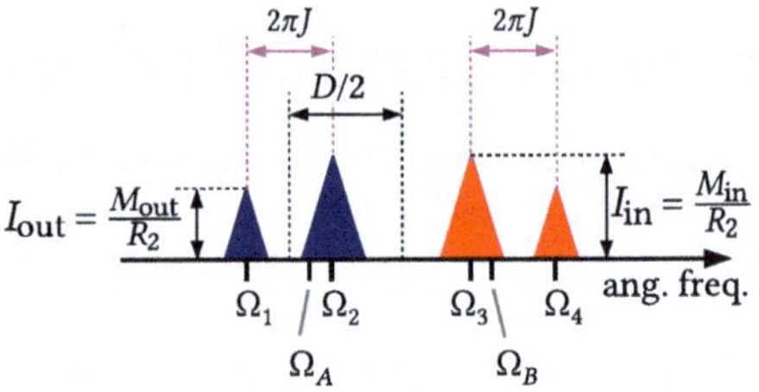

Fig. B.2 Illustration of spectral lineshape two J-coupled spins

where M_0 is equilibrium magnetization due to a single spin. Furthermore, a few other relationships can be derived from the above equations,

$$M_{\text{in}} = \frac{D + 2\pi J}{2D} M_0 \,, \tag{B.18a}$$

$$M_{\text{out}} = \frac{D - 2\pi J}{2D} M_0 \,, \tag{B.18b}$$

$$|\Omega_A - \Omega_B| = \sqrt{D^2 - (2\pi J)^2} = \sqrt{(\Omega_1 - \Omega_4)(\Omega_2 - \Omega_3)} \,. \tag{B.18c}$$

Total area below the curve (in angular frequency scale) is $\pi R_2(2I_{\text{in}} + 2I_{\text{out}}) = 2\pi M_0$, which is the expected peak intensity due to two spins.

The *Mathematica* code is

```
(*J-coupled spins 1/2 lineshape*)
JcoupledSpinFunc[δ_, δ0List_, R2_, J_, M0_] :=
 Module[{α1, α2, α3, α4, δA, δB, δToΩConvFac,
  Ω, ΩA, ΩB, Ω1, Ω2, Ω3, Ω4, D, Min, Mout},
 δToΩConvFac = 2π*500.13;
 {δA, δB} = δ0List;
 Ω = δ*δToΩConvFac; ΩA = δA*δToΩConvFac; ΩB = δB*δToΩConvFac;
 D = Sqrt[(ΩA - ΩB)^2];
 Ω1 = 1/2 (ΩA + ΩB) + 1/2 (D + 2π J);
 Ω2 = 1/2 (ΩA + ΩB) + 1/2 (D - 2π J);
 Ω3 = 1/2 (ΩA + ΩB) - 1/2 (D - 2π J);
 Ω4 = 1/2 (ΩA + ΩB) - 1/2 (D + 2π J);
 α1 = R2 - I (Ω - Ω1); α2 = R2 - I (Ω - Ω2);
 α3 = R2 - I (Ω - Ω3); α4 = R2 - I (Ω - Ω4);
 Min = (D + 2π J)/(2 D) M0; Mout = (D - 2π J)/(2 D) M0;
 Re[Mout/α1 + Min/α2 + Min/α3 + Mout/α4] ];
deltaData = Range[0,2,.002];
absorptionSpectrum = JcoupledSpinFunc[deltaData,{(*δA*).9,(*δB*)1.1},(*R2*)40,
(*J*)12,(*M0*)10];
```

B.2 Rapid Derivation of *N*-State Exchange Lineshape in *Mathematica*

Here we show a "copy-paste" code in the *Mathematica* software to rapidly derive the four- and five-state exchange spectral lineshapes. However, it can be readily extended to an arbitrary number of states. We do not show the actual formulae generated by the code due to the excessive number of terms contained.

```
 (* four-state exchange lineshape *)
Kmatrix = {{-kAB - kAC - kAD, kBA, kCA, kDA}, {kAB, -kBA - kBC - kBD, kCB, kDB},
{kAC, kBC, -kCA - kCB - kCD, kDC}, {kAD, kBD, kCD, -kDA - kDB - kDC}};
Amatrix = DiagonalMatrix[{αA, αB, αC, αD}];
(* three following command lines: keeping together the terms on diagonal (βj)
   of the matrix B=A-K reduces the size of the final symbolic expression *)
Bmatrix = Amatrix - Kmatrix;
betaRule = MapThread[Rule[#1, #2] &,
{{βA, βB, βC, βD}, Diagonal[Bmatrix]}]; (* later restores the diagonal
   values of B *)
```

```
Bmatrix = Bmatrix - DiagonalMatrix[Diagonal[Bmatrix]] +
  DiagonalMatrix[{βA, βB, βC, βD}]; (* replaces diagonal values with βj *)
{fA, fB, fC, fD} = M0 Inverse[Bmatrix].{pA, pB, pC, pD};
fourStateLineshapeComplex = (Total[{fA, fB, fC, fD}] // Together) /. betaRule;
alphaRule = {αA -> RA + I (Ω - ΩA), αB -> RB + I (Ω - ΩB),
αC -> RC + I (Ω - ΩC), αD -> RD + I (Ω - ΩD)};
fourStateLineshapeComplex = fourStateLineshapeComplex /. alphaRule; (* substitute
for αj *)
(* in order to obtain the absorption part of the spectrum,
    numerical values of the parameters must be inserted and then the real part
    taken, see the example in the following lines *)
absorptionSpectrum = fourStateLineshapeComplex /. {Ω -> Range[4000],
    ΩA -> 1000., ΩB -> 1500., ΩC -> 2500., ΩD -> 3000.,
    M0 -> 1., pA -> 1/4, pB -> 1/4, pC -> 1/4, pD -> 1/4,
    kAB -> 30., kBA -> 30., kAC -> 30., kCA -> 30., kAD -> 30., kDA -> 30.,
    kBC -> 30., kCB -> 30., kBD -> 30., kDB -> 30., kCD -> 30., kDC -> 30.,
    RA -> 30., RB -> 30., RC -> 30., RD -> 30.};
absorptionSpectrum = Re[absorptionSpectrum];
```

```
 (* five-state exchange lineshape *)
Kmatrix = {{-kAB - kAC - kAD - kAE, kBA, kCA, kDA, kEA},
{kAB, -kBA - kBC - kBD - kBE, kCB, kDB, kEB},
{kAC, kBC, -kCA - kCB - kCD - kCE, kDC, kEC},
{kAD, kBD, kCD, -kDA - kDB - kDC - kDE, kED},
{kAE, kBE, kCE, kDE, -kEA - kEB - kEC - kED}};
Amatrix = DiagonalMatrix[{αA, αB, αC, αD, αE}];
(* three following command lines: keeping together the terms on diagonal (βj)
    of the matrix B=A-K reduces the size of the final symbolic expression *)
Bmatrix = Amatrix - Kmatrix;
betaRule = MapThread[Rule[#1, #2] &,
{{βA, βB, βC, βD, βE}, Diagonal[Bmatrix]}]; (* later restores the
    diagonal values of B *)
Bmatrix = Bmatrix - DiagonalMatrix[Diagonal[Bmatrix]] +
  DiagonalMatrix[{βA, βB, βC, βD, βE}]; (* replaces diagonal values with βj *)
{fA, fB, fC, fD, fE} = M0 Inverse[Bmatrix].{pA, pB, pC, pD, pE};
fiveStateLineshapeComplex = (Total[{fA, fB, fC, fD, fE}] // Together) /. betaRule;
alphaRule = {αA -> RA + I (Ω - ΩA), αB -> RB + I (Ω - ΩB),
αC -> RC + I (Ω - ΩC), αD -> RD + I (Ω - ΩD), αE -> RE + I (Ω - ΩE)};
fiveStateLineshapeComplex = fiveStateLineshapeComplex /. alphaRule; (* substitute
    for αj *)
(* in order to obtain the absorption part of the spectrum,
    numerical values of the parameters must be inserted and then the real part
    taken, see the example in the following lines *)
absorptionSpectrum = fiveStateLineshapeComplex /. {Ω -> Range[4000],
    ΩA -> 1000., ΩB -> 1500., ΩC -> 2500., ΩD -> 3000., ΩE -> 3300.,
    M0 -> 1., pA -> 1/4, pB -> 1/4, pC -> 1/4, pD -> 1/8, pE -> 1/8,
    kAB -> 30., kBA -> 30., kAC -> 30., kCA -> 30., kAD -> 30., kDA -> 30.,
    kAE -> 30., kEA -> 30., kBC -> 30., kCB -> 30., kBD -> 30., kDB -> 30.,
    kBE -> 30., kEB -> 30., kCD -> 30., kDC -> 30., kCE -> 30., kEC -> 30.,
    kDE -> 30., kED -> 30., RA -> 30., RB -> 30., RC -> 30.,
    RD -> 30., RE -> 30.};
absorptionSpectrum = Re[absorptionSpectrum];
```

B.3 Rapid Derivation of Three-State Populations

Steady-state populations are derived with the following *Mathematica* code

```
(* three-state steady-state populations *)
Kmatrix = {{-kAB - kAC, kBA, kCA}, {kAB, -kBA - kBC, kCB},
  {kAC, kBC, -kCA - kCB}};
pvector = {pA, pB, pC};
equations = MapThread[Equal[#1, #2] &,
  {Kmatrix . pvector, ConstantArray[0, Length@pvector]}];
ssRule = Solve[(And @@ equations)~And~(Total@pvector == 1),
  pvector][[1]] // Factor
```

This code can be extended to higher number of states.

Three-state equilibrium populations are derived from the steady-state populations with additional equilibrium condition $k_{AB} = \frac{k_{AC}k_{BA}k_{CB}}{k_{BC}k_{CA}}$ (Eq. (3.39)). The corresponding code reads

```
(* three-state equilibrium populations *)
eqCondRule = {kAB -> kAC kBA kCB/(kBC kCA)}
ssRule /. eqCondRule // Factor
```

References

1. Březina V, Hanyková L, Velychkivska N, Hill JP, Labuta J (2022) NMR lineshape analysis using analytical solutions of multi-state chemical exchange with applications to kinetics of host-guest systems. Sci Rep.10.1038/s41598-022-20136-4.
2. Levitt MH (2001) Spin dynamics: basics of nuclear magnetic resonance. Wiley.
3. Günthe H (2013) NMR spectroscopy: basic principles, concepts and applications in chemistry. Wiley.

Appendix C
Experimental Details, Models and Data Analysis

C.1 Properties of Used Chemicals

Quantities characterizing the compounds used in this study are given in Table C.1. They were used for calculation of concentrations during the experiments. Molar mass of oxoporphyrinogens was calculated in the *Chem Draw Pro* program. **OxP** and its derivatives were synthesized in National Institute for Materials Science (NIMS), Japan, by Jonathan Hill. Other compounds were purchased from Sigma-Aldrich.

C.2 UV/vis Spectroscopy Measurements

UV/vis absorption spectra were recorded using Jasco J-820, Shimadzu UV-3600, and Hitachi U-2910 spectrophotometers (1 cm or 1 mm quartz cells with screw caps were used). Baseline, determined by calibration measurement with empty cuvette, was subtracted from each spectrum. All UV/vis measurements were conducted at room temperature. Titration experiments (i.e. measurement of spectra at different guest concentrations when the guest is gradually added into the host solution) were performed by the addition of stock solution (acid dissolved in chloroform, in some cases in the presence of oxoporphyrinogen with the same concentration as in the sample to keep the total dye concentration constant) into the sample using a microsyringe. Deuterated spectroscopic chloroform was used as solvent. Spectral calculator [1] was used for the conversion of the measured UV/vis spectra into the corresponding RGB values.

Measurements of solvatochromic effects in **Bz_2OxP** and **Bz_4OxP** in DMF/$CDCl_3$ mixture were started in the pure solution of oxoporphyrinogen in $CDCl_3$. Then, appropriate amount of the sample was gradually removed and replaced from the stock solution containing DMF (stock solution also contained the oxoporphyrinogen

V. Březina, *Supramolecular Complexes of Oxoporphyrinogens with Organic Molecules*, Springer Theses, https://doi.org/10.1007/978-3-031-68646-7

Table C.1 Basic properties of chemicals used in this work

Compound	Molar mass/g.mol^{-1}	Density/g.mL^{-1}	Notes
OxP	1125.57		Powder
Bz$_2$OxP	1463.61		Powder
Bz$_4$OxP	1801.68		Powder
CSA	232.3	1.331	Powder
			Solubility in $CDCl_3$: 2.3 mg/mL
			$[CSA]_{max} \approx 0.01$ M
DFA	96.03	1.526	Liquid
TFA	114.0	1.489	Liquid
$CDCl_3$	120.4	1.5	$\mu^{el} = 1.15$ D
DMF	73.10	0.948	Basic, p$K_a \approx -0.3$ for conjugate acid,
			μ^{el}=3.86 D

Full names of the investigated oxoporphyrinogens are *meso*-tetrakis(3,5-di-*tert*-butyl-4-oxo-2,5-cyclohexadienylidene)porphyrinogen (**OxP**), N_{21},N_{23}-bis(4-bromobenzyl)-*meso*-tetrakis(3,5-di-*tert*-butyl-4-oxo-2,5-cyclohexadienylidene)porphyrinogen (**Bz$_2$OxP**) and N_{21},N_{22},N_{23},N_{24}-tetrakis(4-bromobenzyl)-*meso*-tetrakis(3,5-di-*tert*-butyl-4-oxo-2,5-cyclohexadienylidene)porphyrinogen (**Bz$_4$OxP**). Numbering of the porphyrinogen macrocycle is shown in Fig. 1.2d.

to prevent dilution), total sample volume was kept constant. Prior the sample preparation, the solvent was filtered through potassium carbonate to remove all residual acid molecules (which are sometimes created by solvent decomposition). Deuterated spectroscopic chloroform with silver foil was used to prevent formation of HCl. In these cases, titration is not appropriate because the complete picture of spectral changes should be obtained in the whole vol.f. range (φ(DMF) $= 1$ is inaccessible in titration starting in pure chloroform). Measurements of solvatochromic effects in **OxP** were investigated by titration of **OxP** in $CDCl_3$ with DMF similarly to the previous paragraph (solvatochromic spectral change is saturated at φ(DMF) $= 0.2$, which is accessible in titration).

C.3 RGB Coordinates of UV/vis Spectra

Conversion of UV/vis spectra to corresponding RGB coordinates was done using Excel spreadsheet developed by Bruce Lindbloom (Spectral Calculator) [1]. Input data in this spreadsheet are in the form of transmittance T, which was calculated from the measured absorbance A using $T = 10^{-A}$. Output format was set to Adobe RGB (1998). RGB values of spectra of individual absorbing components obtained by titration of oxoporphyinogens with DFA are listed in Table C.2.

Table C.2 RGB values of protonated oxoporphyrinogen species (listed in Figs. 6.3, 6.4 and 6.5) calculated from measured absorbance using [1]

	R	G	B
OxP	177	92	170
OxP^{2+}	212	152	166
OxP^{4+}	151	186	160
Bz_2OxP	217	129	171
Bz_2OxP^{+}	204	137	197
Bz_2OxP^{2+}	157	169	206
Bz_2OxP^{4+}	147	196	192
Bz_4OxP	244	165	173
Bz_4OxP H-bond	225	135	184
Bz_4OxP^{+}	195	168	194
Bz_4OxP^{2+}	121	195	213

C.4 NMR Spectroscopy Measurements

NMR spectroscopy measurements were performed using Bruker Avance III HD 500 and JEOL AL300BX spectrometers. Tetramethylsilane was used as the internal standard, chloroform with silver foil was used as solvent to prevent the formation of HCl. Titration experiments (i.e. measurement of spectra at different guest concentrations when the guest is gradually added into the host solution) were performed by the addition of stock solution (acid dissolved in chloroform, in some cases in the presence of oxoporphyrinogen with the same concentration as in the sample) into the sample using a microsyringe. All measurements were performed at 25 °C unless otherwise specified. Variable temperature measurements were performed at the Bruker spectrometer, refrigerator was used for cooling up to ca. −40 °C, liquid nitrogen was used to reach lower temperatures.

C.5 NMR Analysis of Host-Guest Binding

In order to obtain useful information presented in Chap. 6, the measured spectra were analyzed in two steps:

Step 1—Lineshape Fitting
Fitting of all relevant peaks in raw data (after proper phase correction and subtraction of baseline with polynomial curve fitted outside peaks) using appropriate formulae: Lorentzian, J-coupled or exchange lineshapes. This type of fitting is illustrated in Fig. C.1. The spectra were not fitted in the whole ppm range, but rather only in the close vicinity of the analyzed peaks. Partially overlapping peaks can be fitted without problems. Fitting of the exchange lineshapes requires careful treatment of δ_j and R_2^j parameters, which is discussed in next sections.

	TB free	TB bound 1	TB bound 2	water	CSA methyl 1a	CSA methyl 1b	CSA methyl 2
I	2.4E+7	1.2E+8	1.3E+8	3.3E+8	9.7E+6	9.9E+6	1.9E+7
delta0 / ppm	1.33	1.38	1.53	1.55	0.56	0.58	0.86
gamma / ppm	0.0096	0.0051	0.0056	0.0053	0.0050	0.0048	0.0061

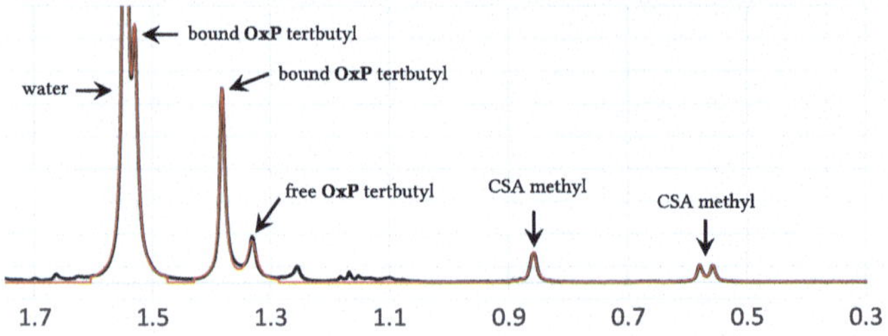

Fig. C.1 Illustration of fitting the measured spectra (**OxP** + (*R*)-CSA). Experimental data (black) are successfully reproduced by the Lorentzian fitting curves (red)

Fitting of *tert*-butyl resonances of the host, methyl group resonances of the guest (averaged signal due to fast exchange between **G** and **HG** in both CSA and DFA), and resonance due to water (averaged signal due to fast exchange between **W** and **HW**) also provided accurate number of equivalents of guest and water (with respect to host) and consequently their total concentrations using $[\mathbf{G}]_t = \text{equiv.}(\mathbf{G}) \times [\mathbf{H}]_t$ and $[\mathbf{W}]_t = \text{equiv.}(\mathbf{W}) \times [\mathbf{H}]_t$. Total concentration of host was calculated from the weight of the sample in powder form and the corresponding volume of solvent (the effect of dilution during the titration was also incorporated). The lineshape fitting procedure was conducted in *Excel*.

Step 2—Fitting of Binding Isotherms

Simultaneous least squares fitting of appropriate binding isotherms to the peak parameters obtained in Step 1 as every titration experiment shows several spectral changes (peak shifts or newly formed isolated peaks). Figure C.2 shows isotherms calculated from shift of averaged peaks in fast exchange that were fitted using Eq. (3.62) (isotherms 2,3,4) and isotherms constructed as ratio of peak areas that were fitted by Eq. (3.61) (isotherms 1,5,6). Proper binding model should be used in order to describe the system correctly. The lineshape fitting procedure was conducted in *Excel* using Solver add-in. Contributions of sum of squares due to particular isotherms to the total sum of squares are weighted to be in the same order as shown in Fig. C.3. This figure also shows all parameters used to describe the binding isotherms within the 1:4 host-guest binding model. The parameters with blue background were fitted, those with yellow background were fixed and obtained directly from the spectra. The least-square fitting procedure was conducted in *Excel* using Solver add-in and VBA programming environment to implement calculations of binding isotherms. The fitted values are listed in Table 6.1. In order to obtain the corresponding error, the fitting procedure was repeated several times with different fixed values of K_j. Acceptable

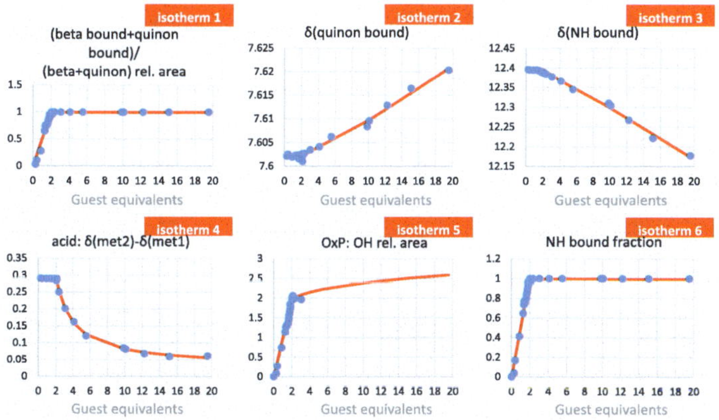

Fig. C.2 Illustration of fitted binding isotherms

	isotherm 1	isotherm 2	isotherm 3	isotherm 4	isotherm 5	isotherm 6
sums of squares	6.9E-2	7.0E-6	9.4E-4	3.8E-4	4.0E+1	2.7E-2
weights	250	2000000	20000	50000	1	200
weighted SSq	**17.24**	**14.06**	**18.77**	**18.85**	**40.48**	**5.38**
TOTAL SSq	114.8					

						isotherm 4	
OxP + R-CSA NMR		isotherm 2		isotherm 3		δG(CSA met2)-δG(CSA met1)	0.021
K1	4.8E+5	δHG(quin bound)	7.60	δHG(NH bound)	12.4	δHG(CSA met2)-δHG(CSA met1)	0.27
K2	1.1E+8	δHG2(quin bound)	7.60	δHG2(NH bound)	12.4	δHG2(CSA met2)-δHG2(CSA met1)	0.29
K3	131	δHG3(quin bound)	7.61	δHG3(NH bound)	12.2	δHG3(CSA met2)-δHG3(CSA met1)	0.24
K4	2.8	δHG4(quin bound)	8.41	δHG4(NH bound)	3.82	δHG4(CSA met2)-δHG4(CSA met1)	1.46

Fig. C.3 Illustration of simultaneous weighted fitting of binding isotherms (**OxP** + (*R*)-CSA) in *Excel*. Parameters with blue background were fitted, the other were obtained directly from the measured spectra

Fig. C.4 Plot of total sum of squares for repeated fitting procedure with different fixed K_1 values (**Bz$_4$OxP** + DFA)

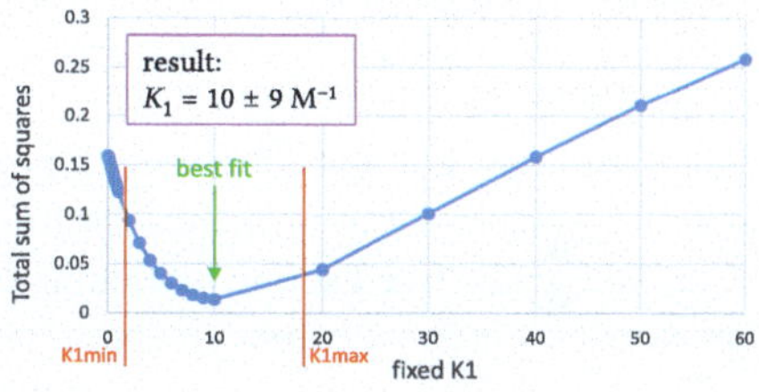

fits were determined by visual inspection of fitted curves and subsequently confidence intervals were constructed from minimum and maximum values of these fitted parameters. This procedure is illustrated in Fig. C.4 for the titration of **Bz$_4$OxP** with DFA.

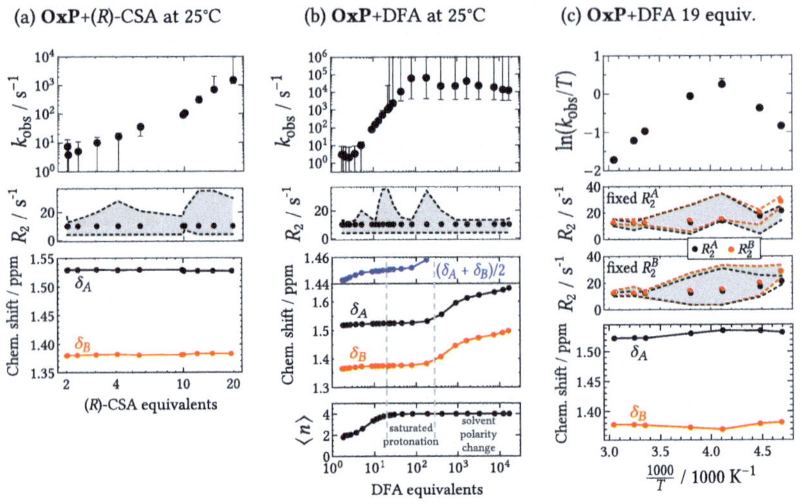

Fig. C.5 Fitting of **OxP** *tert*-butyl signals with symmetric two-state exchange lineshape. (**a**) Titration of **OxP** with (*R*)-CSA (full spectra in Fig. E.3). (**b**) Titration of **OxP** with DFA (full spectra in Fig. E.2). (**c**) Variable temperature measurements at 18 equiv. of DFA (full spectra in Fig. E.9). The quantity $\langle n \rangle$ stands for mean protonation number

C.6 Fitting of Exchange Lineshapes

This appendix describes describes fitting exchanging of *tert*-butyl-H signals with symmetric two-state exchange lineshape (Eq. (B.11)). Dependence on guest concentration or on temperature of the transition rate coefficient k_{obs} was determined. The fitting procedure was realized in *Mathematica* (using in-house scripts with `FindFit` command for the minimization procedure). Chemical shifts δ_A and δ_B and transverse relaxation rates R_2^A and R_2^B (including the inhomogeneous broadening effect, see Sect. 3.1.3) have a big influence on the resulting exchange coefficients. However, they could not be fixed at one particular value throughout the whole experiment since they might change with concentration or temperature.

Parameters δ_A and δ_B were obtained from the positions of peak maxima in the slow regime (at low guest concentration or at low temperature). During the titration experiment the parameters δ_A and δ_B might change with changing polarity of the solvent (e.g. upon addition of acid) or when new species form in the sample (which must not change the number of states). In this situation, the difference $|\delta_A - \delta_B|$ often remains nearly constant and only the center $(\delta_A + \delta_B)/2$ shifts with increasing guest concentration. On the other hand, variable temperature experiments imply significant change in δ_A and δ_B parameters including their difference, which makes the fitting procedure more difficult. Due to these reasons, change of δ_A and δ_B during the whole experiment is also given.

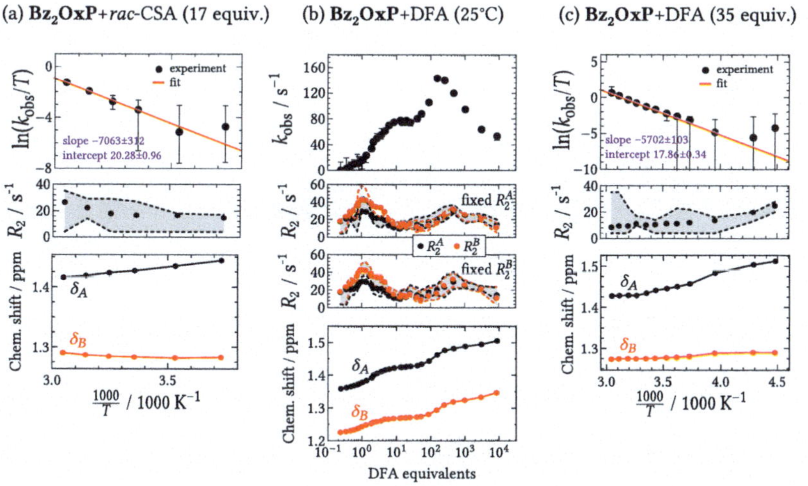

Fig. C.6 Fitting of **Bz₂OxP** *tert*-butyl signals with symmetric two-state exchange lineshape. (**a**) Variable temperature measurements at 17 equiv. of (*rac*)-CSA (full spectra in Fig. E.15). (**b**) Titration of **Bz₂OxP** with DFA (full spectra in Fig. E.4). (**c**) Variable temperature measurements at 35 equiv. of DFA (full spectra in Fig. E.11)

Relaxation rates R_2^A and R_2^B (which include homogeneous broadening) were also obtained at the slow regime from peak width 2γ (R_2 [s^{-1}] $= 2\pi\nu_0$ [MHz]γ [ppm]). In some cases, $R_2^A = R_2^B$ was assumed. However, they might also vary with concentration or with temperature. During the fitting procedure, one of the relaxation rates were fixed. The parameters adjusted during the fitting procedure are: k_{obs}, one of transverse relaxation rates R_2^A or R_2^B (only if they are not equal) and total magnetization in relative units M_0 (here, M_0 plays the role of arbitrary proportionality constant). To obtain the confidence intervals, fitting was repeated for every measured spectrum for different fixed R_2 values, which were varied in a reasonable range of $R_2 \in [10\ \mathrm{s}^{-1}, 40\ \mathrm{s}^{-1}]$. Quality of the fits to the experimental data was determined upon visual inspection, fixed R_2 parameters providing a good fit are denoted by gray zone in Figs. C.5 and C.6. Error bars of fitted k_{obs} in these figures correspond to maximum or minimum R_2 inside the gray zone. Mean values of k_{obs} correspond to best fit (with minimal sum of squares) or to the fixed R_2 obtained from peak width at slow exchange.

Analysis of **OxP** + acid systems is shown in Fig. C.5. Fitting of titration of **OxP** with (*R*)-CSA in Fig. C.5a assumed $R_2^A = R_2^B$, δ_A and δ_B were almost constant during the whole experiment. Fitting of titration of **OxP** with DFA in Fig. C.5b also assumed $R_2^A = R_2^B$, $|\delta_A - \delta_B|$ was kept constant during the whole experiment. However, the center $(\delta_A + \delta_B)/2$ (blue curve) slightly shifted due to formation of $\mathbf{HG_4}$ complex at 2–20 equiv. of DFA until saturation, furthermore, the center is shifted by a greater extent above 300 equiv. due to solvent polarity change. Fitting of variable temperature

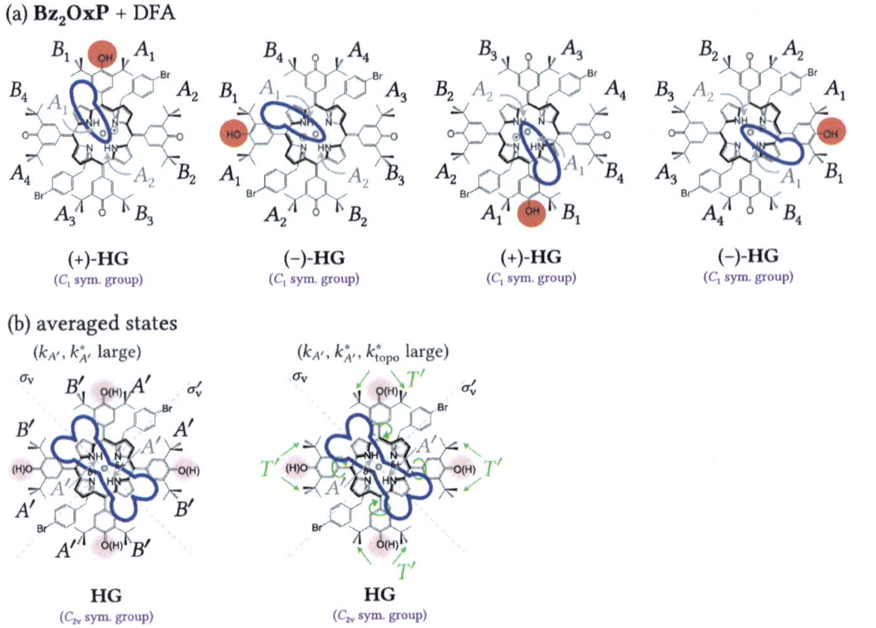

Fig. C.7 Structures of **Bz2OxP** + DFA complex, corresponding states of *tert*-butyl-H spins (black labels; also valid for *ortho*-H and β-H spins) and states of NH spins (gray labels). (**a**) Spin states in **Bz2OxP**$^+$·DFA complex, (**b**) Fast prototropic tautomerization processes cause averaging of chiral field in **Bz2OxP** + DFA system (consists of averaged states from (a)). Slow topomerization process (left) implies two states (i.e. A' and B') of *tert*-butyl-H and *ortho*-H spins, fast topomerization process (right) induces one averaged state (i.e. T'). In the latter case, topomerization does not influence states of β-H spins. Geometric shapes with proper symmetry represent chiral field created by the acid anion. Dashed straight lines denote symmetry elements σ_v (vertical mirror plane). Transition rate coefficients $k_{A'}$ and $k^*_{A'}$ correspond to reaction rate coefficients κ_{taut} and κ^*_{taut}, respectively. Fast exchange implies (Eq. (3.62)) in *tert*-butyl-H peaks: $\delta_{A'} = \frac{1}{4}(\delta_{A_1} + \delta_{A_2} + \delta_{A_3} + \delta_{A_4})$, $\delta_{B'} = \frac{1}{4}(\delta_{B_1} + \delta_{B_2} + \delta_{B_3} + \delta_{B_4})$ and $\delta_{T'} = \frac{1}{2}(\delta_{A'} + \delta_{B'})$. Fast exchange implies in NH peaks: $\delta_{A'} = \frac{1}{2}(\delta_{A_1} + \delta_{A_2})$

measurements of **OxP** with 18 equiv. of DFA in Fig. C.5c showed different values of R_2^A or R_2^B. Therefore, in the first part of the analysis, R_2^A was fixed and R_2^B was fitted together with k_{obs} and M_0, which is shown in the second panel. Successful fits correspond to R_2^A within the gray zone, the fitted values of R_2^B are enclosed by red dashed line. Mean values of R_2^A (black circles) correspond to mean values of R_2^B (red circles). In the second part of the analysis, the roles of R_2^A and R_2^B were swapped as shown in the third panel. Confidence intervals in the Eyring plot in the first panel correspond to intersection of k_{obs} confidence intervals obtained in the second and third panel. Values of δ_A and δ_B varied during the experiment.

Analysis of **Bz2OxP** + acid systems is shown in Fig. C.6. Fitting of variable temperature mesurement of **Bz2OxP** with 17 equiv. of (*rac*)-CSA in Fig. C.6a assumed $R_2^A = R_2^B$, δ_A and δ_B significantly varied during the whole experiment. Linear Eyring plot (see Sect. 2.1.4) in first panel yielded (using weighted linear least squares) the energy barrier parameters listed in Table 7.1, the following *Mathematica* code has been used

```
(* Eyring plot ln(k/T)=a 1/T + b *)
ΔH‡ = -R a/1000 /. R -> 8.3145 (* in kJ/mol *)
ΔS‡ = R (b + Log[h/kBoltz])/1000 /. {R -> 8.3145,
kBoltz -> 1.3806*10^-23, h -> 6.6261*10^-34} (* in kJ/mol/K *)
```

Fitting of titration of **Bz₂OxP** with DFA in Fig. C.6b showed different values of R_2^A or R_2^B. Fitting was performed for range of fixed R_2^A values and then for fixed R_2^B values as shown in second and third panel, respectively. However, the "hillocks" in R_2^A and R_2^B values below 20 equiv. of DFA (corresponding to $\langle n \rangle < 1$) in Fig. C.6b indicate that two-state fitting model is not appropriate, which causes the failed linear dependence in Fig. 7.9. In this concentration range, population of free host is not negligible and broadens the signals. Three-state exchange fitting is needed to obtain corrected k_{obs} values. Parameters δ_A and δ_B significantly varied but their difference remained almost constant. Structures and corresponding states of the **Bz$_2$OxP** + DFA system are shown in Fig. C.7. Fitting of variable temperature mesurement of **Bz$_2$OxP** with 35 equiv. of DFA in Fig. C.6c assumed $R_2^A = R_2^B$, δ_A and δ_B significantly varied during the whole experiment. Linear Eyring plot (see Sect. 2.1.4) in first panel yielded (using weighted linear least squares) the energy barrier parameters listed in Table 7.1.

C.7 Fitting of Experimental Data in Bz$_2$OxP + (*R*)-CSA System

The following text describing data analysis from Sect. 7.2.3 is taken from our paper [2], where files with implementation in *Excel* and *Mathematica* can also be found. The fitting procedure comprises of two consecutive steps:

Step 1—Lineshape Fitting

Fitting of tertiary butyl (TB) resonances of the host, methyl (MET) group resonances of the guest (averaged signal due to fast exchange between **G** and **HG**), and resonance due to water (averaged signal due to fast exchange between **W** and **HW**) with Lorentzian lineshapes was performed in order to provide accurate number of equivalents of guest and water (with respect to host) and consequently their total concentrations. This fitting procedure was realized in *Excel* (with Solver add-in using iterative generalized reduced gradient (GRG) nonlinear method for minimization). Total concentration of host was calculated from the weight of the sample in powder form and the corresponding volume of solvent (the effect of dilution during the titration was also incorporated). Values of $[\mathbf{G}]_t$, $[\mathbf{W}]_t$ and $[\mathbf{H}]_t$ were interpolated and served for calculation of continuous values of all transition rate coefficients and concentrations shown in Fig. 7.12 by solid lines.

Symmetric two-state exchange lineshape (Eq. (B.11)) and half-symmetric three-state exchange (Eqs. (3.52) and (3.37)) lineshapes were fitted to the NH peaks of the host to obtain the concentration dependence of transition rate coefficients. Parameters $\delta_{A'_{\mathrm{I}}}$, $\delta_{A'_{\mathrm{II}}}$ and $\delta_{C'}$ were obtained from the positions of peak maxima in the slow regime. In the intermediate and fast regime the Larmor frequencies shifted due to variation in solvent polarity, so the difference $|\delta_{A'_{\mathrm{I}}} - \delta_{A'_{\mathrm{II}}}|$ was fixed, while the center $(\delta_{A'_{\mathrm{I}}} + \delta_{A'_{\mathrm{II}}})/2$ was adjusted during the fitting, see gray dashed lines in Fig. 7.12a, b. Relaxation rates were assumed to be equal, i.e. $R_2^{A'_{\mathrm{I}}} = R_2^{A'_{\mathrm{II}}} = R_2^{C'}$. During the fitting procedure, the relaxation rates were fixed. The parameters adjusted during the fitting procedure are: $k_{A'}$ for two-state exchange, $k_{A'}$ and $k_{A'C'}$ for three-state exchange and total magnetization in relative units M_0. This fitting procedure was realized in *Mathematica* (using in-house scripts with `FindFit` command for the minimization procedure). The excellent match of fits is illustrated in Fig. C.8a,c. The fitting procedure was repeated for every measured spectrum (at particular $[\mathbf{G}]_{\mathrm{t}}$) for different R_2 values, which were varied in a reasonable range of $R_2 \in [10\ \mathrm{s}^{-1}, 40\ \mathrm{s}^{-1}]$, gray zone in Fig. C.8b, d (bottom). Upon visual inspection, the fitted lineshapes provided an excellent match to the experimental data in almost the whole range of the R_2 parameter (gray zone in Fig. C.8b, d (bottom)). Maximum and minimum values of the fitted transition rate coefficients $k_{A'}$, $k_{A'C'}$ and $k_{C'A'}$ form the error bars in Fig. C.8b, d (top). Mean $k_{A'}$, $k_{A'C'}$ and $k_{C'A'}$ values correspond to $R_2 = 30\ \mathrm{s}^{-1}$. Figure 7.12c in the main manuscript shows that the transition rate coefficients in the intermediate exchange regime are determined with the highest accuracy since the spectral lineshape at this point is most sensitive to the change of the transition rate coefficients (see also Sect. S7.4 in [2]).

Step 2—Binding Model Fitting

Simultaneous fitting of $k_{A'}$ concentration dependence together (using Eq. (C.8a)) with shift of fast-exchanging peaks of **H/HW** (denoted as C' in Fig. 7.12b), two host TB resonances and two guest MET resonances (Fig. C.9) was performed. This fit provides equilibrium binding constant $K_{\mathbf{HG}}$, reaction rate coefficient κ_{taut} and Larmor frequencies of the peaks involved. The following binding isotherms were fitted simultaneously

$$k_{A'} = 2\kappa_{\mathrm{taut}}[\mathbf{G}]\,,$$

$$\delta_{C'} = \delta_C \frac{[\mathbf{H}]}{[\mathbf{H}] + [\mathbf{HW}]} + \delta_D \frac{[\mathbf{HW}]}{[\mathbf{H}] + [\mathbf{HW}]}\,,$$

$$\delta(\mathbf{G}\mathit{MET}1\ \text{observed}) = \delta(\mathbf{G}\mathit{MET}1\ \text{free})\frac{[\mathbf{G}]}{[\mathbf{G}]_{\mathrm{t}}} + \delta(\mathbf{G}\mathit{MET}1\ \text{bound})\frac{[\mathbf{HG}]}{[\mathbf{G}]_{\mathrm{t}}}\,,$$

$$\delta(\mathbf{G}\mathit{MET}2\ \text{observed}) = \delta(\mathbf{G}\mathit{MET}2\ \text{free})\frac{[\mathbf{G}]}{[\mathbf{G}]_{\mathrm{t}}} + \delta(\mathbf{G}\mathit{MET}2\ \text{bound})\frac{[\mathbf{HG}]}{[\mathbf{G}]_{\mathrm{t}}}\,,$$

$$\delta(\mathbf{H}\mathit{TB}1\ \text{observed}) = \delta(\mathbf{H}\mathit{TB}1\ \text{free})\frac{[\mathbf{H}]}{[\mathbf{H}]_{\mathrm{t}}} + \delta(\mathbf{H}\mathit{TB}1\ \text{bound})\frac{[\mathbf{HG}]}{[\mathbf{H}]_{\mathrm{t}}}\,,$$

$$\delta(\mathbf{H}\mathit{TB}2\ \text{observed}) = \delta(\mathbf{H}\mathit{TB}2\ \text{free})\frac{[\mathbf{H}]}{[\mathbf{H}]_{\mathrm{t}}} + \delta(\mathbf{H}\mathit{TB}2\ \text{bound})\frac{[\mathbf{HG}]}{[\mathbf{H}]_{\mathrm{t}}}\,.$$

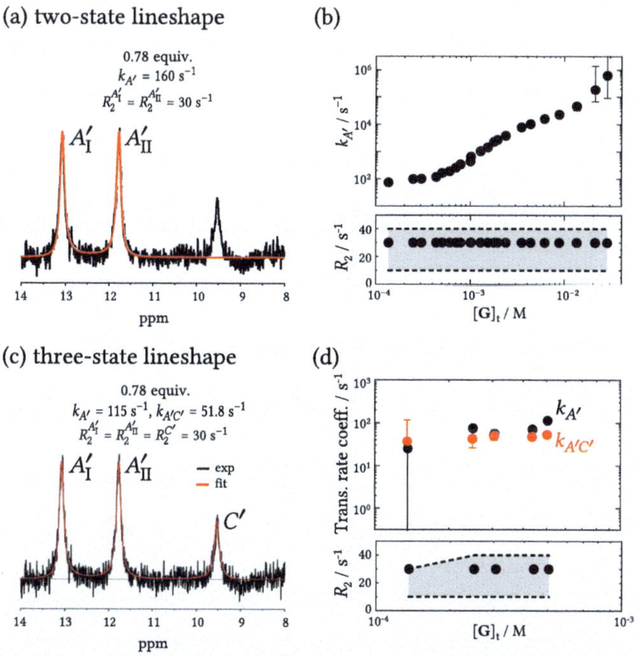

Fig. C.8 Fitting of two- and three-state exchange spectral lineshape and determination of transition rate coefficients in Step 1. (**a**) Example of two-state fitting (spectrum at 0.78 equiv. of guest). Parameters $\delta_{A'_I}$, $\delta_{A'_{II}}$ and $R_2^{A'_I} = R_2^{A'_{II}}$ were fixed during the fitting of one particular spectrum. Values of $k_{A'}$ and M_0 are fitted parameters (value of M_0 differs from the real total magnetization since it depends on technical details of the experimental setup and data processing). (**b**) Resulting values of $k_{A'}$ from two-state fitting (top) and the range of the R_2 parameter, where the fit was successful (gray zone, bottom). Full circles (top and bottom) correspond to $R_2 = 30\ \mathrm{s}^{-1}$, error bars denote maximum and minimum $k_{A'}$ values from repeated fitting procedure at different R_2 values in the range $R_2 \in [10\ \mathrm{s}^{-1}, 40\ \mathrm{s}^{-1}]$ (gray zone, bottom). (**c**) Example of three-state fitting. Parameters $\delta_{A'_I}$, $\delta_{A'_{II}}$, $\delta_{C'}$ and $R_2^{A'_I} = R_2^{A'_{II}} = R_2^{C'}$ were fixed during the fitting of one particular spectrum. Values of $k_{A'}$, $k_{A'C'}$ and M_0 are fitted parameters. (**d**) Resulting values of $k_{A'}$ and $k_{A'C'}$ from three-state fitting (top) and the range of the R_2 parameter (gray zone, bottom). Full circles (top and bottom) correspond to $R_2 = 30\ \mathrm{s}^{-1}$, error bars denote maximum and minimum $k_{A'C'}$, $k_{C'A'}$ values from the repeated fitting procedure at different R_2 values in the range $R_2 \in [10\ \mathrm{s}^{-1}, 40\ \mathrm{s}^{-1}]$ (gray zone, bottom)

Concentrations of the chemical species were obtained from numerical solution of the competitive host-ligand binding model, summarized in Eqs. (A.2) and (A.3). The fitting procedure was realized in *Mathematica* (using in-house scripts). In order to obtain error of all fitted parameters, the fitting procedure was repeated several times with different fixed values of parameters $K_{\mathbf{HG}}$, κ_A, δ_C or δ_D. Acceptable fits were determined by visual inspection of fitted curves and subsequently confidence intervals were constructed from minimum and maximum values of these fitted parameters. The parameter κ_{taut} was calculated as a mean, hence its error was deter-

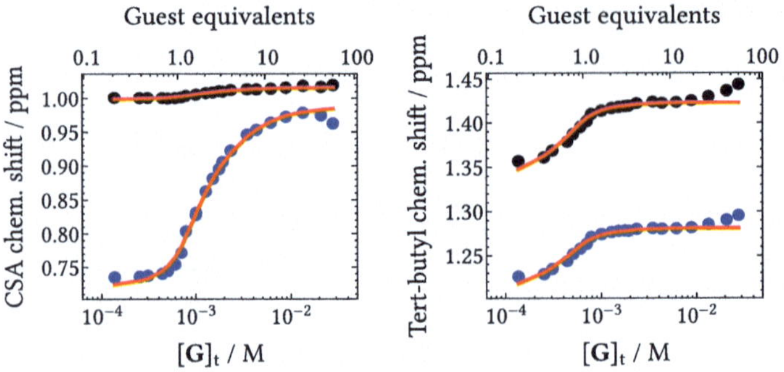

Fig. C.9 Result of simultaneous fitting of chemical shifts of guest's MET resonances (left panel) and host's TB resonances (right panel) during the Step 2

mined as standard deviation. Furthermore, errors of other parameters, denoted here as $\sigma_\bullet$, were obtained from propagation of uncertainty. In particular, error of κ_{CA} (where $\kappa_{\text{ass}} = K_{\mathbf{HG}}\kappa_{\text{diss}}/4$) was calculated as $\sigma_{\kappa_{\text{diss}}} = \sqrt{(\frac{\kappa_{\text{diss}}}{4})^2\sigma^2_{K_{\mathbf{HG}}} + (\frac{K_{\mathbf{HG}}}{4})^2\sigma^2_{\kappa_{\text{diss}}}}$, errors of Gibbs energy barriers $\Delta G_j^\ddagger$ (where $\Delta G_j^\ddagger = -RT\ln\frac{h\kappa_j}{k_B T}$) were calculated as $\sigma_{\Delta G_j^\ddagger} = \sqrt{(\frac{RT}{\kappa_j})^2\sigma^2_{\kappa_j}}$ and errors of standard reaction Gibbs energies ΔG_j° (where $\Delta G_j^\circ = -RT\ln K_j$) were calculated as $\sigma_{\Delta G_j^\circ} = \sqrt{(\frac{RT}{K_j})^2\sigma^2_{K_j}}$. All parameters used during the fitting procedures in Step 1 and Step 2 are listed in Table C.3.

C.8 Kinetic Models of Bz$_2$OxP + (*R*)-CSA System

Our model for the system of di-bromobenzylated oxoporphyrinogen (**H**) in the presence of (*R*)-camphorsulfonic acid (**G**) and water (**W**) combines the competitive host-ligand binding scheme for chemical kinetics (described in Sect. 2.2.3) and the half-symmetric three-state exchange model for the observed spin kinetics. The spin kinetics corresponds to the central NH nuclear spins located at the host.

There are multiple levels of description of the molecular kinetics in our system. The most elementary description is a simplified chemical kinetics Scheme C.1a, assuming the host molecule in three possible chemical states, **H**, **HG** and **HW**. The simplified chemical kinetics is governed by the following equations

$$\frac{\mathrm{d}[\mathbf{HG}]}{\mathrm{d}t} = -\kappa_{\text{diss}}[\mathbf{HG}] + 4\kappa_{\text{ass}}[\mathbf{G}][\mathbf{H}]\,, \tag{C.1a}$$

$$\frac{\mathrm{d}[\mathbf{H}]}{\mathrm{d}t} = -4\kappa_{\text{ass}}[\mathbf{G}][\mathbf{H}] - \kappa_{CD}[\mathbf{W}][\mathbf{H}] + \kappa_{\text{diss}}[\mathbf{HG}] + \kappa_{DC}[\mathbf{HW}]\,, \tag{C.1b}$$

$$\frac{\mathrm{d}[\mathbf{HW}]}{\mathrm{d}t} = -\kappa_{DC}[\mathbf{HW}] + \kappa_{CD}[\mathbf{W}][\mathbf{H}]\,. \tag{C.1c}$$

Table C.3 Overview of fitted and fixed parameters in the fitting procedures in Step 1 and Step 2

Parameter	Value	Comment
$k_{A'C'}$	$47 \pm 6\ s^{-1}$	Mean from raw data fit (Step 1)
$k_{A'}$	10–$10^5\ s^{-1}$	Fitted (Step 1), changing with concentration
$K_{\mathbf{HG}}$	$(7.4 \pm 3.0) \times 10^4\ M^{-1}$	Fitted (Step 2),
		Analysis from Sect. 6.9 yields $(8.0 \pm 5.0) \times 10^4\ M^{-1}$
$K_{\mathbf{HW}}$	$240 \pm 35\ M^{-1}$	Reported value [3], fixed during Step 2
κ_{taut}	$(10 \pm 1) \times 10^5\ M^{-1}s^{-1}$	Fitted (Step 2), binding isotherm $k_{A'} = 2\kappa_{\text{taut}}[\mathbf{G}]$
κ_{diss}	$47 \pm 6\ s^{-1}$	$\kappa_{\text{diss}} = k_{A'C'}$, mean from raw data fit (Step 1),
		Fixed during Step 2
κ_{ass}	$(9 \pm 4) \times 10^5\ M^{-1}s^{-1}$	Calculated, $\kappa_{\text{ass}} = K_{\mathbf{HG}}\kappa_{\text{diss}}/4$,
$\delta_{A'_{\text{I}}}$	13.08 ppm	At low $[\mathbf{G}]_t$; fixed (Step 1), obtained from peak maxima,
	< 13.08 ppm	At high $[\mathbf{G}]_t$; fitted with fixed $\delta_{A'_{\text{I}}} - \delta_{A'_{\text{II}}}$ (Step 1)
$\delta_{A'_{\text{II}}}$	11.76 ppm	At low $[\mathbf{G}]_t$; fixed (Step 1), obtained from peak maxima,
	< 11.76 ppm	At high $[\mathbf{G}]_t$; fitted with fixed $\delta_{A'_{\text{I}}} - \delta_{A'_{\text{II}}}$ (Step 1)
$\delta_{C'}$	9.19–9.65 ppm	Fixed, changing with concentration, obtained from the peak maxima (Step 1)
δ_C	7.8 ± 0.2 ppm	Fitted (Step 2), reported value [3] 7.86 ± 0.40 ppm
δ_D	10.4 ± 0.2 ppm	Fitted (Step 2), reported value [3] 10.09 ± 0.03 ppm
δ(**G***MET*1 free)	1.02 ± 0.01 ppm	Fitted (Step 2)
δ(**G***MET*1 bound)	1.00 ± 0.01 ppm	Fitted (Step 2)
δ(**G***MET*2 free)	0.99 ± 0.01 ppm	Fitted (Step 2)
δ(**G***MET*2 bound)	0.71 ± 0.01 ppm	Fitted (Step 2)
δ(**H***TB*1 free)	1.33 ± 0.01 ppm	Fitted (Step 2)
δ(**H***TB*1 bound)	1.42 ± 0.01 ppm	Fitted (Step 2)
δ(**H***TB*2 free)	1.20 ± 0.01 ppm	Fitted (Step 2)
δ(**H***TB*2 bound)	1.28 ± 0.01 ppm	Fitted (Step 2)

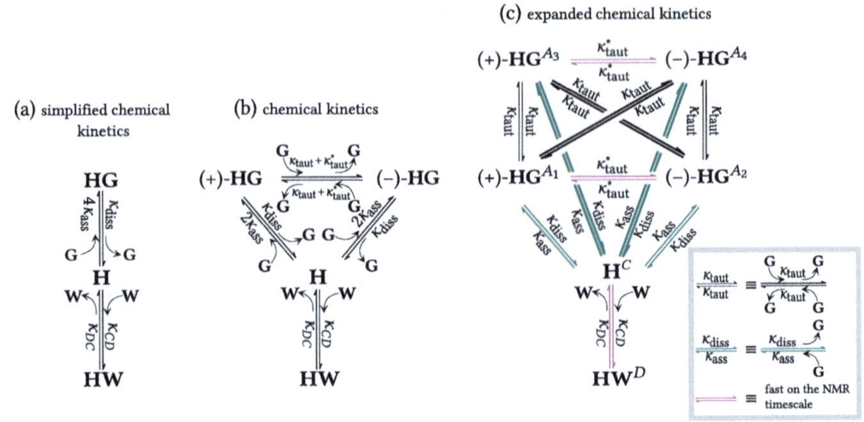

Scheme C.1 Chemical kinetics schemes for the multi-state system of di-bromobenzylated oxoporphyrinogen (host **H**) in the presence of two potential ligands, (R)-camphorsulfonic acid (ligand **G**) and water (ligand **W**). (**a**) Simplified chemical kinetics scheme corresponding to 1:1 **H**:**G** binding with competitive 1:1 **H**:**W** binding described by Eq. (C.1). (**b**) Chemical kinetics scheme described by Eq. (C.4). (**c**) Expanded chemical kinetics scheme described by Eq. (C.5). All relevant molecular processes and their reaction rate coefficients are shown. Processes denoted by magenta arrows are fast on the NMR timescale

The factor of four in the term $4\kappa_{\text{ass}}[\mathbf{G}][\mathbf{H}]$ is caused by the presence of four possible protonation sites, increasing the rate of **HG** formation. The reaction rate coefficients are related to the equilibrium constants through the following formulae (obtained for the time derivatives in Eqs. (C.1a) and (C.1c) equal to zero)

$$K_{\mathbf{HG}} = 4K_{\mathbf{HG}}^{\text{micro}} = 4\frac{\kappa_{\text{ass}}}{\kappa_{\text{diss}}}\,, \tag{C.2a}$$

$$K_{\mathbf{HW}} = \frac{\kappa_{CD}}{\kappa_{DC}}\,. \tag{C.2b}$$

The stepwise binding constant, defined as $K_{\mathbf{HG}} = [\mathbf{HG}]/([\mathbf{H}][\mathbf{G}])$, describes protonation of any of the four possible protonation sites (hence the factor of four in Eq. (C.2a)). On the other hand, the microscopic binding constant $K_{\mathbf{HG}}^{\text{micro}} = \kappa_{\text{ass}}/\kappa_{\text{diss}}$ describes binding to one particular binding site [4]. Thus, this constant corresponds to the "microscopic" standard reaction Gibbs energy $\Delta G_{\mathbf{HG}}^{\circ\text{micro}}$, which can be obtained from quantum chemistry calculations [5] as $\Delta G_{\mathbf{HG}}^{\circ\text{micro}} = \mu_{\mathbf{HG}}^{\circ} - \mu_{\mathbf{H}}^{\circ} - \mu_{\mathbf{G}}^{\circ}$ ($\mu_{\bullet}^{\circ}$ represents the standard chemical potential of species indicated in subscript). The quantity $\Delta G_{\mathbf{HG}}^{\circ\text{micro}}$ is higher than the standard reaction Gibbs energy $\Delta G_{\mathbf{HG}}^{\circ} = -RT \ln K_{\mathbf{HG}}$ (the quantity $\Delta G_{\mathbf{HG}}^{\circ}$ is always defined by the latter equation from the stepwise equilibrium constant, but it should be correctly interpreted [6, 7]),

$$\Delta G_{\mathbf{HG}}^{\circ\text{micro}} = -RT \ln K_{\mathbf{HG}}^{\text{micro}} = -RT \ln \frac{K_{\mathbf{HG}}}{4} = \Delta G_{\mathbf{HG}} + RT \ln 4\,. \tag{C.3}$$

The numerical value is $\Delta G_{\mathbf{HG}}^{\text{omicro}} = \Delta G_{\mathbf{HG}} + RT \ln 4 = -27.76 + 3.43$ kJ.mol^{-1} = -24.33 kJ.mol^{-1} at $T = 298$ K.

In fact, the protonated species **HG** are not identical as they form two enantiomers (+)-**HG**) and (−)-(**HG**). The chemical kinetics is described by Scheme C.1b, which is an expansion of the Scheme C.1a. This scheme is governed by the following equations

$$\frac{\mathrm{d}[(+)\text{-}\mathbf{HG})]}{\mathrm{d}t} = -(\kappa_{\text{taut}} + \kappa^{*}_{\text{taut}})[\mathbf{G}][(+)\text{-}\mathbf{HG})] - \kappa_{\text{diss}}[(+)\text{-}\mathbf{HG})] + (\kappa_{\text{taut}} + \kappa^{*}_{\text{taut}})[\mathbf{G}][(-)\text{-}(\mathbf{HG})] + 2\kappa_{\text{ass}}[\mathbf{G}][\mathbf{H}], \tag{C.4a}$$

$$\frac{\mathrm{d}[(-)\text{-}(\mathbf{HG})]}{\mathrm{d}t} = -(\kappa_{\text{taut}} + \kappa^{*}_{\text{taut}})[\mathbf{G}][(-)\text{-}(\mathbf{HG})] - \kappa_{\text{diss}}[(-)\text{-}(\mathbf{HG})] + (\kappa_{\text{taut}} + \kappa^{*}_{\text{taut}})[\mathbf{G}][(+)\text{-}\mathbf{HG})] + 2\kappa_{\text{ass}}[\mathbf{G}][\mathbf{H}], \tag{C.4b}$$

$$\frac{\mathrm{d}[\mathbf{H}]}{\mathrm{d}t} = -4\kappa_{\text{ass}}[\mathbf{G}][\mathbf{H}] - \kappa_{CD}[\mathbf{W}][\mathbf{H}] + \kappa_{\text{diss}}\left([(+)\text{-}\mathbf{HG})] + [(-)\text{-}(\mathbf{HG})]\right) + \kappa_{DC}[\mathbf{HW}], \tag{C.4c}$$

$$\frac{\mathrm{d}[\mathbf{HW}]}{\mathrm{d}t} = -\kappa_{DC}[\mathbf{HW}] + \kappa_{CD}[\mathbf{W}][\mathbf{H}]. \tag{C.4d}$$

These equations are compatible with the simplified chemical kinetics (Eq. (C.1)) by summing Eq. (C.4a) and (C.4b), and substituting [(+)-**HG**)] + [(−)-(**HG**)] = [**HG**].

In order to establish the connection between chemical kinetics and spin kinetics, Scheme C.1b must be expanded to incorporate all relevant microstates, see Scheme C.1c. The states (+)-**HG**)^{A_1}, (+)-**HG**)^{A_3}, (−)-(**HG**)^{A_2} and (−)-(**HG**)^{A_4} correspond to protonation of different C=O sites (see Fig. 7.10 for molecular structures). The expanded chemical kinetics is described by the following equations

$$\begin{aligned}\frac{\mathrm{d}[(+)\text{-}\mathbf{HG})^{A_1}]}{\mathrm{d}t} &= -(2\kappa_{\text{taut}} + \kappa^{*}_{\text{taut}})[\mathbf{G}][(+)\text{-}\mathbf{HG})^{A_1}] - \kappa_{\text{diss}}[(+)\text{-}\mathbf{HG})^{A_1}] \\ &+ \kappa_{\text{taut}}[\mathbf{G}][(+)\text{-}\mathbf{HG})^{A_3}] + \kappa^{*}_{\text{taut}}[\mathbf{G}][(-)\text{-}(\mathbf{HG})^{A_2}] + \kappa_{\text{taut}}[\mathbf{G}][(-)\text{-}(\mathbf{HG})^{A_4}] + \kappa_{\text{ass}}[\mathbf{G}][\mathbf{H}^{C}],\end{aligned} \tag{C.5a}$$

$$\begin{aligned}\frac{\mathrm{d}[(+)\text{-}\mathbf{HG})^{A_3}]}{\mathrm{d}t} &= -(2\kappa_{\text{taut}} + \kappa^{*}_{\text{taut}})[\mathbf{G}][(+)\text{-}\mathbf{HG})^{A_3}] - \kappa_{\text{diss}}[(+)\text{-}\mathbf{HG})^{A_3}] \\ &+ \kappa_{\text{taut}}[\mathbf{G}][(+)\text{-}\mathbf{HG})^{A_1}] + \kappa_{\text{taut}}[\mathbf{G}][(-)\text{-}(\mathbf{HG})^{A_2}] + \kappa^{*}_{\text{taut}}[\mathbf{G}][(-)\text{-}(\mathbf{HG})^{A_4}] + \kappa_{\text{ass}}[\mathbf{G}][\mathbf{H}^{C}],\end{aligned} \tag{C.5b}$$

$$\begin{aligned}\frac{\mathrm{d}[(-)\text{-}(\mathbf{HG})^{A_2}]}{\mathrm{d}t} &= -(2\kappa_{\text{taut}} + \kappa^{*}_{\text{taut}})[\mathbf{G}][(-)\text{-}(\mathbf{HG})^{A_2}] - \kappa_{\text{diss}}[(-)\text{-}(\mathbf{HG})^{A_2}] \\ &+ \kappa^{*}_{\text{taut}}[\mathbf{G}][(+)\text{-}\mathbf{HG})^{A_1}] + \kappa_{\text{taut}}[\mathbf{G}][(+)\text{-}\mathbf{HG})^{A_3}] + \kappa_{\text{taut}}[\mathbf{G}][(-)\text{-}(\mathbf{HG})^{A_4}] + \kappa_{\text{ass}}[\mathbf{G}][\mathbf{H}^{C}],\end{aligned} \tag{C.5c}$$

$$\begin{aligned}\frac{\mathrm{d}[(-)\text{-}(\mathbf{HG})^{A_4}]}{\mathrm{d}t} &= -(2\kappa_{\text{taut}} + \kappa^{*}_{\text{taut}})[\mathbf{G}][(-)\text{-}(\mathbf{HG})^{A_4}] - \kappa_{\text{diss}}[(-)\text{-}(\mathbf{HG})^{A_4}] \\ &+ \kappa_{\text{taut}}[\mathbf{G}][(+)\text{-}\mathbf{HG})^{A_1}] + \kappa^{*}_{\text{taut}}[\mathbf{G}][(+)\text{-}\mathbf{HG})^{A_3}] + \kappa_{\text{taut}}[\mathbf{G}][(-)\text{-}(\mathbf{HG})^{A_2}] + \kappa_{\text{ass}}[\mathbf{G}][\mathbf{H}^{C}],\end{aligned} \tag{C.5d}$$

$$\begin{aligned}\frac{\mathrm{d}[\mathbf{H}^{C}]}{\mathrm{d}t} &= -4\kappa_{\text{ass}}[\mathbf{G}][\mathbf{H}^{C}] - \kappa_{CD}[\mathbf{W}][\mathbf{H}^{C}] \\ &+ \kappa_{\text{diss}}\left([(+)\text{-}\mathbf{HG})^{A_1}] + [(+)\text{-}\mathbf{HG})^{A_3}] + [(-)\text{-}(\mathbf{HG})^{A_2}] + [(-)\text{-}(\mathbf{HG})^{A_4}]\right) + \kappa_{DC}[\mathbf{HW}^{D}],\end{aligned} \tag{C.5e}$$

$$\frac{\mathrm{d}[\mathbf{HW}^{D}]}{\mathrm{d}t} = -\kappa_{DC}[\mathbf{HW}^{D}] + \kappa_{CD}[\mathbf{W}][\mathbf{H}^{C}]. \tag{C.5f}$$

The above equations are compatible with the chemical kinetics scheme governed by Eq. (C.4) as shown by summing Eq. (C.5a) with (C.5b) and Eq. (C.5c) with (C.5d), and substituting $[(+)\text{-}\mathbf{HG})^{A_1}] + [(+)\text{-}\mathbf{HG})^{A_3}] = [(+)\text{-}\mathbf{HG})]$ and $[(-)\text{-}(\mathbf{HG})^{A_2}] + [(-)\text{-}(\mathbf{HG})^{A_4}] = [(-)\text{-}(\mathbf{HG})]$.

This expanded chemical kinetics scheme accounts for all relevant molecular processes and microstates (i.e. there is absence of integer prefactors or sums of reaction rate coefficients accounting for degeneracy) and explains all the prefactors or sums of reaction rate coefficients used in previous Schemes C.1a, b. The corresponding energy barriers can be calculated employing the Eyring equation (Eq. (2.20)). In order to simplify the model, the guest-mediated prototropic tautomerization between states $(+)\text{-}\mathbf{HG})^{A_1} \leftrightarrow (+)\text{-}\mathbf{HG})^{A_3}$, $(+)\text{-}\mathbf{HG})^{A_1} \leftrightarrow (-)\text{-}(\mathbf{HG})^{A_4}$, $(-)\text{-}(\mathbf{HG})^{A_2} \leftrightarrow (+)\text{-}\mathbf{HG})^{A_3}$ and $(-)\text{-}(\mathbf{HG})^{A_2} \leftrightarrow (-)\text{-}(\mathbf{HG})^{A_4}$ is characterized by single effective reaction rate coefficient κ_A, although the real energy barriers might differ. On the other hand, prototropic tautomerization between states $(+)\text{-}\mathbf{HG})^{A_1} \leftrightarrow (-)\text{-}(\mathbf{HG})^{A_2}$ and $(+)\text{-}\mathbf{HG})^{A_3} \leftrightarrow (-)\text{-}(\mathbf{HG})^{A_4}$ is characterized by the reaction rate coefficient κ_A^*.

At this point, fast exchange between states $(+)\text{-}\mathbf{HG})^{A_1} \leftrightarrow (-)\text{-}(\mathbf{HG})^{A_2}$, $(+)\text{-}\mathbf{HG})^{A_3} \leftrightarrow (-)\text{-}(\mathbf{HG})^{A_4}$ and $\mathbf{H}^C \leftrightarrow \mathbf{H}^D$ should be taken into account. The equivalent reduced spin kinetics scheme is constructed from the expanded chemical kinetics Scheme C.1c according to the procedure explained in Sect. 3.3.8. In this case, Eq. (C.5a) and (C.5c), Eq. (C.5b) and (C.5d), and Eq. (C.5e) and (C.5f), respectively, are added up, expansion $[\mathbf{H}^C] = \frac{[\mathbf{H}^C]}{[\mathbf{H}^C]+[\mathbf{H}^D]}([\mathbf{H}^C] + [\mathbf{H}^D])$ is made and substitutions $[(+)\text{-}\mathbf{HG})^{A_1}] + [(-)\text{-}(\mathbf{HG})^{A_2}] = [\mathbf{HG}^{A'_{\mathrm{I}}}]$, $[(+)\text{-}\mathbf{HG})^{A_3}] + [(-)\text{-}(\mathbf{HG})^{A_4}] = [\mathbf{HG}^{A'_{\mathrm{II}}}]$ and $[\mathbf{H}^C] + [\mathbf{H}^D] = [\mathbf{H}^{C'}]$ are used. This procedure yields governing equations for three-state spin kinetics corresponding to the spin kinetics Scheme C.2a, in particular

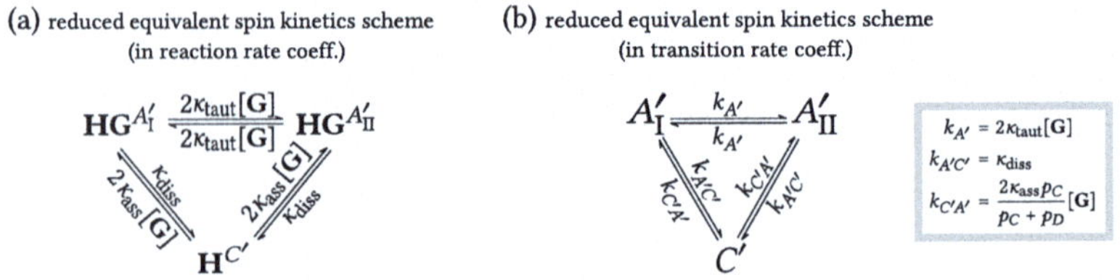

Scheme C.2 Reduced equivalent spin kinetics schemes for the multi-state system of dibromobenzylated oxoporphyrinogen (host **H**) in the presence of two potential ligands, (*R*)-camphorsulfonic acid (ligand **G**) and water (ligand **W**). The schemes refer to central NH protons of the host molecule. (**a**) Spin kinetics in terms of reaction rate coefficients as obtained from contraction of Scheme C.1c. It is described by Eq. (C.6). (**b**) Corresponding spin kinetics in terms of transition rate coefficients described by Eq. (C.7). It has the form of half-symmetric three-state exchange. Comparison with (a) gives the the relationship between transition and reaction rate coefficients in Eq. (C.8). This model was used for lineshape fitting, see Sect. C.7 for details

$$\frac{d[\mathbf{HG}^{A'_{\mathrm{I}}}]}{dt} = -2\kappa_{\mathrm{taut}}[\mathbf{G}][\mathbf{HG}^{A'_{\mathrm{I}}}] - \kappa_{\mathrm{diss}}[\mathbf{HG}^{A'_{\mathrm{I}}}] + 2\kappa_{\mathrm{taut}}[\mathbf{G}][\mathbf{HG}^{A'_{\mathrm{II}}}] + 2\kappa_{\mathrm{ass}}[\mathbf{G}][\mathbf{H}^{C}] , \tag{C.6a}$$

$$\frac{d[\mathbf{HG}^{A'_{\mathrm{II}}}]}{dt} = -2\kappa_{\mathrm{taut}}[\mathbf{G}][\mathbf{HG}^{A'_{\mathrm{II}}}] - \kappa_{\mathrm{diss}}[\mathbf{HG}^{A'_{\mathrm{II}}}] + 2\kappa_{\mathrm{taut}}[\mathbf{G}][\mathbf{HG}^{A'_{\mathrm{I}}}] + 2\kappa_{\mathrm{ass}}[\mathbf{G}][\mathbf{H}^{C}] , \tag{C.6b}$$

$$\frac{d[\mathbf{H}^{C'}]}{dt} = -4\kappa_{\mathrm{ass}}\frac{[\mathbf{G}][\mathbf{H}^{C}]}{[\mathbf{H}^{C}] + [\mathbf{H}^{D}]}[\mathbf{H}^{C'}] + \kappa_{\mathrm{diss}}\left([\mathbf{HG}^{A'_{\mathrm{I}}}] + [\mathbf{HG}^{A'_{\mathrm{II}}}]\right) . \tag{C.6c}$$

These equations can be further compared with a spin kinetics scheme formulated in terms of transition rate coefficients as shown in Scheme C.2b, which is governed by the following set of differential equations

$$\frac{dp_{A'_{\mathrm{I}}}}{dt} = -(k_{A'} + k_{A'C'})p_{A'_{\mathrm{I}}} + k_{A'}p_{A'_{\mathrm{II}}} + k_{C'A'}p_{C'} , \tag{C.7a}$$

$$\frac{dp_{A'_{\mathrm{II}}}}{dt} = -(k_{A'} + k_{A'C'})p_{A'_{\mathrm{II}}} + k_{A'}p_{A'_{\mathrm{I}}} + k_{C'A'}p_{C'} , \tag{C.7b}$$

$$\frac{dp_{C'}}{dt} = -2k_{C'A'}p_{C'} + k_{A'C'}p_{A'_{\mathrm{I}}} + k_{A'C'}p_{A'_{\mathrm{II}}} . \tag{C.7c}$$

Relationships between the transition and reaction rate coefficients are established by comparison of Eqs. (C.6) and (C.7) using the appropriate relations for spin state populations, i.e. $p_{A'_{\mathrm{I}}} = \frac{[\mathbf{HG}^{A'_{\mathrm{I}}}]}{[\mathbf{H}]_{\mathrm{t}}}$, $p_{A'_{\mathrm{II}}} = \frac{[\mathbf{HG}^{A'_{\mathrm{II}}}]}{[\mathbf{H}]_{\mathrm{t}}}$ and $p_{C'} = \frac{[\mathbf{H}^{C'}]}{[\mathbf{H}]_{\mathrm{t}}}$. As a result, we also obtain the dependencies of transition rate coefficients on guest concentration,

$$k_{A'} = 2\kappa_{A}[\mathbf{G}] , \tag{C.8a}$$

$$k_{A'C'} = \kappa_{\mathrm{diss}} , \tag{C.8b}$$

$$k_{C'A'} = \frac{2\kappa_{\mathrm{ass}}p_{C}}{p_{C} + p_{D}}[\mathbf{G}] . \tag{C.8c}$$

C.9 Prototropic Tautomerization Processes in Bz_2OxP

Prototropic tautomerization consists of several steps, which are indistinguishable in NMR (due to fast exchange regime). Here, we suggest the corresponding reaction schemes for tautomerization processes. Initial state (−)-(**HG**) can undergo the tautomerization process in three different ways as illustrated in Fig. C.10 (final state (+)-**HG**), characterized by κ_{A}^{*}), Fig. C.11 (final state (−)-(**HG**), characterized by κ_{A}) and Fig. C.12 (final state (+)-**HG**), characterized by κ_{A}). All three reaction paths

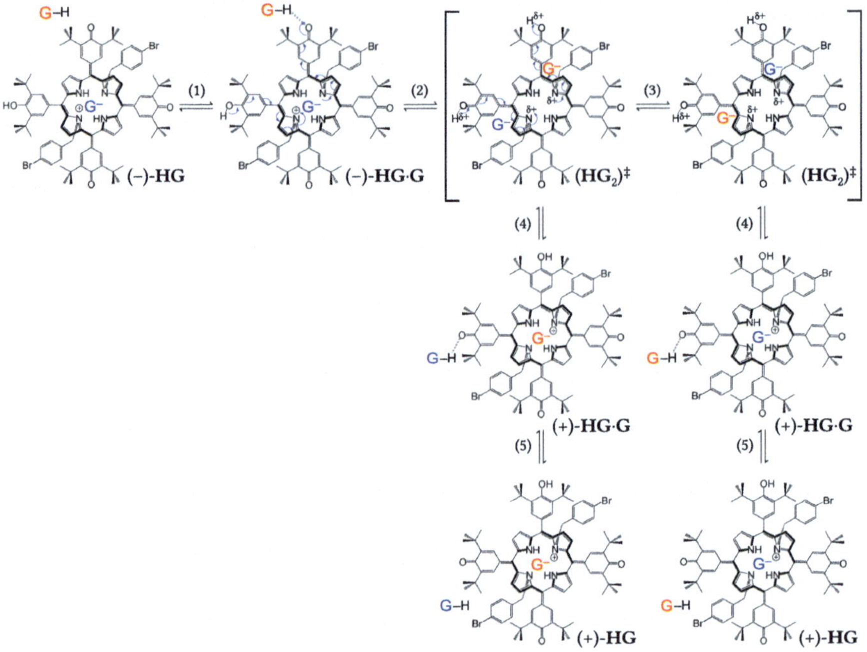

Fig. C.10 Suggested transition structures for prototropic tautomerization from (−)-(**HG**) to (+)-**HG**) characterized by κ_A^*. Dotted line in (+)-**HG**·**G** and (−)-**HG**·**G** represents hydrogen bond, dashed lines in $(\mathbf{HG}_2)^\ddagger$ represent partially formed bonds in transition state

are reversible and consist of analogous steps. In step (1), an incoming guest (red) forms a hydrogen bond with host's carbonyl group thus forming hydrogen-bonded structure (−)-**HG**·G. The next step (2) shows the suggested transition state $(\mathbf{HG}_2)^\ddagger$, which is created by formal redistribution of electron density (denoted by blue arrows in (−)-**HG**·**G** structure and by dashed bonds in $(\mathbf{HG}_2)^\ddagger$ structure). Formally, partial charge $+\delta$ is located at the central alkylated amines and at the protonated carbonyls although in reality, the charge is delocalized. During this step, guest anions change positions and are located close to partially charged atoms of the host in $(\mathbf{HG}_2)^\ddagger$. This allows for (blue and red) guest anions exchange, as shown in step (3). Hence, prototropic tautomerization can, in principle, proceed in two branches, i.e. the initial guest counteranion (blue) either remains at the porphyrinogen center or it is replaced by the incoming acid anion (red). Both branches lead to hydrogen bonded species (+)-**HG**·**G** or (−)-**HG**·**G** in step (4) (redistribution of electron density is denoted by blue arrows in $(\mathbf{HG}_2)^\ddagger$ structure). Finally, monoprotonated host-guest complex (+)-**HG**) or (−)-(**HG**) is formed in step (5). Although the tautomerization processes consist of several steps, in our chemical kinetics scheme (Scheme 7.2c), they are described by effective barriers $\Delta G_A^\ddagger$ (corresponding to κ_A)

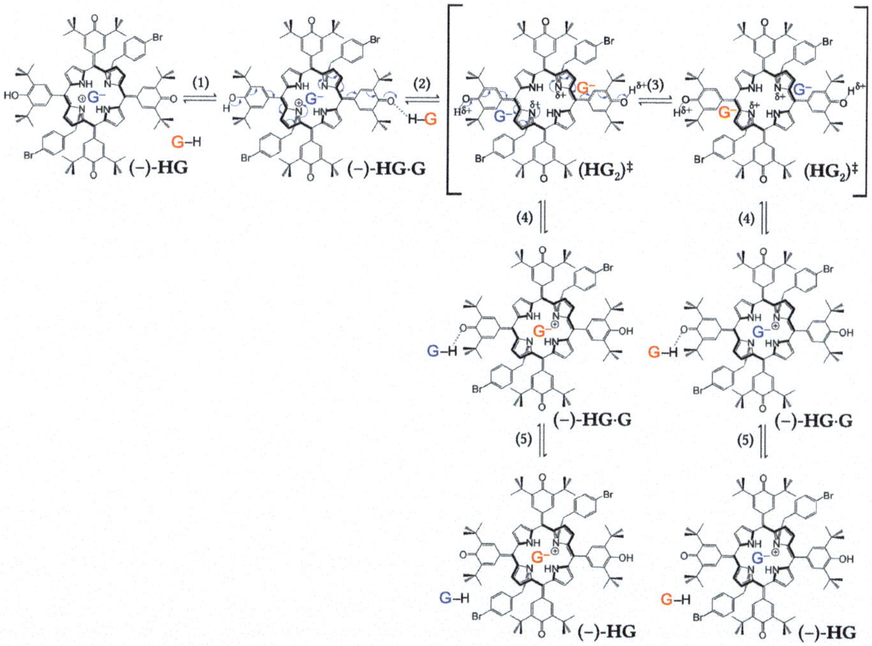

Fig. C.11 Suggested transition structures for prototropic tautomerization from (−)-(**HG**) to (−)-(**HG**) characterized by κ_A. Dotted line in (−)-**HG·G** represents hydrogen bond, dashed lines in $(\mathbf{HG}_2)^{\ddagger}$ represent partially formed bonds in transition state

and $\Delta G^{\ddagger}_{A*}$ (corresponding to κ^*_A). Step (1) is a second-order reaction, hence the effective forward reaction rate (κ_A[**G**][**HG**] or κ^*_A[**G**][**HG**]) depends linearly on [**G**] and [**HG**]. Similarly, step (5) is also a second-order reaction, hence the effective backward reaction rate (κ_A[**G**][**HG**] or κ^*_A[**G**][**HG**]) depends linearly on [**G**] and [**HG**]. Note that we assume equal effective barriers $\Delta G^{\ddagger}_A$ for protonation transfers "left-to-right" (Fig. C.11) and "left-to-down" and (Fig. C.12).

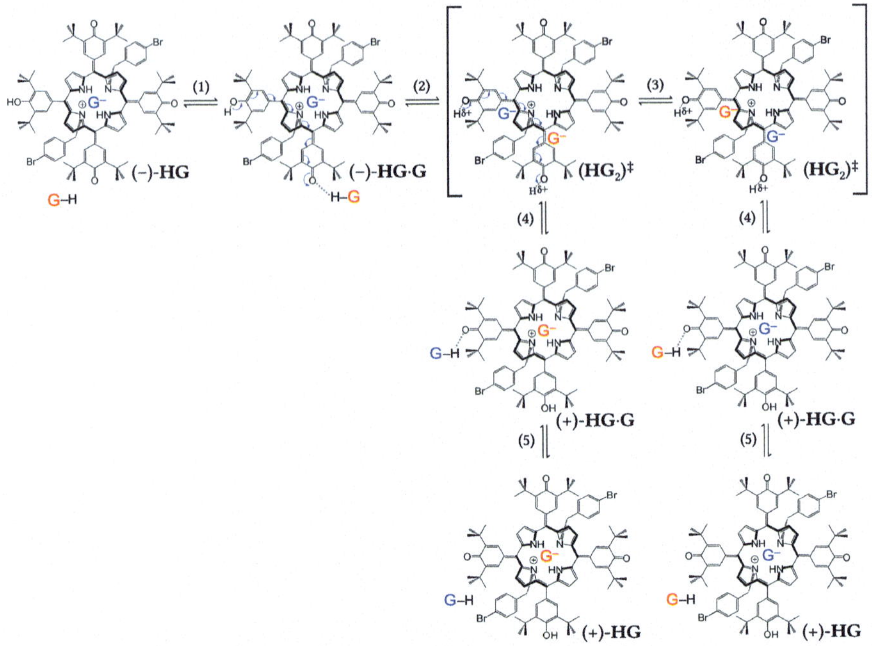

Fig. C.12 Suggested transition structures for prototropic tautomerization from (−)-(**HG**) to (+)-**HG**) characterized by κ_A. Dotted line in (+)-**HG·G** and (−)-**HG·G** represents hydrogen bond, dashed lines in $(\mathbf{HG}_2)^{\ddagger}$ represent partially formed bonds in transition state

References

1. Lindbloom B Spectral calculator. http://www.brucelindbloom.com/.
2. Březina V, Hanyková L, Velychkivska N, Hill JP, Labuta J (2022) NMR lineshape analysis using analytical solutions of multi-state chemical exchange with applications to kinetics of host-guest systems. Sci Rep. 10.1038/s41598-022-20136-4.
3. Labuta J, Ishihara S, Šikorský T, Futera Z, Shundo A, Hanyková L, Burda JV, Ariga K, Hill JP (2013) NMR spectroscopic detection of chirality and enantiopurity in referenced systems without formation of diastereomers. Nat Commun 4:2188.
4. Connors KA (1987) Binding constants: the measurement of molecular complex stability. Wiley.
5. Muddana HS, Gilson MK (2012) Calculation of host-guest binding affinities using a quantum-mechanical energy model. J Chem Theory Comput 8:2023–2033.
6. Atkins PW, Paula JD, Keeler J (2018) Atkins' physical chemistry. Oxford University Press.
7. Ercolani G (2003) Assessment of cooperativity in self-assembly. J Am Chem Soc 125:16097–16103.

Appendix D
SVD Analysis

SVD of UV/vis spectra (in the matrix **A**) was conducted in *Mathematica* using:

```
{U, S, V} = SingularValueDecomposition[A];
```

Several graphs showing quality of DSIS and SVD analysis were adopted from our publication [1] with permission of Wiley-VCH Verlag.

D.1 $\mathbf{Bz_2OxP}$ + (*R*)-CSA Titration

Total host concentration changed during the titration due to dilution by stock solution (see Appendix C.2). Therefore, the measured spectra shown in Fig. D.1a were rescaled to the initial concentration (see footnote 7), plot of the absorbance at 700 nm in Fig. D.1b illustrates the extent of rescaling. SVD analysis following Sect. 5.2.3 was performed on the original as well as rescaled spectra. Figures D.1c–f prove little effect of spectra rescaling on SVD except for the first amplitude vector $\mathbf{v}_1$ (note that $\mathbf{v}_1$ is most sensitive to concentration fluctuations also in other UV/vis titration experiments) and consequently the first combination coefficient $s_1\mathbf{v}_1$. Mutual plot of combination coefficients (e) forms a straight line segment confirming the presence of two absorbing species – free and protonated $\mathbf{Bz_2OxP}$ host. Spectrum of free host $\mathbf{z}_1$ is identified as the first titration measurement $\mathbf{a}_1$ (black bold line in (a)) yielding the $\mathbf{t}_1$ coordinates in combination coefficients space (black point in (e)) using $\mathbf{t}_1 = \mathbf{U}_2^{\ddagger}\mathbf{a}_1$ (from Eqs. (5.21) and (5.14)). However, spectrum of protonated host $\mathbf{z}_2$ was not measured in isolation (the spectral change was not saturated) and its $\mathbf{t}_2$ coordinates were obtained by minimization of the function in Eq. (5.26) using the 1:1 host-guest binding model (see Sect. 2.2.1) for the model populations $\mathbf{P}_{\mathrm{mod}}$. Fitting parameters were the vector $\mathbf{t}_2$ (i.e. T_{12} and T_{22} elements of the transformation matrix **T**) and K_1. Resulting value is $K_{\mathbf{HG}} = (5.0 \pm 1.0) \times 10^4\ \mathrm{M}^{-1}$, uncertainties were obtained by repeated fitting with fixed K_1 value.

V. Březina, *Supramolecular Complexes of Oxoporphyrinogens with Organic Molecules*, Springer Theses, https://doi.org/10.1007/978-3-031-68646-7

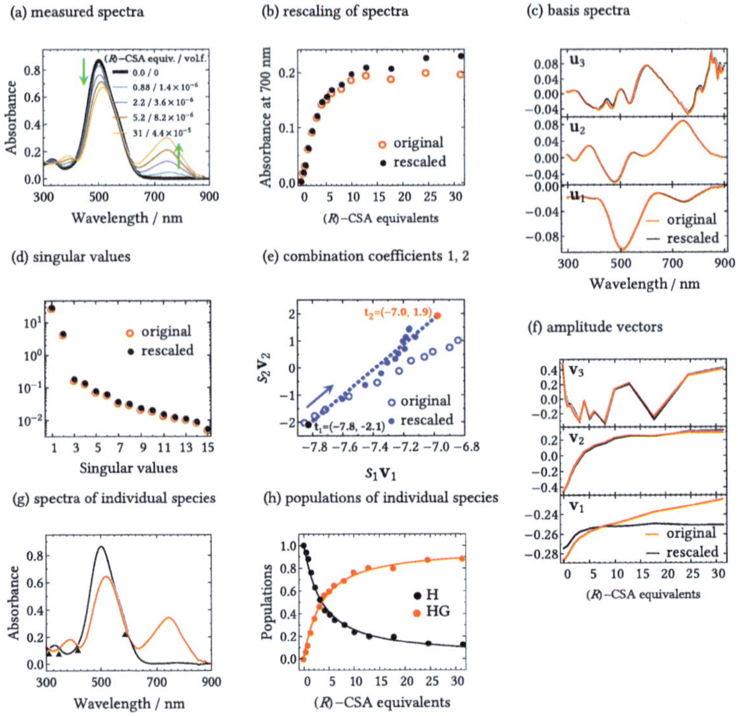

Fig. D.1 UV/vis titration of **Bz_2OxP** (7×10^{-6} M, dilution during measurement compensated by rescaling to initial dye concentration, $CDCl_3$) with (*R*)-CSA and corresponding analyses. (**a**) Rescaled titration spectra. (**b**) Absorbance at 700 nm before and after rescaling. SVD analysis before and after rescaling yielded (**c**) basis spectra, (**d**) singular values, (**e**) mutual plot of combination coefficients 1,2 with coordinates $\mathbf{t}_1$ and $\mathbf{t}_2$ of the absorbing species, and (**f**) amplitude vectors. The fitting procedure, described in Sect. 5.2.3, yielded (**g**) spectra of free (identified as bold line in panel (a)) and protonated host (obtained from SVD analysis), black triangles denote intersections which produce isosbestic points as explained in Sect. 4.4; and (**h**) populations $\mathbf{P}_{\text{exp}}$ (points) and $\mathbf{P}_{\text{mod}}$ (lines) of these species

D.2 OxP + TFA Titration

The presence of three absorbing components and two consecutive spectral changes is confirmed by singular values in Fig. D.2a and mutual plot of combination coefficients in Fig. D.2c–f showing two straight line segments. Coordinates of **OxP^{2+}** in Fig. D.2c, d do not lie exactly at the intersection of the two lines as would be expected. This is probably because the first combination coefficient $s_1\mathbf{v}_1$ in Fig. D.2b shows a "jump" since it is sensitive to the total dye concentration, which slightly changed (by a few percent as estimated from the ratio of $s_1\mathbf{v}_1$ values before and after the jump) when different stock solution was used to observe the second spectral change.

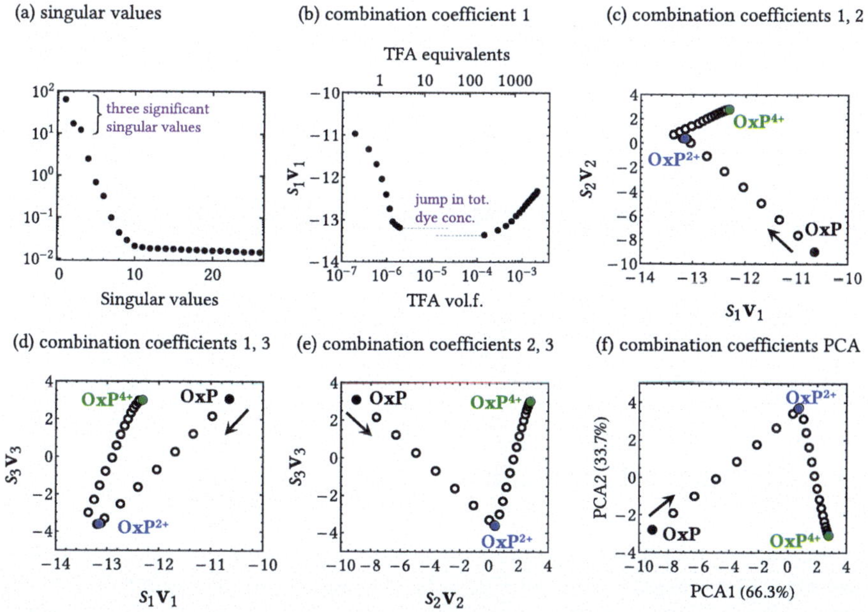

Fig. D.2 SVD analysis of UV/vis titration of **OxP** (10^{-5} M, CH_2Cl_2, data first published in [2]) with TFA, (**a**) singular values, (**b**) first combination coefficients shows 'jump' due to total dye concentration change, (**c, d, e**) mutual plot of combination coefficients. (**f**) PCA on the combination coefficients enables 2D plot of 3D data lying on a plane

Since the combination coefficients from three species lie on a 2D plane in 3D combination coefficients space, they can be conveniently visualized when the plane is properly rotated. As shown in Sect. 5.3, the PCA method is suited for such a rotation. The PCA of combination coefficients almost coincides with the mutual plot of the second and third combination coefficients (the plane has been originally oriented almost perpendicularly to the $s_1\mathbf{v}_1$ dimension), where the effect of total dye concentration change is filtrated out (**OxP**$^{2+}$ lies at the intersection of straight lines). See Appendix D.5 for the detailed implementation of PCA.

Populations were obtained by DSIS (see Sect. 5.2.5) by minimization of Eq. (5.28) with constraints $\sum \mathbf{p}_j = 1$ and $0 \leq \mathbf{p}_j \leq 1$. These populations were fitted using the 1:4 host-guest binding model from Appendix A.4. Furthermore, fitting of the host-guest model was performed in the 3D SVD approximation as described in Sect. 5.2.3, see [3] for details. Figure D.3 shows the high quality of these fits.

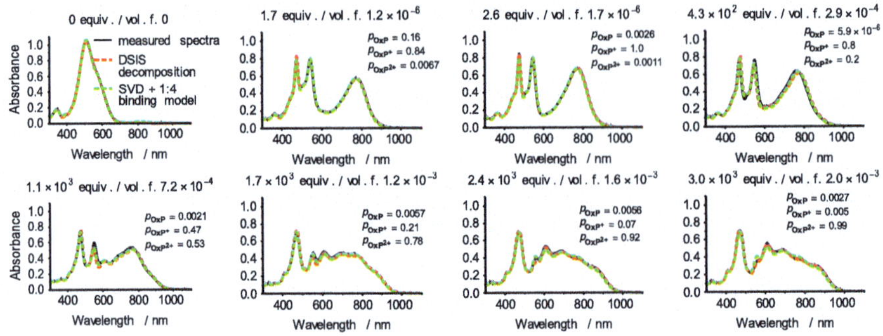

Fig. D.3 Quality check of DSIS and 3D SVD approximation of UV/vis titration of **OxP** with TFA

D.3 Bz_2OxP + DFA Titration

The presence of four absorbing components and three consecutive spectral changes is confirmed by singular values in Fig. D.2a and mutual plot of combination coefficients in Fig. D.2c showing three line segments (Fig. D.4). Because of experimental difficulties (unwanted photoreaction at high acid contents) the titration was repeated several times at different concentration ranges. Finally, the spectra were combined together and rescaled to the initial dye concentration (see footnote 7). Imperfections in rescaling are visible as small "jumps" in the first combination coefficient in Fig. D.2b. Mutual plot of combination coefficients 2 and 3 is not sensitive to these concentration fluctuations.

Populations were obtained by DSIS (see Sect. 5.2.5) by minimization of Eq. (5.28) with constraints $\sum \mathbf{p}_j = 1$ and $0 \leq \mathbf{p}_j \leq 1$. These populations were fitted using the

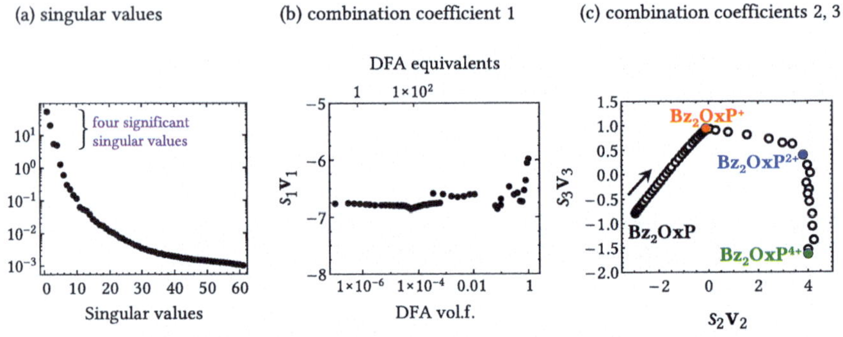

Fig. D.4 SVD analysis of UV/vis titration of **Bz_2OxP** (7×10^{-6}M, $CDCl_3$, diluted during measurement, rescaled to initial concentration) with DFA, **a** singular values, **b** first combination coefficients shows "jumps" due to total dye concentration change, **c** mutual plot of combination coefficients 2, 3

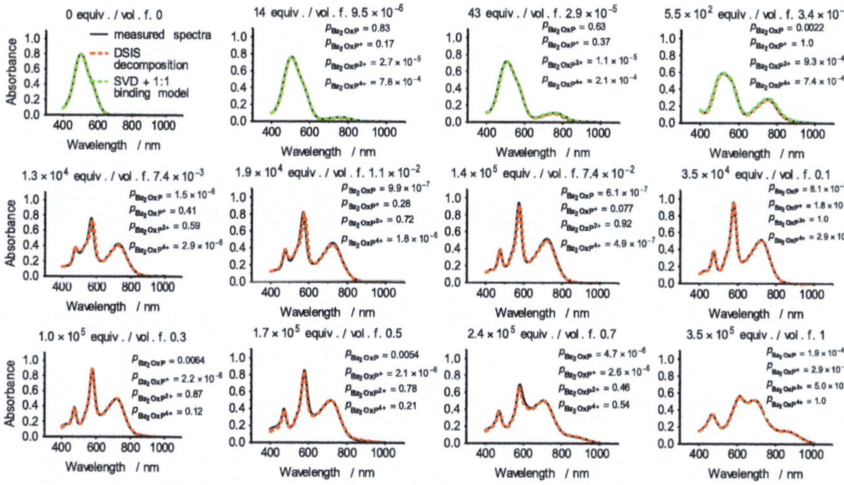

Fig. D.5 Quality check of DSIS and 4D SVD approximation of UV/vis titration of $\mathbf{Bz_2OxP}$ with DFA

1:1 host-guest binding model from Appendix A.1. Furthermore, fitting of the host-guest model was performed in the 4D SVD approximation as described in Sect. 5.2.3, see [3] for details. Figure D.5 shows the high quality of these fits.

D.4 $\mathbf{Bz_4OxP}$ + DFA Titration

Data excluding the first solvatochromic shift were analyzed by SVD. The presence of three absorbing components and two consecutive spectral changes is confirmed by singular values in Fig. D.6a and mutual plot of combination coefficients in Fig. D.6c showing two straight line segments. Rescaling of the experimental spectra produced on jumps (only one outlier) as deduced from the first combination coefficient in Fig. D.6b.

Populations were obtained by DSIS (see Sect. 5.2.5) by minimization of Eq. (5.28) with constraints $\sum \mathbf{p}_j = 1$ and $0 \leq \mathbf{p}_j \leq 1$. Figure D.7 shows the high quality of this decomposition.

D.5 OxP + DMF Titration

Plot of singular values in Fig. D.8a indicates three significant components (capturing 99.975% of the signal), which indicates presence of three absorbing species. However, the structure of combination coefficients plot in Fig. D.8b is nontrivial. The **OxP**

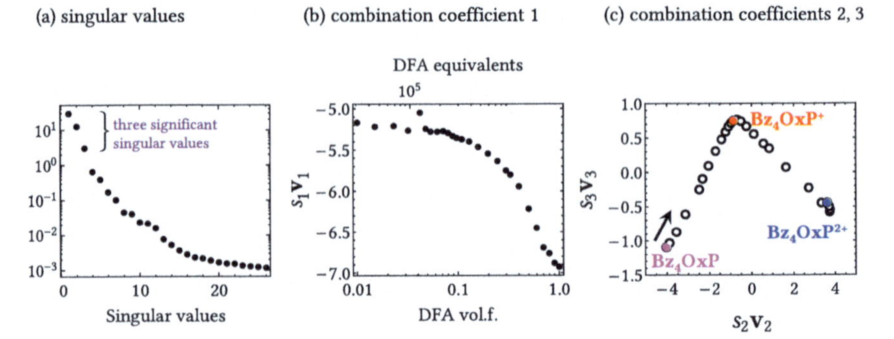

Fig. D.6 SVD analysis of UV/vis titration of **Bz_4OxP** (5×10^{-6}M, $CDCl_3$) with DFA, diluted during measurement, measured spectra rescaled to initial concentration. (**a**) Singular values, (**b**) first combination coefficients shows only one outlier, so the rescaling was very good, (**c**) mutual plot of combination coefficients 2, 3

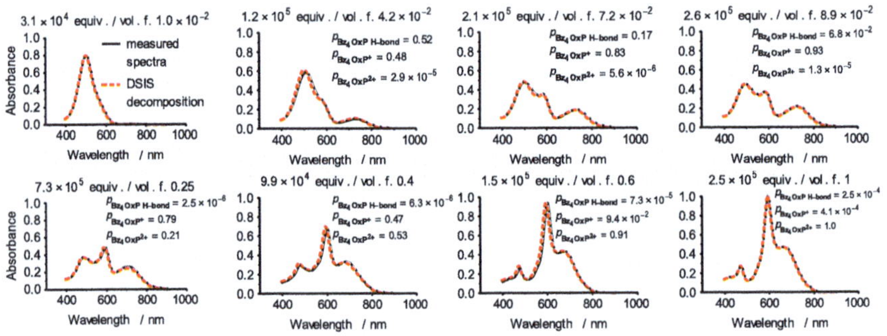

Fig. D.7 Quality check of DSIS of UV/vis titration of **Bz_4OxP** with DFA

and **OxP H-bond** species lie on a straight line but before saturation of the spectral change, there is a "digression" to the **porphodimethene form**. In our publication [3], we concluded the existence of two hydrogen-bonded species (very similar to each other and thus consistent with three significant singular values), however, here we propose this simpler solution with only three absorbing species on the basis of the combination coefficients plot (not present in [3]).

Figure D.8b shows PCA of the combination coefficients. Since PCA identifies directions of maximal variance in a data set (see Sect. 5.3), the first two PCs project the combination coefficients from 3D to 2D plane so that they are maximally "stretched out". This projection accounts for 99.95% of the signal variance, so the combination coefficients indeed lie on a 2D plane in their 3D space, further supporting the presence of three absorbing components (N_{spc} components form an $(N_{spc} - 1)$-dimensional object in the combination coefficients space, although small overlap with other dimensions occurs due to the noise). Coordinates of **OxP** and the **porphodimethene form** in combination coefficients plot were determined from the measured spectra at the endpoints of the spectral changes, where they appear in iso-

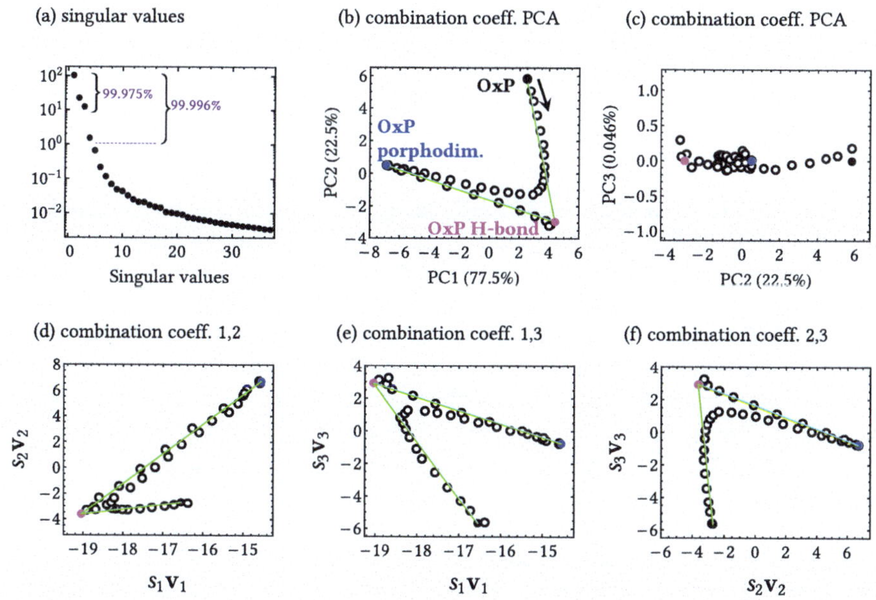

Fig. D.8 SVD analysis of UV/vis titration of **OxP** with DMF, (**a**) singular values show three significant species, (**b**, **c**) PCA plot of combination coefficients, (**d**, **e**, **f**) mutual plot of combination coefficients 2, 3. Colored points show coordinates of the absorbing components. Coordinates in PCA3 are set to zero in **c**

lation (black and blue points in Fig. D.8b). The coordinates of **OxP H-bond** were determined as intersection of the lines delineating the first and third spectral changes in the PCA plot (green lines in Fig. D.8b) while setting the PC3 coordinate of all three species to zero as shown in Fig. D.8c.

For calculation of PCA, the combination coefficients were arranged as columns $\mathbf{x}_j$ of the row-wise data matrix **B** according to Eq. (5.29). This calculation was implemented in *Mathematica* using SVD (Sect. 5.3.2) with the following code

```
B = Transpose@{s1 v1, s2 v2, s3 v3 }; (* row-wise data matrix *)
Bbar = B - Outer[Times,ConstantArray[1,Length@B],
Mean[B]]; (* mean-centering *)
{U, S, V} = SingularValueDecomposition[Bbar];
V1 = V[[All,1]]; V2 = V[[All,2]]; V3 = V[[All,3]];
PC1 = Bbar.V1; (* = U1 s1 *)
PC2 = Bbar.V2; (* = U2 s2 *)
PC3 = Bbar.V3; (* = U3 s3 *)
```

Consequently, PCs are transformed to coordinates $\mathbf{t}_j$ (using inversion of Eq. (5.31)) with

$$\underbrace{\begin{pmatrix} \dots \mathbf{t}_1 \dots \\ \dots \mathbf{t}_2 \dots \\ \dots \mathbf{t}_3 \dots \end{pmatrix}}_{\mathbf{T}^\mathsf{T}} = \begin{pmatrix} \vdots & \vdots & & \vdots \\ \boldsymbol{\alpha}_1 & \boldsymbol{\alpha}_2 & \dots & \boldsymbol{\alpha}_n \\ \vdots & \vdots & & \vdots \end{pmatrix} \begin{pmatrix} \mathrm{PC1(comp.1)} & \mathrm{PC2(comp.1)} & 0 \\ \mathrm{PC1(comp.2)} & \mathrm{PC2(comp.2)} & 0 \\ \mathrm{PC1(comp.3)} & \mathrm{PC2(comp.3)} & 0 \end{pmatrix} + \overline{\mathbf{B}}^\mathsf{T}, \quad \text{(D.1)}$$

where $\boldsymbol{\alpha}_j$ are PCA loadings. As the data during PCA become mean-centered, the mean $\overline{\mathbf{B}}$ should be added in order to obtain true coordinates. The coordinates obtained through PCA nicely fit into the three original combination coefficients planes, Fig. D.8d–f. It would be very tedious if not impossible to find the proper positions of the **OxP H-bond** coordinate directly in these plots without the use of PCA since bad placement results in negative spectra or forbidden values of populations.

When all coordinates (columns of the transformation matrix **T**) are known, the individual spectra are reconstructed using $\mathbf{Z} = \mathbf{U}_{N_\mathrm{spc}}\mathbf{T}$ and populations using $\mathbf{P}_\mathrm{exp}^\mathsf{T} = \mathbf{T}^{-1}\mathbf{S}_{N_\mathrm{spc}}\mathbf{V}_{N_\mathrm{spc}} = \mathbf{Z}^\dagger\mathbf{A}$. The reconstruction of spectra and populations can be implemented as follows

```
{U, S, V} = SingularValueDecomposition[A];
T = Transpose@{t1, t2, t3};
Z = U[[All,1;;3]].T;
PT = PseudoInverse[Z].A;
(* alternatively: PT=Inverse[T].(S.Transpose[V])[[1;;3,All]];*)
```

D.6 Bz_2OxP + DMF at Different vol.f.

Singular values in Fig. D.9a indicate the presence maximum three absorbing components (99.99% of the signal), however, interpretation of mutual plot of combination coefficients in Fig. D.9b requires three components and two straight line segments. The analysis was conducted using PCA in similar manner as in the previous section. Coordinates of **Bz_2OxP** in (b) were obtained directly from the measured spectra. The second spectral change is unsaturated, thus, only interval for reasonable location for PCA coordinates of **form 2** (transparent green line segment) was determined, it lies on the straight line corresponding to the second spectral change (orange line). PCA coordinates of **form 1** lie at the intersection of the two spectral changes (orange lines in Fig. D.9b). PC3 component of the coordinates was set to zero and the coordinates in Fig. D.9d–f were calculated. Then individual spectra and their populations were obtained analogously to the previous appendix.

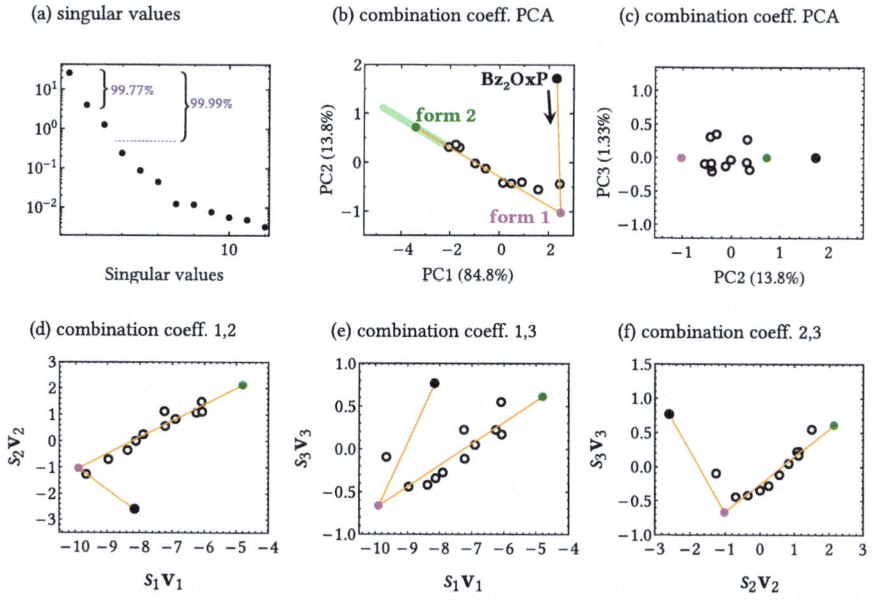

Fig. D.9 SVD analysis of UV/vis titration of **Bz_2OxP** with DMF, **a** singular values show three significant species, **b**, **c** PCA plot of combination coefficients, **d**, **e**, **f** mutual plot of combination coefficients 2, 3. Colored points show coordinates of the absorbing components. Coordinates in PCA3 are set to zero in **c**

D.7 Bz_4OxP + DMF at Different vol.f.

Since the measured spectral change is very small, it cannot be decided if the solvatochromism is caused by formation of new species or by shift of absorption bands (see Sect. 4.5). Here, the hypothetical formation of new species is investigated. The SVD analysis is fully analogous to the previous appendix. The hypothetical individual spectra in Fig. D.10g resemble the hydrogen bonded species of **Bz_2OxP** in Fig. 6.16b. However, no absorption bands arise, which suggests that no new species are formed.

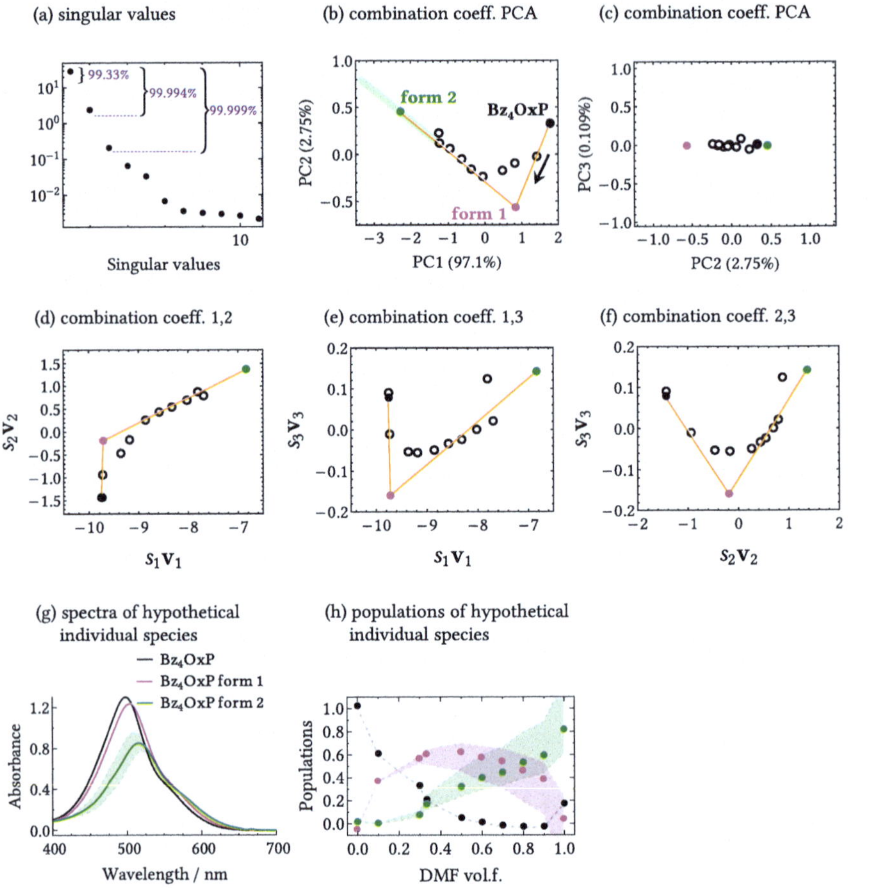

Fig. D.10 SVD analysis of UV/vis titration of **Bz4OxP** with DMF, **a** singular values, **b**, **c** PCA plot of combination coefficients, **d**, **e**, **f** mutual plot of combination coefficients 2, 3. Colored points show coordinates of the absorbing components. Coordinates in PCA3 are set to zero in **c**

References

1. Březina V, Hanyková L, Velychkivska N, Hill JP, Labuta J (2022) NMR lineshape analysis using analytical solutions of multi-state chemical exchange with applications to kinetics of host-guest systems. Sci Rep. 10.1038/s41598-022-20136-4.
2. Shundo A, Ishihara S, Labuta J, Onuma Y, Sakai H, Abe M, Ariga K, Hill JP (2013) Colorimetric visualization of acid-base equilibria in non-polar solvent. Chem Commun 49:6870–6872.
3. Březina V, Ishihara S, Lang J, Hanyková L, Ariga K, Hill JP, Labuta J (2018) Structural modulation of chromic response: effects of binding-site blocking in a conjugated Calix[4]pyrrole chromophore. ChemistryOpen 7:323–335.

Appendix E
NMR Spectra

NMR spectra analyzed in this work are shown here in the full ppm range. Several figures were adopted from our publication [1] with modifications with permission of Wiley-VCH Verlag. Experimental details about NMR measurements can be found in Appendix C.4. Note that NMR assignment of **Bz_2OxP** in our study [1] and in the master's thesis [2] was partially incorrect, the correct assignment can be found here or in [3].

E.1 Spectra of Acids

See Fig. E.1.

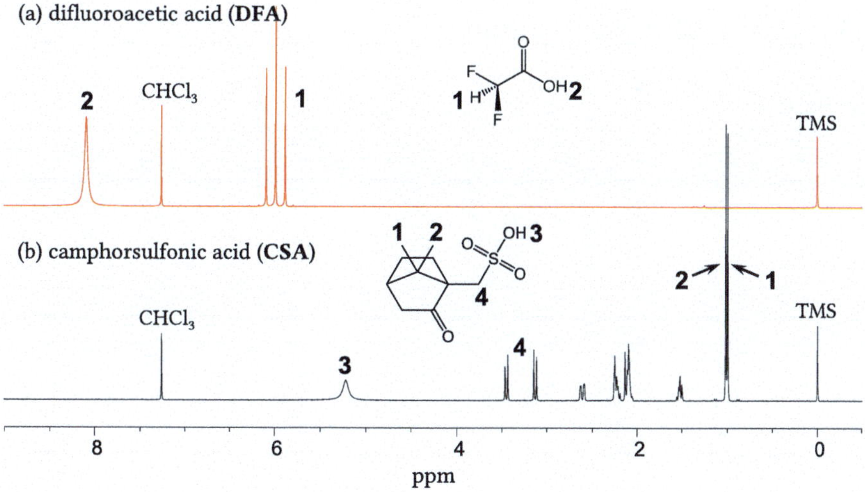

Fig. E.1 NMR spectra of **a** difluoroacetic acid and **b** camphorsulfonic acid, both in $CDCl_3$ at 298 K

V. Březina, *Supramolecular Complexes of Oxoporphyrinogens with Organic Molecules*, Springer Theses, https://doi.org/10.1007/978-3-031-68646-7

E.2 OxP + DFA Titration

See Fig. E.2.

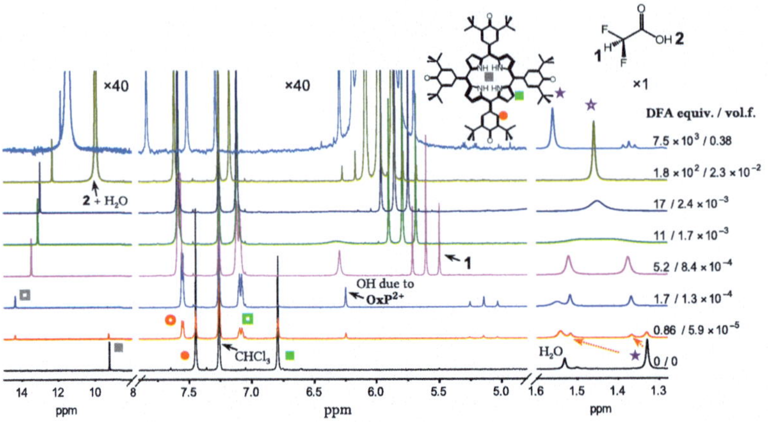

Fig. E.2 Titration of **OxP** with DFA (in $CDCl_3$ at 298 K, host concentration 0.002 M decreased upon addition of stock solution). Free and protonated/complexed species are denoted by full and empty symbols, respectively. Red dotted arrows denote splitting of *tert*-butyl signal due to symmetry breaking after protonation and anion binding. Some regions of the spectra are magnified for clarity. The factor of magnification is given above each region (see factors above each region)

E.3 OxP + (*R*)-CSA Titration

See Fig. E.3.

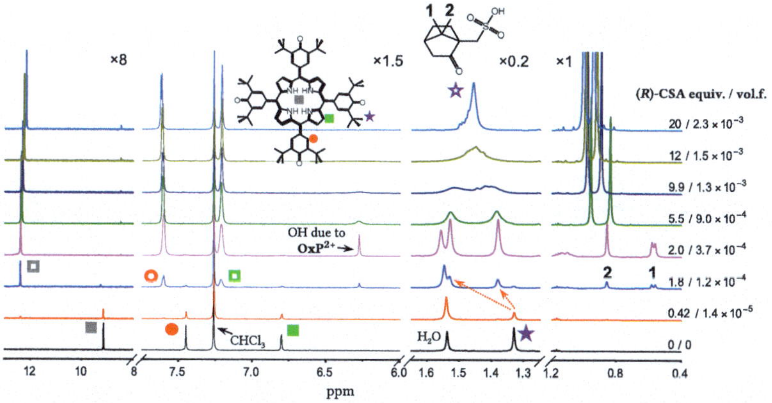

Fig. E.3 Titration of **OxP** with (*R*)-CSA (in $CDCl_3$ at 298 K, host concentration 9×10^{-4} M decreased upon addition of stock solution). Free and protonated/complexed species are denoted by full and empty symbols, respectively. Red dotted arrows denote splitting of *tert*-butyl signal due to symmetry breaking after protonation and anion binding. Certain regions of spectra have been magnified for clarity (see factors above each region). Splitting of methyl signals due to (*R*)-CSA at 0.57 ppm below 2 equiv. is caused by formation of (+)-**H**(*R*)-$\mathbf{G}_2$ and (*R*)-**H**(*R*)-$\mathbf{G}_2$ diastereomers

E.4 Bz$_2$OxP + DFA Titration

See Fig. E.4.

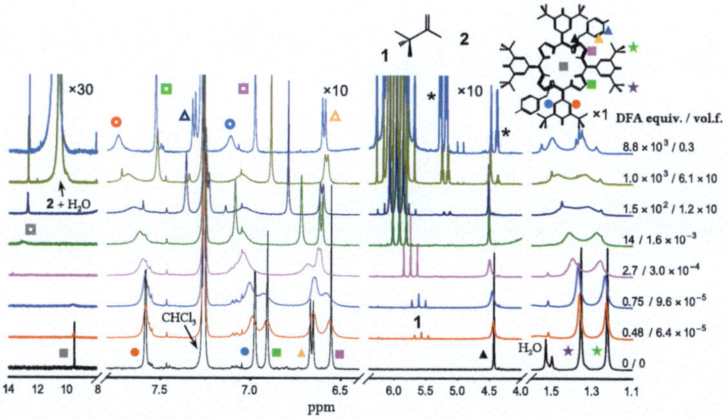

Fig. E.4 Titration of **Bz$_2$OxP** with DFA (in $CDCl_3$ at 298 K, host concentration 7×10^{-4} M decreased upon addition of stock solution). Free and protonated/complexed species are denoted by full and empty symbols, respectively. Asterisk (*) denotes impurities contained in acid. Some regions of the spectra are magnified for clarity (see factors above each region). The factor of magnification is given above each region

E.5 Bz$_2$OxP + (*R*)-CSA Titration

See Fig. E.5.

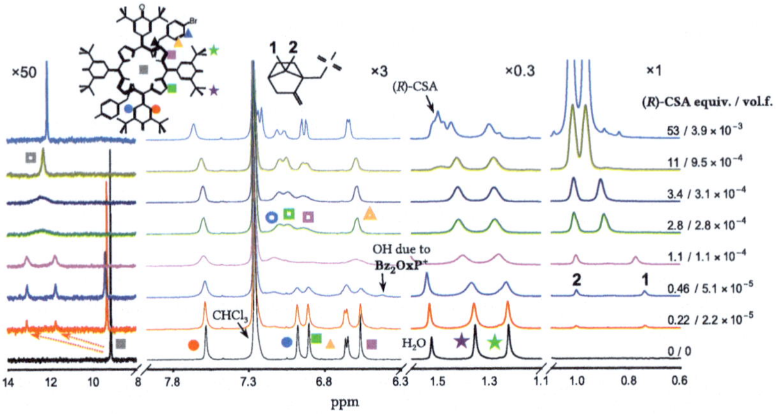

Fig. E.5 Titration of **Bz$_2$OxP** with (*R*)-CSA (in $CDCl_3$ at 298 K, host concentration 6×10^{-4} M decreased upon addition of stock solution). Free and protonated/complexed species are denoted by full and empty symbols, respectively. Red dotted arrows denote splitting of *tert*-butyl signal due to symmetry breaking after protonation and anion binding. Certain regions of spectra have been magnified for clarity (see factors above each region)

E.6 Bz$_2$OxP + (*rac*)-CSA Titration

See Fig. E.6.

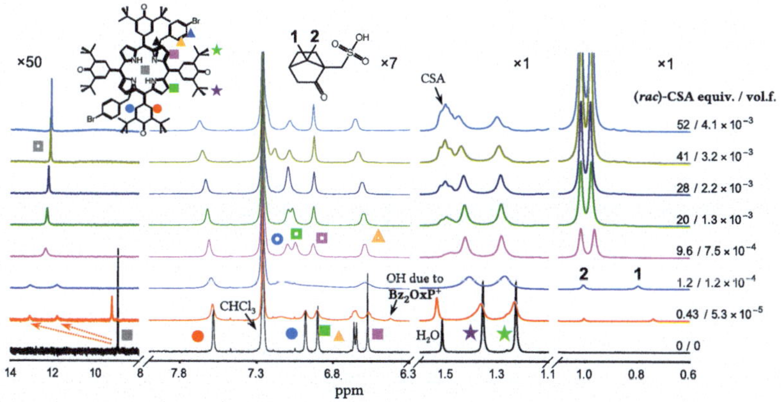

Fig. E.6 Titration of **Bz$_2$OxP** with (*rac*)-CSA (in $CDCl_3$ at 298 K, host concentration 6×10^{-4} M decreased upon addition of stock solution). Free and protonated/complexed species are denoted by full and empty symbols, respectively. Red dotted arrows denote splitting of *tert*-butyl signal due to symmetry breaking after protonation and anion binding. Certain regions of spectra have been magnified for clarity (see factors above each region). Structure of the protonated form is confirmed by COSY in [2]

E.7 Bz$_4$OxP + DFA Titration

See Fig. E.7.

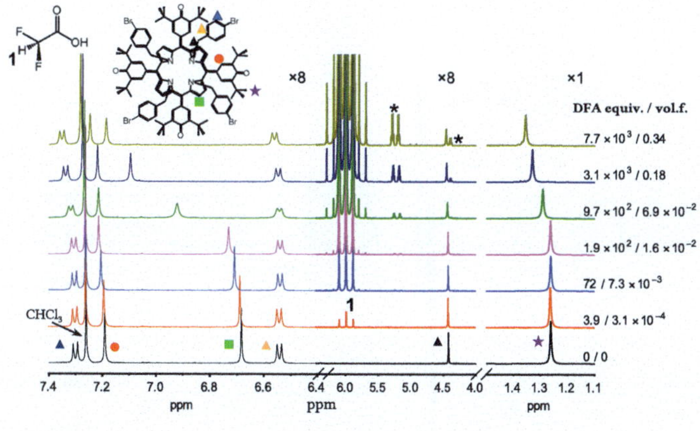

Fig. E.7 Titration of **Bz$_4$OxP** with DFA (in $CDCl_3$ at 298 K, host concentration 8×10^{-4} M decreased upon addition of stock solution). Asterisk (*) denotes impurities contained in acid. Some regions of the spectra are magnified for clarity (see factors above each region)

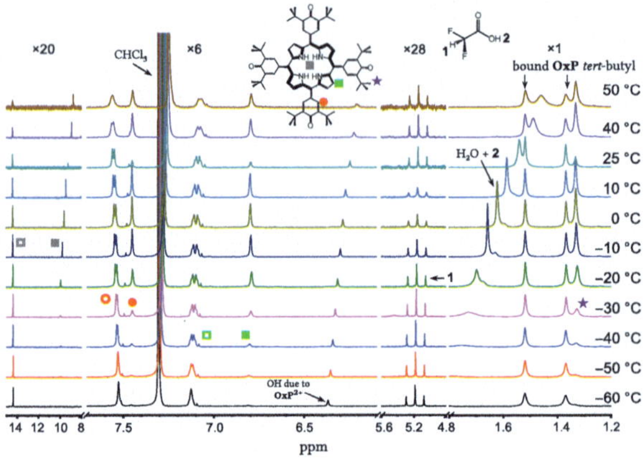

Fig. E.8 Variable temperature measurement of **OxP** with 1.1 equiv. DFA (in $CDCl_3$, host concentration 8.9×10^{-4} M). Full and empty symbols denote free and bound species, respectively. The spectrum at 25 °C shows decrease of free host peak intensity (probably due to insufficiently dissolved host since this spectrum was measured first

E.8 OxP + DFA 1.1 equiv. Variable Temperature

See Fig. E.8.

E.9 OxP + DFA 18 equiv. Variable Temperature

See Fig. E.9.

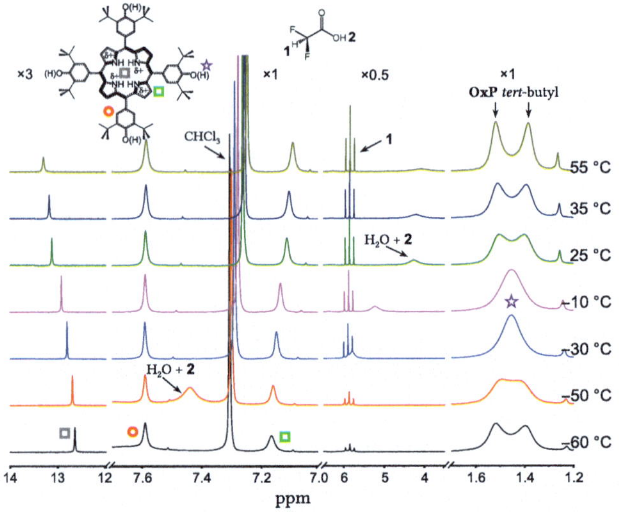

Fig. E.9 Variable temperature measurement of **OxP** with 18 equiv. DFA (in $CDCl_3$, host concentration 8.4×10^{-4} M)

E.10 Bz_2OxP Variable Temperature

See Fig. E.10.

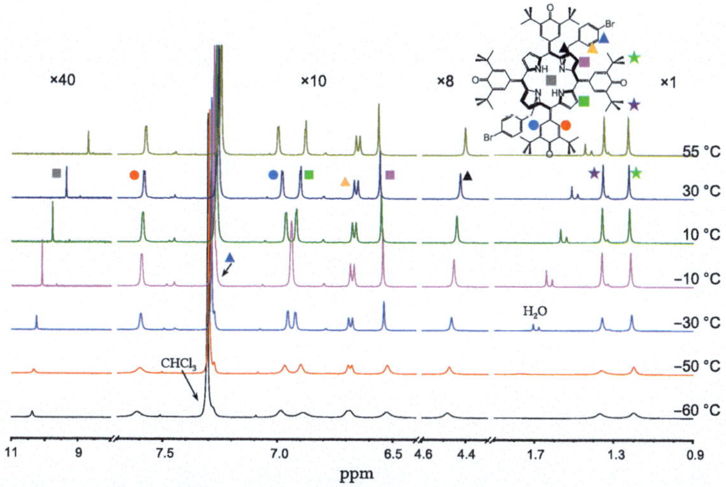

Fig. E.10 Variable temperature measurement of pure $\mathbf{Bz_2OxP}$ (in $CDCl_3$)

E.11 Bz_2OxP + DFA 35 equiv. Variable Temperature

See Fig. E.11.

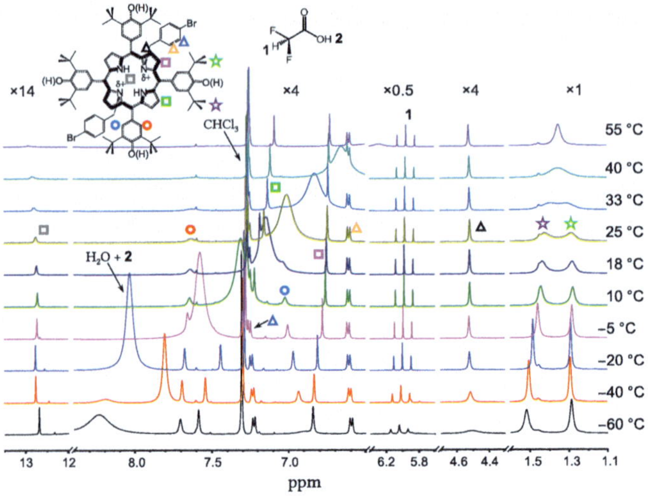

Fig. E.11 Variable temperature measurement of **Bz$_2$OxP** with 35 equiv. DFA (in $CDCl_3$, host concentration 7×10^{-4} M)

E.12 Bz$_2$OxP + (*R*)-CSA 0.59 equiv. Variable Temperature

See Fig. E.12.

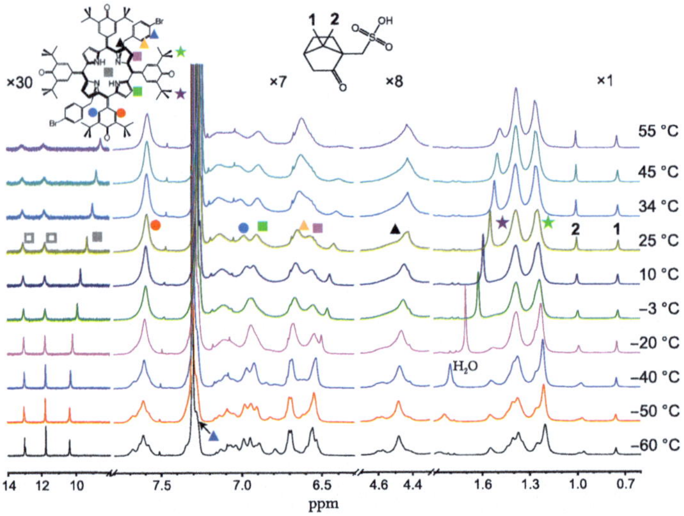

Fig. E.12 Variable temperature measurement of **Bz$_2$OxP** with 0.59 equiv. (*R*)-CSA (in $CDCl_3$)

E.13 Bz$_2$OxP + (*R*)-CSA 14 equiv. Variable Temperature

See Fig. E.13.

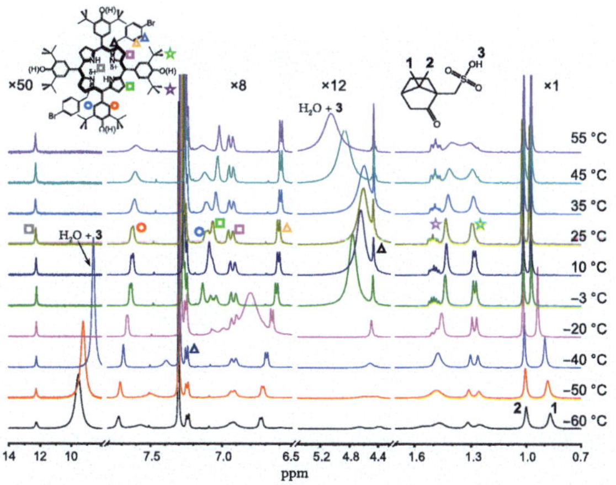

Fig. E.13 Variable temperature measurement of **Bz$_2$OxP** with 14 equiv. (*R*)-CSA (in $CDCl_3$, host concentration 8.4×10^{-4} M). Structure of the protonated form is confirmed by COSY in [2]

E.14 Bz$_2$OxP + (*rac*)-CSA 0.5 equiv. Variable Temperature

See Fig. E.14.

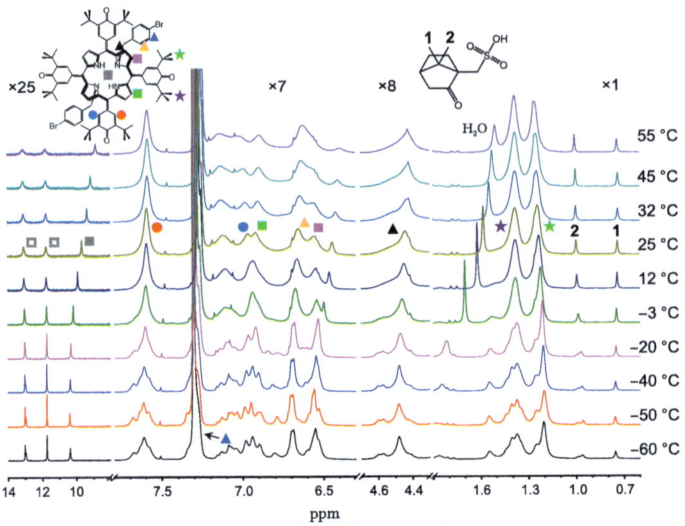

Fig. E.14 Variable temperature measurement of **Bz$_2$OxP** with 0.5 equiv. (*rac*)-CSA (in $CDCl_3$)

E.15 Bz$_2$OxP + (*rac*)-CSA 17 equiv. Variable Temperature

See Fig. E.15.

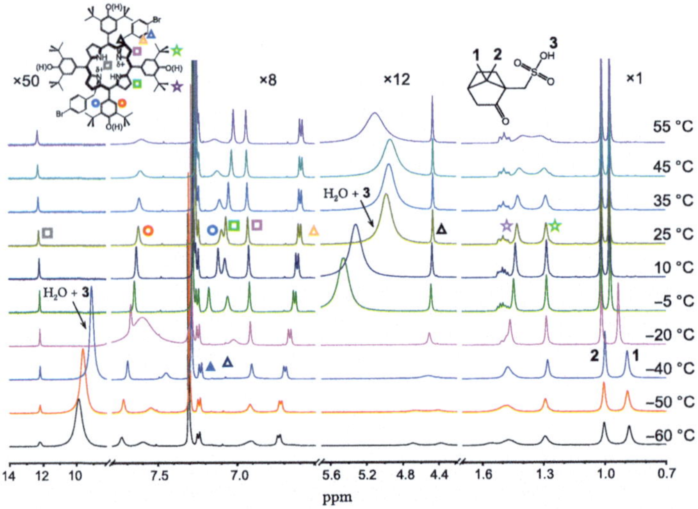

Fig. E.15 Variable temperature measurement of **Bz$_2$OxP** with 17 equiv. (*rac*)-CSA (in $CDCl_3$)

E.16 OxP + DMF Titration

See Fig. E.16.

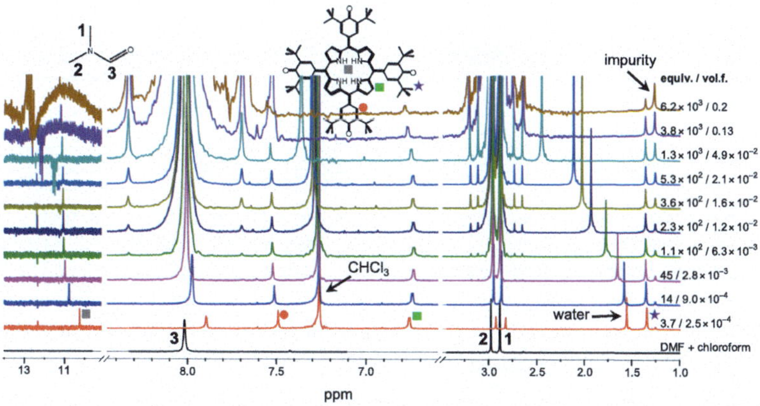

Fig. E.16 NMR titration of **OxP** with DMF (in $CDCl_3$ at 298 K, host concentration 0.001 M decreased to 5×10^{-4} M upon addition of stock solution). Spectra were rescaled to constant **OxP** concentration. Spectrum of DMF in $CDCl_3$, which was used as stock solution, is included. Peak of free water (1.56 ppm) is shifting probably due to the solvent polarity change

E.17 Detection of OH Signal in Bz_2OxP

Hydrogen atoms from the host OH groups exchange with hydrogen atoms of water present in the sample. This exchange is mediated by the acid OH groups. Hence, the presence of an OH signal in the host NMR spectrum can be confirmedby addition of small amount of D_2O into the sample. Then, the host OH signal should vanish since the deuterium is not detectable in ^{1}H NMR measurement. Figure E.17a confirms the presence of OH resonance (red arrow) in **OxP**·DFA complex. Also, Fig. E.17b suggests assignment of peak at 6.5 ppm as OH (red arrow), although this assignment is not as unequivocal as in the previous case.

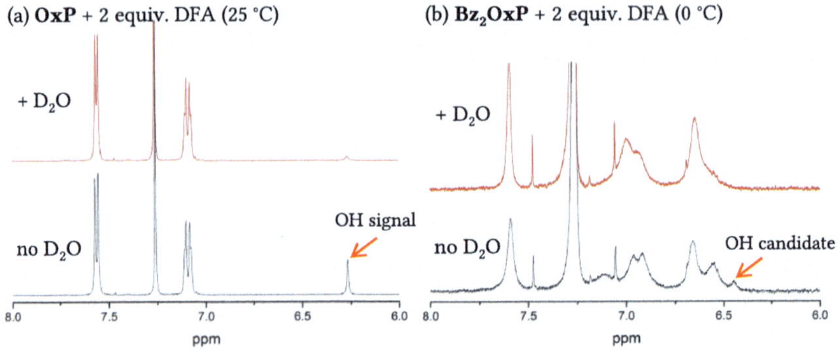

Fig. E.17 Confirmation of presence of exchangeable phenolic OH peak (**a**) bottom: **OxP** + 2 equiv. of DFA (25 °C), top: after addition of D_2O, (**b**) bottom: $\mathbf{Bz_2OxP}$ + 0.5 equiv. of CSA (0 °C), top: after addition of D_2O. The OH peak denoted by red arrow disappears in both cases

E.18 Bz_4OxP Assignment—HMBC

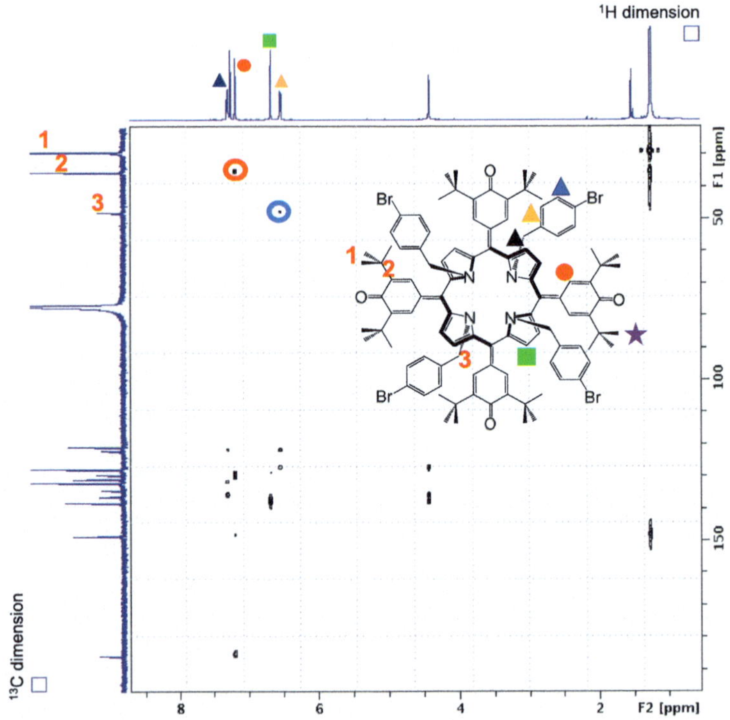

Fig. E.18 NMR HMBC measurement of $\mathbf{Bz_4OxP}$. Geometric shapes denote assignment of hydrogen nuclei, numbers denote assignment of carbon nuclei

Two-dimensional technique Heteronuclear Multiple Bond Correlation (HMBC) detected correlations via chemical bonds between ^{1}H and ^{13}C nuclei (Fig. E.18). The measurement shows correlations (**i**) between *tert*-butyl carbon and hemiquinonoid hydrogen (**2** ↔ ●, red circle), (**ii**) between CH_2 carbon and inner benzyl hydrogen (**3** ↔ ▲, blue circle). Peaks **1**, **2**and **3**were assigned according to they chemical shift and peak area, the HMBC correlation was used to assign ●and ▲.

References

1. Březina V, Ishihara S, Lang J, Hanyková L, Ariga K, Hill JP, Labuta J (2018) Structural modulation of chromic response: effects of binding-site blocking in a conjugated Calix[4]pyrrole chromophore. ChemistryOpen 7:323–335.
2. Březina V (2014) Spectroscopic and theoretical study of supramolecular complexes of symmetric porphyrins with chiral guests, MA thesis. Univerzita Karlova, Matematicko-fyzikální fakulta.
3. Labuta J, Ishihara S, Šikorský T, Futera Z, Shundo A, Hanyková L, Burda JV, Ariga K, Hill JP (2013) NMR spectroscopic detection of chirality and enantiopurity in referenced systems without formation of diastereomers. Nat Commun 4:2188.

Curriculum Vitae

Basic Information

Family name	Březina
Given name	Václav
Year of birth	1990
Nationality	Czech
Affiliation	Faculty of Mathematics and Physics, Charles University in Prague
E-mail address	vaclav.brezina@matfyz.cuni.cz

Education in Physics

2009–2012	Bachelor, *General Physics*, thesis *Diffusion of interacting particles*, supervisor Dr. Artem Ryabov
2012–2014	Master, *Physics of Condensed Matter and Materials*, thesis *Spectroscopic and theoretical study of supramolecular complexes of symmetric porphyrins with chiral guests*, supervisor Dr. Lenka Hanyková
2014–2022	Doctoral, *Biophysics, chemical and macromolecular physics*, thesis *Supramolecular complexes of oxoporphyrinogens with chiral organic molecules*, supervisor Dr. Lenka Hanyková

Publications

1 V. Březina, S. Ishihara, J. Lang, L. Hanyková, K. Ariga, J. P. Hill, and J. Labuta, "Structural Modulation of Chromic Response: Effects of Binding-Site Blocking in a Conjugated Calix[4]pyrrole Chromophore", ChemistryOpen **7**, 323–335 (2018).

2 V. Březina, L. Hanyková, N. Velychkivska, J. P. Hill, and J. Labuta, "NMR lineshape analysis using analytical solutions of multi-state chemical exchange with applications to kinetics of host–guest systems", Sci. Rep., 10.1038/s41598-022-20136-4 (2022).

V. Březina, *Supramolecular Complexes of Oxoporphyrinogens with Organic Molecules*, Springer Theses, https://doi.org/10.1007/978-3-031-68646-7

3 N. Velychkivska, L. Starovoytova, V. Březina, L. Hanyková, J. P. Hill, and J. Labuta, "Improving the Colloidal Stability of Temperature-Sensitive Poly(N-isopropylacrylamide) Solutions Using Low Molecular Weight Hydrophobic Additives", ACS Omega **3**, 11865–11873 (2018).
4 M. K. Chahal, J. Labuta, V. Březina, P. A. Karr, Y. Matsushita, W. A. Webre, D. T. Payne, K. Ariga, F. D'Souza, and J. P. Hill, "Knock-on synthesis of tritopic calix[4]pyrrole host for enhanced anion interactions", Dalton Trans., 10.1039/C9DT02365H (2019).
5 D. T. Payne, M. K. Chahal, V. Březina, W. A. Webre, K. Ariga, F. D'Souza, J. Labuta, and J. P. Hill, "Diporphyrin tweezer for multichannel spectroscopic analysis of enantiomeric excess", Front. Chem. Sci. Eng., 10.1007/s11705-019-1869-1 (2020).
6 K. Takimoto, S. Ishihara, J. Labuta, V. Březina, D. T. Payne, J. P. Hill, K. Ariga, M. Sumita, S. Mori, and H. Sato, "Enantiomeric Excess Dependent Splitting of NMR Signal through Dynamic Chiral Inversion and Coligand Exchange in a Coordination Complex", J. Phys. Chem. Lett., 8164–8169 (2020).
7 D. T. Payne, J. Labuta, Z. Futera, V. Březina, L. Hanyková, M. K. Chahal, and J. P. Hill, "Molecular rotor based on an oxidized resorcinarene", Organic Chemistry Frontiers **9**, 39–50 (2022).

The manufacturer's authorised representative in the EU is Springer Nature Customer Service Centre GmbH, Europaplatz 3, 69115 Heidelberg, Germany. If you have any concerns regarding our products, please contact ProductSafety@springernature.com

Printed and bound by CPI Group (UK) Ltd, Croydon, CR0 4YY
01/10/2026
02225970-0001